FOCUSED ION BEAM SYSTEMS

Basics and Applications

The focused ion beam (FIB) system is an important tool for understanding and manipulating the structure of materials at the nanoscale. Combining this system with an electron beam creates a TwoBeam system – a single system that can function as an imaging, analytical, and sample modification tool. Presenting the principles, capabilities, challenges, and applications of the FIB technique, this edited volume comprehensively covers the state-of-the-art in ion beam technology including the TwoBeam system.

The basic principles of ion beam and two-beam systems, their interaction with materials, etching, and deposition are all covered, as well as *in-situ* materials characterization, sample preparation, three-dimension reconstruction, and applications in biomaterials and nanotechnology.

With nanostructured materials becoming increasingly important in micromechanical, electronic, and magnetic devices, this self-contained working knowledge of the full range of ion beam methods, their advantages, and when best to implement them is a valuable resource for researchers in materials science, electrical engineering, and nanotechnology.

NAN YAO holds several positions at Princeton University, New Jersey. He is director of the Imaging and Analysis Center, Senior Research Scholar at the Institute for the Science and Technology of Materials, and a senior lecturer in Materials Science and Engineering.

FOCUSED ION BEAM SYSTEMS

Basics and Applications

Edited by

NAN YAO
Princeton University, New Jersey

CAMBRIDGE
UNIVERSITY PRESS

CAMBRIDGE UNIVERSITY PRESS
Cambridge, New York, Melbourne, Madrid, Cape Town, Singapore,
São Paulo, Delhi, Dubai, Tokyo, Mexico City

Cambridge University Press
The Edinburgh Building, Cambridge CB2 8RU, UK

Published in the United States of America by Cambridge University Press, New York

www.cambridge.org
Information on this title: www.cambridge.org/9780521158596

© Cambridge University Press 2007

First published 2007
First paperback edition 2010

A catalogue record for this publication is available from the British Library

ISBN 978-0-521-83199-4 Hardback
ISBN 978-0-521-15859-6 Paperback

Contents

Contributors

Tatsuya Asahata
SII Nano Technology Inc., Japan

Clive Chandler
FEI Company, 5350 NE Dawson Creek Drive, Hillsboro, OR 97124-5793

Derren Dunn
IBM Microelectronics, Hopewell Junction, NY 12533

Toshiaki Fujii
SII NanoTechnology Inc., Japan

Lucille A. Giannuzzi
FEI Company, 5350 NE Dawson Creek Drive, Hillsboro, OR 97124-5793

Hyoung Ho (Chris) Kang
IBM Microelectronics, East Fishkill, New York, NY

Kirk Hou
Princeton University, Princeton Institute for the Science and Technology of Materials, 70 Prospect Avenue, Princeton, NJ 08540

Robert Hull
University of Virginia, Department of Materials Science and Engineering, 116 Engineers Way, Charlottesville, VA 22904-4745

Nobutsugu Imanishi
Kyoto University, Department of Nuclear Engineering, 3-1-14 Ayameike-minami, Nara, 631-0033, Japan

Tohru Ishitani
Naka division, Hitachi High-technologies Corporation, Ichige 882, Hitachi-naka, Ibaraki, 312-8504, Japan

Takashi Kaito
SII NanoTechnology Inc., Japan

T. Kamino
Hitachi Science Systems, Ltd., 11-1 Ishikawa-cho, Hitachinaka-shi, Ibaraki-ken, 312-0057, Japan

Alan J. Kubis
University of Virginia, Department of Materials Science, 116 Engineers Way, Charlottesville, VA 22904

Richard Langford
University of Manchester, Materials Science Centre, School of Materials, Grosvenor Street, Manchester M1 7HS, UK

T. Ohnishi
Hitachi High-technologies Corporation, 882, Ichige, Hitachinaka-shi, Ibaraki-ken, 312-8504, Japan

Kaoru Ohya
The University of Tokushima, Department of Electrical and Electronic Engineering, Faculty of Engineering, Minamijosanjima 2-1, Tokushima, 770-8506, Japan

E. L. Principe
Carl Zeiss SMT Inc., Nano Technology Systems Division, 555 Twin Dolphin Drive, Redwood City, CA 94065

Daniel Recht
Princeton University, Princeton Institute for the Science and Technology of Materials, 70 Prospect Avenue, Princeton, NJ 08540

Steve Reyntjens
FEI Company, Achtseweg Noord 5, 5651 GG Eindhoven, The Netherlands

Ampere A. Tseng
Arizona State University, Department of Mechanical and Aerospace Engineering, Tempe, Arizona 85287-6106

Mark Utlaut
University of Portland, Department of Physics, Portland, OR 97203

Matthew Weschler
FEI Company, 5350 NE Dawson Creek Drive, Hillsboro, OR 97124-5793

T. Yaguchi
Hitachi Science Systems, Ltd., 11-1 Ishikawa-cho, Hitachinaka-shi, Ibaraki-ken, 312-0057 Japan

Nan Yao
Princeton University, Princeton Institute for the Science and Technology of Materials, 70 Prospect Avenue, Princeton, NJ 08540

Preface

In the past few years, scientists have begun to gain the exquisite of controlling the arrangement of matter on the nanometer scale ($1\,nm = 10^{-9}\,m$), a new field called nanotechnology, consequently, has started to emerge. As the foundation of nanotechnology, nanostructured materials take on an enormously richer variety of properties and promise exciting new advances in micromechanical, electronic, and magnetic devices as well as in molecular fabrications. The structure–composition–processing–property relationships for these sub-100 nm-sized materials can only be understood by employing the new generation microscopes such as the focused ion beam system in corporate with simultaneous operation of electron beam and in-situ analysis. This book will highlight the principles and vast capabilities of this technique and their applications in this fast-growing nanotechnology field and the challenges in the twenty-first century.

1

Introduction to the focused ion beam system

NAN YAO

Princeton University

1.1 Introduction

The frontier of today's scientific and engineering research is undoubtedly in the realm of nanotechnology: the imaging, manipulation, fabrication, and application of systems at the nanometer scale. To maintain the momentum of current research and industrial progress, the continued development of new state of the art tools for nanotechnology is a clear necessity. In addition, knowledge and innovative application of these tools is in increasingly high demand as greater numbers of them come into use. The interdisciplinary field of materials science, in particular, perpetually seeks imaging and analysis on a smaller and smaller scale for a more complete understanding of materials structure–composition–processing–property relationships. Moreover, the ability to conduct material fabrication via precise micro- and nano-machining has become imperative to the progress of materials science and other fields relying on nanotechnology.

An important tool that has successfully met these challenges and promises to continue to meet future nanoscale demands is the focused ion beam (FIB) system. The technology offers the unsurpassed opportunities of direct micro- and nano-scale deposition or materials removal anywhere on a solid surface; this has made feasible a broad range of potential materials science and nano-technology applications. There has naturally been great interest in exploring these applications, recently spurring the development of the two-beam FIB system, often also called DualBeam or CrossBeam, a new and more powerful tool that has advanced hand in hand with the complexity of new materials.

A focused ion beam system combines imaging capabilities similar to those of a scanning electron microscope (SEM) with a precision machining tool.

Focused Ion Beam Systems: Basics and Applications, ed. N. Yao.
Published by Cambridge University Press. © Cambridge University Press 2007.

It was developed as the result of research on liquid-metal ion sources (LMIS) for use in space, conducted by Krohn in 1961 [1,2]. Liquid-metal ion sources found novel applications in the areas of semiconductors and materials science, and the FIB was commercialized in the 1980s as a tool mainly geared toward the growing semiconductor industry [3]. In the development of semiconductor fabrication, there is a constant struggle to improve the resolution and speed of the lithographic technique. The use of photoresist and masking improved the speed and reproducibility of the result, but not the resolution, due to the fundamental and practical limitations imposed by the wavelengths of the light used. Electron beam lithography was a marked improvement in this area [4], due to the much smaller wavelength of a high energy electron, often on the order of one to two hundredths of a nanometer compared to the hundreds of nanometers associated with light. However, electron beam (or e-beam) lithography is a comparatively slow process, and often has difficulty penetrating harder materials without suffering from considerable distortion effects due to local charge buildup. Electrons, though easy to produce and accelerate, simply did not have the mass to penetrate materials and remove atoms from a lattice quickly, and so e-beams have stayed primarily in the realm of imaging, except in certain very specific environments. Thus the demand for a lithographic method with the advantage of short wavelengths, allowing higher resolution, but without the drawbacks presented by the low mass of electrons, has found an answer in the use of focused ion beams.

Fundamentally, a focused ion beam system produces and directs a stream of high-energy ionized atoms of a relatively massive element, focusing them onto the sample both for the purpose of etching or milling the surface and as a method of imaging. The ions' greater mass allows them to easily expel surface atoms from their positions and produces secondary electrons from the surface, allowing the ion beam to image the sample before, during, and after the lithography process. The ion beam has a number of other uses as well, including the deposition of material from a gaseous layer above the sample. The ions in the beam strike atoms or molecules down onto the surface of the sample, where intermolecular attractions fix them, and the implantation of ions into a surface [5,6].

Today's focused ion beam system utilizes a liquid-metal ion source at the top of its column to produce ions, usually Ga^+. The ions are then pulled out and focused into a beam by an electric field. They subsequently pass through apertures and are scanned over the sample surface. The ion–atom collision is either elastic or inelastic. Whereas elastic collisions result in the excavation of surface atoms, a technique called sputtering or milling, inelastic collisions

Figure 1.1 A typical SEM image showing the simultaneous milling and deposition capabilities of a two-beam FIB system. (*Courtesy of Fibics Incorporated.*)

transfer some of the ions' energy to either the surface atoms or electrons, resulting in the emission of secondary electrons (those that become excited enough to escape from their shell). Secondary ions are also emitted from the surface following the secondary electrons.

The FIB system has four basic functions: milling, deposition, implantation, and imaging; each will be discussed in detail in the following chapters. Milling is a process that allows digging into the sample surface as a result of the use of relatively heavy ions in the beam. It can also be easily converted into a deposition system simply by adding a gas delivery device that allows the application of certain materials, usually metals, to the surface of the material where the beam strikes. When combined with milling, FIB deposition can create almost any microstructure. Figure 1.1 represents a typical example of such capability. Ion implantation is another important component of surface modification that is available using the FIB. In addition to these three variations of material surface adaptation, the FIB system also has extensive imaging capabilities. The large size of the ions provides advantages that are not available with scanning electron microscopes or other imaging tools.

The FIB's unique properties allow it to isolate specific sample regions so that it only makes the necessary modifications without affecting the integrity

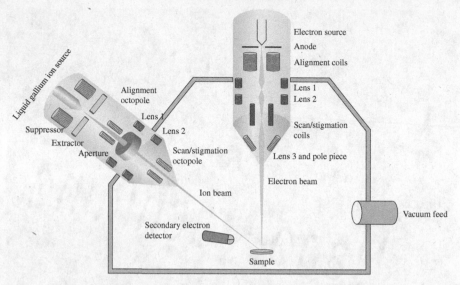

Figure 1.2 A schematic diagram showing the configuration of a two-beam focused ion beam system.

of the whole sample. With this technology the FIB can perform simple techniques such as making probe holes as well as more complicated procedures such as cutting a precise three-dimensional cross section of a sample. The FIB, with its combination of drilling and deposition capabilities, is also ideal for failure analysis and repair.

Using only an FIB system has some disadvantages, however, including that it often causes some undesired damage to the sample. Obstacles associated with the FIB, as well as the growing complexity of materials, has fueled the development of a two-beam focused ion beam system: a system that combines both electron beam and ion beam in a single microscope. Though the FIB system by itself has a wide range of functions and applications, combining the FIB's precise imaging and machining abilities with the scanning electron microscope's high resolution, nondestructive imaging leads new and invaluable applications to emerge that were previously impossible. The two-beam FIB excels at high resolution structural, chemical, and geometric analyses of cross sections of layers of material, a necessary feature for the examination of complex materials and their synthetic analogs as well as for the analysis of phenomena that may affect performance, durability, and reliability of many new materials. The combination of SEM and FIB in a two-beam system, as shown schematically in Figure 1.2, allows the electron and ion beams to work symbiotically to achieve tasks beyond the limitations of either individual system.

In this chapter, we present a basic introduction of the two-beam FIB system. Since ion beam and electron beam are the two key components of the system, we start with a discussion of them first, followed by a discussion of ion and electron sources used in the two-beam system. It is important to explain the essential differences between electrons and ions in order to understand how the properties of each affect the structure and functionality of the FIB and the SEM. Following the discussion of properties of the ion beam and electron beam and their emission sources, we will look at the ion optics and electron optics responsible for focusing ions and electrons from the source onto the sample in the column of a microscope. The detection of secondary and backscattered charged particles from the sample to form images will also be examined. Finally, we will introduce the two-beam system and discuss its advantages versus a standalone SEM or FIB platform, and how its enhanced capabilities open new channels for materials science and nanotechnology.

1.2 Ion beam versus electron beam

All emissions can be sources of information, depending on the capabilities of the instrument. The ejected signals from a focused ion beam or electron beam can be collected, amplified, and then displayed to show detailed information of the sample surface. When the ion beam is focused on one area for an extended length of time, the continuous sputtering process gives the machine another added use, that of removing surface material, which opens the door for probing and milling applications. The FIB system can also be a deposition tool by injecting an organometallic gas in the path of the ion beam, just above the sample surface. This technique allows for many kinds of material fabrication at the micro- and nano-scales.

Since ions are significantly more massive than electrons (Table 1.1), the FIB system has many more applications than a conventional imaging instrument. The collision between the large primary ions of the beam and the surface atoms causes surface alteration of various levels determined by the dosage, overlap, dwell time, and many other ion beam variables. Such surface alteration could not be achieved at the same level with electrons.

The ion beam and electron beam are based on the same principle and serve many of the same purposes. They both consist of a stream of charged particles that is focused by a series of lenses and apertures onto a sample and both employ similar methods to produce and accelerate the particles from their source. Both systems can be used to image a sample, as well as to perform etching and deposition.

Table 1.1 *Quantitative comparison of FIB ions and SEM electrons*

Particle	FIB	SEM	Ratio
Type	Ga^+ ion	Electron	
Elementary charge	$+1$	-1	
Particle size	0.2 nm	0.00001 nm	20 000
Mass	1.2×10^{-25} kg	9.1×10^{-31} kg	130 000
Velocity at 30 kV	2.8×10^5 m/s	1.0×10^8 m/s	0.0028
Velocity at 2 kV	7.3×10^4 m/s	2.6×10^7 m/s	0.0028
Velocity at 1 kV	5.2×10^4 m/s	1.8×10^7 m/s	0.0028
Momentum at 30 kV	3.4×10^{-20} kg m/s	9.1×10^{-23} kg m/s	370
Momentum at 2 kV	8.8×10^{-21} kg m/s	2.4×10^{-23} kg m/s	370
Momentum at 1 kV	6.2×10^{-21} kg m/s	1.6×10^{-23} kg m/s	370
Beam			
Size	nm range	nm range	
Energy	up to 30 kV	up to 30 kV	~
Current	pA to nA range	pA to µA range	~
Penetration depth			
In polymer at 30 kV	60 nm	12000 nm	0.005
In polymer at 2 kV	12 nm	100 nm	0.12
In iron at 30 kV	20 nm	1800 nm	0.11
In iron at 2 kV	4 nm	25 nm	0.16
Average signal per 100 particles at 20 kV			
Secondary electrons	100–200	50–75	1.33–4.0
Backscattered electron	0	30–50	0
Substrate atom	500	0	infinite
Secondary ion	30	0	infinite
X-ray	0	0.7	0

The fundamental difference between the use of an ion beam and that of an electron beam lies in their unique characteristics. The ion is much larger and more massive than the electron and can be positively charged, whereas electrons are always negatively charged. Since ions travel more slowly and require greater fields to focus and control than electrons, different methods are required to control massive ions versus electrons.

Size and mass can appreciably alter the interactions between the beam and the sample (Table 1.1). When a beam of energetic particles, whether ions or electrons, strikes a solid surface several interactions occur. Some particles are backscattered from the surface layers; others are slowed down within the solid. Unlike electrons, the relatively large ions have a hard time penetrating the surface of a sample because it is much harder for them to pass through individual atoms. Instead, their size increases their probability of interactions with atoms, causing a rapid loss of energy. As a result, atomic ionization of the surface atoms and breaking of the chemical bonds between these atoms – both processes involving mainly surface electrons – occur as the main ion–atom interactions. Emission of secondary electrons usually accompanies these processes as well as a change in the chemical state of the material. Unlike in the case of an incident electron beam, however, the inner electrons cannot be reached or excited by an ion beam and characteristic X-rays are therefore unlikely to be generated.

The total length that the ion travels is known as its "penetration depth," a term which also applies to electrons, which often penetrate much deeper into the sample than ions (Table 1.1). Because of the statistical nature of the atomic collision, the penetration depth adheres to a symmetric Gaussian distribution around the mean value. In the process of material modification, the moving ion recoils one or more atoms in the sample, which results in the recoiling of constituent atoms, leading to the creation of atomic defects along the path of the ion beam.

The other difference between the two beam types, of course, is that the ion beam has a much greater direct effect on its target, causing localized heating and removing atoms beneath the focus, as well as implanting ions into the surface and depositing atoms located above the sample onto it. Electron beams generally cause little or no surface damage, have greater difficulty causing deposition, and generally do not change the internal structure of the sample, as the electrons left by the beam's passage dissipate through conduction [7].

Ions are many times more massive than electrons and therefore carry hundreds of times more momentum than electrons. In the ion–atom collision, this momentum is transferred to the atoms on the surface of the material, disturbing them from their aligned positions in a sputtering effect that has important milling applications. The ion beam, as a direct result of the large size and mass of the ion, surpasses the range of capabilities of the electron beam by being able to remove atoms from the surface of a material in a precise and controlled manner.

Gallium (Ga^+) ions are usually used in FIB systems for a number of reasons. First, because of its low melting point, gallium only requires limited heating and can conveniently be in liquid phase during operation; the lower operating temperature also minimizes interdiffusion with the tungsten needle substrate. Second, its mass is heavy enough to allow milling of the heavier elements, but it is not so heavy that a sample is immediately destroyed. Third, its low volatility at the melting point conserves the supply of metal and yields a long source life of about 400 μA-hours/mg. Fourth, its low vapor pressure allows Ga to be used in its pure form instead of in the form of an alloy source, which would require an E × B mass separator in the optics column.

Finally, gallium can be easily distinguished from other elements, so if implantation occurs, the gallium ions will not interfere with the analysis of the sample. Although other ions can be used, gallium has become the ion of choice for the focused ion beam system.

Ultimately, the purpose of precision engineering and attention to detail in FIB design is to produce a focused ion beam that impacts the surface at a desired point. It is important to consider the implications of this impact. A beam of light incident upon a surface causes local temperature increase, small electromagnetic fluctuations, and photoelectron emission. A beam of charged particles does all of this and more. The incident particles raise the temperature of the area they impact, although generally not to a significant degree; they change the local charge densities of the region, resulting in a temporary charge imbalance; the addition of their kinetic energy to the energy of the sample causes secondary electrons to be emitted, which can be captured and used to image the surface of the sample; they produce a degree of characteristic X-ray emission, which can be used for spectroscopy purposes; and, of course, they cause damage to the surface structure at different levels depending on the physical nature of charged particles.

The emission of secondary electrons from the surface is what gives both beam types their imaging abilities. Adding the energy of the particles in the beam to that of the electrons in the sample allows some of those electrons to escape from within the material, depending on the penetration depth of the beam and the conductivity and work function of the sample. These electrons come from a roughly spherical region around and beneath the beam spot, with increased numbers of electrons exiting from the sides of topographical features that would not have escaped from a flat surface otherwise. In addition to the secondary electrons, there is a degree of radiation produced by the rapid deceleration of the charged particles, generally in the X-ray spectrum. This radiation can later be used for X-ray spectroscopy. Between this emitted radiation and the radiation from the accelerating potentials in the sources, the chamber requires good shielding in order to preserve the safety of the operator. All the modern electron and ion instruments have handled this issue satisfactorily.

The ion beam is capable of efficiently and precisely depositing material. By creating a cloud of, for example as shown in Figure 1.3, platinum atoms above the sample, platinum can be deposited by letting the ion beam strike these atoms, imparting some kinetic energy to them and causing them to impact the surface, on which they remain, adsorbed onto the sample by Van der Waals forces. This technique can be used to deposit both conductive and resistive materials, since the type of atom in the suspended

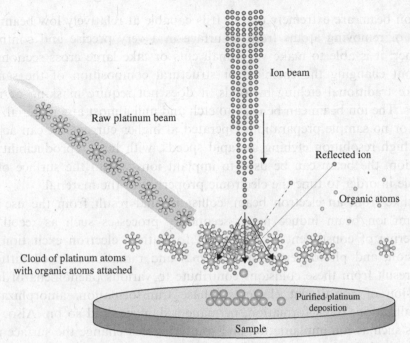

Ion beam

Raw platinum beam

Reflected ion

Organic atoms

Cloud of platinum atoms
with organic atoms attached

Purified platinum
deposition

Sample

Figure 1.3 A Schematic diagram of an ion beam induced platinum deposition process.

material does not matter, making the ion beam quite useful for nano-fabrication purposes. This technique can also be used to improve electron-beam imaging, by depositing a thin layer of conductor over the surface of a feature to be imaged. Similar to sputter- or deposition-coating of a sample, this method preserves more detail at a small scale by using a precisely controlled, very thin layer of conductor to reduce charging effects and define the surface [8].

The electron beam can be used, to some degree, to perform deposition. However, due to the low mass of the electrons, the deposition occurs slowly, and it is more feasible if the beam is very intense and focused precisely, to increase the probability of electrons intersecting atoms of the deposition material. In general, electron-beam deposition is not done with the beams found in smaller scanning electron microscopes, instead requiring a larger and more powerful apparatus. Until the advent of the focused ion beam, this type of precision deposition was extremely difficult to achieve.

While the electron beam barely affects the surface, the heavy particles of the ion beam penetrate deeper within the lattice, kicking out atoms as they go and embedding them in the sample. Therefore, the lithographic abilities of the ion beam are extremely useful. It is capable at relatively low beam currents of removing atoms from a surface in a very precise and controlled manner; it is able to make very small cuts or take large cross sections, all without changing the chemical or structural composition of the sample. Unlike traditional etching methods, it does not require masking or resist stages. The ion beam can be used to etch and mill almost any material, with little or no sample preparation. Operated at higher currents, it can achieve very high resolution etching at rapid speeds, with high reproducibility. In addition, the beam can be used to implant ions within the surface of the sample in order to tune the electronic properties of the material.

Unlike from an electron beam, collisions that result from the use of a gallium ion beam induce many secondary processes such as recoil and sputtering of constituent atoms, defect formation, electron excitation and emission, and photon emission. Thermal and radiation-induced diffusion that result from these collisions contribute to various phenomena of inter-diffusion of constituent elements, phase transformation, amorphization, crystallization, track formation, permanent damage, and so on. Also, processes such as ion implantation and sputtering will change the surface morphology of the sample, possibly creating craters, facets, grooves, ridges, pyramids, blistering, exfoliation, or a spongy surface.

Because of the interrelatedness of these processes, no single phenomenon can be understood without the discussion of several others. Therefore, it is imperative that one possesses a quantitative understanding of the experimental observations as well as creativity in design so that new and more sophisticated combinations of these versatile processes can be applied in the field of nanotechnology. With it, we can aim at more advanced material modification, deposition techniques, implantation, erosion, nano-fabrication, surface analysis, and many other applications.

1.3 The ion source and electron source

In order to properly understand the focused ion beam, it is necessary to consider the source of the beam itself, as illustrated in Figure 1.4. In almost all focused ion beam systems, a reservoir of heavy-metal atoms (typically gallium for the aforementioned reasons) is heated to near evaporation, after which it flows and wets a sharp, heat-resistant tungsten needle with a tip radius of 2–5 μm. Once heated, the Ga can remain liquid for weeks without further heating due to its super-cooling properties. The Ga atoms then flow to the very end of the needle, drawn there by an annular electrode concentric with the tip of the needle and positioned close to it, called the extractor.

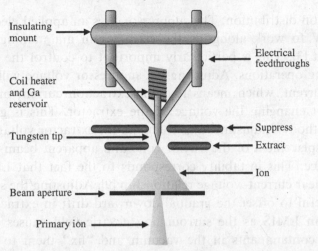

Figure 1.4 A cross-sectional diagram of a liquid-metal ion source (LMIS) found in FIB systems.

A potential difference between needle tip and aperture creates an electric field on the order of 10^{10} V/m and causes energetic ions in the region immediately above the tip to accelerate toward the extractor. These ions exist in a region where the balance between electrostatic and surface tension forces has drawn the liquid-metal into a sharp "Taylor cone" whose apex is only about 5 nm in diameter. The cone tip is small enough such that the extraction voltage of the aperture can pull Ga from the needle tip and efficiently ionize it by field evaporation of the Ga at the end of the Taylor cone, after which it is accelerated by a potential down the ion column. The current density of ions that may be extracted from such an LMIS system is on the order of 10^8 A/cm^2, as the evaporated ions are continuously replaced by a flow of liquid Ga to the cone. The source is generally operated at low emission currents of 1–3 μA to reduce the energy spread of the beam and to yield a stable beam. At higher emission currents, the probability of the formation of dimers, trimers, and droplets increases, which both increases energy spread and decreases source lifetime.

The tip of the tungsten needle is situated just above the extractor, which is held at a voltage on the order of −6 kV relative to the source. The resulting intense electric field at the source tip draws the liquid Ga into a Taylor cone and creates a tiny cusp at its end (called the "incipient jet"). After field evaporation causes ion emission to occur, the ions begin to accelerate down the column. The current emitted from the tip is known as the extraction current. It is regulated by both the suppressor and the extractor, as shown in Figure 1.4, which roughly correspond to "fine" and "coarse" controls of the emitted ion distribution. The suppressor uses an applied electric field of up to $+2\,\text{kV}$ to work alongside the extractor in maintaining a constant beam current [3]. This is particularly important to control the etching rate during milling operations. Adjusting the suppressor voltage will change the extraction current, which means that the extraction current may be regulated without changing the voltage of the extractor. This is generally the preferred method of adjustment, as changing the extractor voltage can result in spatial displacement of the Taylor cone and apparent beam drift on the sample surface. This instability corresponds to the fact that LMIS have a highly nonlinear current–voltage relationship [9]. Adjusting the suppressor is also very useful to offset the gradual downward drift in extraction current that occurs in LMIS as the surrounding electric field causes electrons to collide with contaminants in the vacuum and "fix" them to the source. Thus, the ability of the suppressor to maintain constant extraction voltage without altering the source tip is an essential requirement for FIB system stability.

Field evaporation, the process responsible for ion production in the LMIS, is a physically complex phenomenon, and complete treatment is beyond the scope of this chapter. Fundamentally, however, field evaporation takes place when the potential barrier preventing evaporation has been lowered by the presence of a field and can only be crossed by the ionization of the evaporating atom on the surface of the field emitter. It can be described analytically by first calculating the field needed to produce ions in free space, that is, in the absence of any other field. The energy Q_0 needed to produce an ion in free space in this case is given by $Q_0 = H_a + I_n - n\phi$, where H_a denotes the heat of atomic desorption, I_n refers to the ionization energy to produce an n-fold ion, and ϕ is the work function of the field emitter. The term $n\phi$ corresponds to the released energy when n electrons return to the metal. Taking into account the presence of a field E, which lowers the potential barrier for field ionization by $\sqrt{2^3 e^3 E}$, the required energy for evaporation of the liquid Ga is

$$Q_E = Q_0 - \sqrt{n^3 e^3 E}. \tag{1.1}$$

Alternatively, the atoms in the Taylor cone will ionize when

$$I - \phi = eEx_c, \tag{1.2}$$

where I is the ionization potential near a metal with work function ϕ in the presence of an electric field E. It has been found that the critical distance x_c from the tungsten needle tip required for ion production is approximately $0.2\,\text{nm}$ at a field strength of 10^{10} V/m.

For comparison, we can consider one type of source of electrons in an electron beam column. In the most common configuration, a tungsten filament is heated by a large current, causing it to emit a spectrum of radiation accompanied by a number of loose electrons that have gained sufficient energy to overcome the work function of the metal and escape. These electrons are then accelerated away from the tip by a set of electrostatic fields generated by large coils, and reduced to a relatively clean beam by an aperture below the source. The filament itself is formed into a sharp point, since this shape causes charge to cluster at the tip, giving a greater output current from the source.

The basics of the electron gun revolve around raising an electron's energy above the Fermi level of the cathode material. In a standard thermal emission gun, the energy of the electrons in the emission region is raised by the addition of heat in the cathode and a potential from the anode, resulting in a current density j_c that follows Richardson's Law:

$$j_c = ATc^2 e^{\frac{-\phi_w}{kT_c}}, \tag{1.3}$$

where A is a constant related to the cathode material, T_c is the temperature of the emissive tip, ϕ_w is the work function (related to the Fermi level) of the cathode, and k is Boltzmann's constant. This current density produces a roughly Gaussian beam profile:

$$j(r) = j_0 e^{-\left(\frac{r}{r_0}\right)^2}, \tag{1.4}$$

where r_0 is the radius of the emissive region on the cathode tip, usually ranging from 10 to 50 micrometers.

The tungsten filament is also usually zirconated (ZrO) to increase the thermionic emission of electrons by means of the Schottky effect, in which an accelerating field for electrons exists at the surface of the filament due to an external applied electrostatic field. Zirconated tungsten exhibits a decreased work function (2.5 eV versus 4.5 eV) and provides adequate electron emission at only about 1600 K, compared with 2500–3000 K.

Two other classes of enhanced electron guns are worth mentioning. The potential of rare-earth hexaboride materials, especially lanthanum hexaboride (LaB$_6$), for thermionic electron emission was first reported in 1951 by J. M. Lafferty [10]. The work function of LaB$_6$ is lower than that of W (2.7 eV compared to 4.5 eV), which, combined with its low vapor pressure at high temperature, makes it a superior thermionic electron source for the electron microscopes (both scanning and transmission). LaB$_6$ crystals can provide considerably higher current densities, and LaB$_6$ offers improved coherence and a smaller energy spread [11]. Extensive research in the 1970s and 1980s

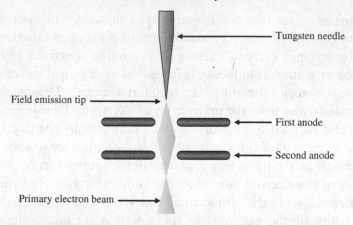

Figure 1.5 A cross-sectional diagram of a cold field emission electron source found in high-resolution SEM systems.

followed Lafferty's work in an attempt to optimize the performance of this class of materials. The drawback to this source is mainly on its relative larger energy spread and low beam intensity for an electron probe of nanometer size. Cold field emission guns (CFEGs) are another recent development in SEM source technology. Unlike thermionic guns, they operate at relatively low temperatures of about 300 K and offer superior resolution and performance. FEGs are simpler in their operation – they use a pair of anodes below the tungsten tip to generate intense electric fields that extract electrons by enabling them to tunnel from the extremely sharp tip (Figure 1.5). As a result of the small tip size and low operating temperature, the electron beam is highly spatially coherent and experiences almost no energy spread, which limits the deleterious effects of chromatic aberration; its current density is also remarkably high [12]. The weakness of FEGs is that they can only be under ultra high vacuum ($< 10^{-10}\tau$), which requires more expensive machinery.

Both ion and electron sources are subject to a pair of limitations. First, the total current in the beam leaving the extraction aperture is only a fraction of that produced by the sources; the majority of the particles generated are blocked by the sides of the aperture, as their velocity vectors are not pointed along the direction of the beam. This means that the "brightness" of the resulting beam depends on both aperture size and source current. A larger aperture means a brighter beam with better imaging and milling characteristics; however, the beam may not be as uniform or as precisely focused as one traveling through a smaller aperture. A higher source current means more charged particles produced, but only a fraction of those will go into the beam itself, so a substantial increase in current is needed for a marked increase in beam brightness. The increased current can cause source instability and lower the lifetime of the source element.

A second limitation is that of uniformity. While the electrons emitted from the zirconated tungsten are relatively evenly distributed in energy, the ions emitted by a liquid-metal ion source (LMIS) as described above tend to follow a Gaussian energy distribution, sometimes asymmetrical, which leads to chromatic aberration. The energy required to evaporate ions from a liquid is dependent on local temperature, field strength, and surface tensions in the area around each atom, which can vary from point to point within the emissive region [13]. The ion emission current is strongly dependent on the tip radius and on the tip surface condition. The sharper the tip is, the higher the field; and the higher the field, the stronger the ion emission. However, the inter-particle repulsion effect at high emission and liquid flow rates with capillary characteristics need to be balanced out in order to maintain a stable, consistent ion beam emission. Electron emission is determined by the work function of the metal used and the field strength at the source, with electrons escaping as soon as they have reached a well-defined critical energy.

Both sources experience an additional degree of energy spread due to mutual repulsion between the charges just beyond the source. The like charges of the ions and the electrons repel each other, imparting a small but significant random velocity that changes the overall energy profile, causing further chromatic aberration and forming the fundamental limit for the focusing ability of the system. Variations in emission characteristics can be controlled to a degree, but before the limit of that control is reached, it is overshadowed by the aberrations arising from these mutually repulsing charged particles.

1.4 Ion optics and electron optics

After a beam of charged particles is produced, whether of ions or electrons, it must be focused to the desired spot size by a series of lenses, usually with one or more apertures along the beam path as well to help control aberrations. A lens for an ion or electron beam can be thought of in the abstract as being almost identical to a lens for light, with a similar function and similar parameters, such as focal length and refractive index. Standard light optics concerns thus enter the picture, including chromatic aberration, spherical aberration, and astigmatism. An additional concern specific to electron and ion beam applications is that of apparent source size, which is related to the inter-particle repulsion mentioned above.

Electron and ion lenses function very much like light optics, but their construction is quite different. Instead of using a material with a certain geometry and index of refraction to bend the path of the light, electron and ion optics use magnetic and electrostatic fields to change the paths of the particles.

Electron beams, consisting of fast-moving, low-mass particles, are generally focused using only magnetic fields. The advantage of a magnetic field is that it is relatively easy to produce a uniform field over a region, and that magnetic fields do no net work on objects within them, so that the kinetic energy of the electrons is not changed. This makes it easier to determine the results of the electron's impact on the surface and to process and understand the resulting data. The magnetic lenses in an electron beam column are generally washer-shaped coils with small central holes, so that the field is intensified by being compressed into a smaller space. The deflection of the beam is proportional to the distance from the axis of the lens, as in light optics, so that the field behaves just as a glass lens would.

Charged particles, however, require stronger fields to focus higher-energy particles because charged particles have kinetic energy vectors that must be diverted by an applied force, while light can be focused by changing the nature of the medium through which it propagates. The higher the kinetic energy, the higher the force needed and the higher the lens fields must be. Higher kinetic energies correspond to shorter wavelengths, which provide higher resolution. In order to achieve this higher resolution, however, ever-stronger fields must be produced in the focusing column.

Ion beams, in contrast to electron beams, use electrostatic lenses almost exclusively. The reason for this stems from the fact that the force on a charged particle due to an electric field E, $F_e = qE$ is independent of the particle's velocity, whereas the force exerted by a magnetic field $\mathbf{B}$, $\mathbf{F} = q\mathbf{v} \times \mathbf{B}$, depends directly on the velocity. A particle accelerated by a potential drop ΔV will gain a speed $\sqrt{2q\Delta V/m}$, where $q\Delta V$ is the energy and m is the particle's mass. Given the higher mass of ions, their velocity is 0.0028 that of an electron accelerated by the same potential, while their momentum is 370 times higher. Magnetic optics would need to be impractically large to provide enough focusing power for an ion beam. Electrostatic lenses, however, can be made extremely small and are capable of producing much faster response for

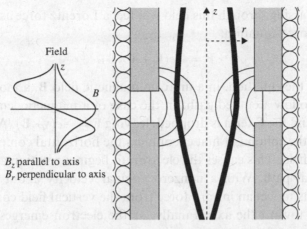

Figure 1.6 A cross section side view of the cylindrical magnetic lens within an SEM column. Magnetic field lines in a magnetic lens are shown in thin lines, while the thick light lines represent two possible electron paths. The smaller inset to the left gives an approximation of the magnetic field strength along the optical axis and perpendicular to the optical axis. The narrowing helical rotation of electrons within the lens cannot be faithfully represented in two dimensions but can be visualized.

beam deflections, an important capability also used for beam blanking (a toggling of the beam's incidence on the sample by deflecting it from the optical axis upstream from an aperture).

1.4.1 The lenses

Immediately after emission in an SEM, the electrons are accelerated by the voltage V between cathode and anode to a kinetic energy of $E = eV$, where e is the electron charge. After the beam is reduced to a relatively coaxial column by an aperture, it enters the top of the electron optics column and passes into the magnetic field of the first electron lens, an axial field with rotational symmetry about the axis of the column. This field B_z has a bell-shaped distribution along the axis where it swells out from the center of the lens coil (see Figure 1.6). B_r, the radial component of the magnetic field, obeys the following equation:

$$B_r = \left(\frac{-r}{2}\right)\frac{\partial B_r}{\partial z}.\tag{1.5}$$

An electron passing through this field will feel a Lorentz force as described by the Lorentz force equation:

$$\mathbf{F} = -e(\mathbf{E} + \mathbf{v} \times \mathbf{B}).\tag{1.6}$$

According to the equation, in a uniform magnetic field $\mathbf{B}$, a moving charged particle will follow a curved path. In the case of a magnetic lens, there is no electric field and the $\mathbf{E}$ term is dropped, leaving $\mathbf{F} = -e(\mathbf{v} \times \mathbf{B})$. As an electron travels down the column, it first encounters the horizontal component (B_r) of the magnetic field. This causes the electron to begin a rotation about the axis along a helical path. With a nonzero angular velocity about the axis, the electron begins to feel an inward force from the vertical field component (B_z) that draws it toward the axis. Finally, as the electron emerges through the bottom horizontal field component, it receives a reverse angular impulse that cancels its rotation about the axis. The constricted beam then continues to narrow toward the focal point.

The lens equation (relevant here since the fields are relatively weak compared to the velocities of the electrons) is

$$\frac{1}{f} = \frac{1}{p} + \frac{1}{q},\tag{1.7}$$

where f is the focal length of the lens, and p and q the distances between the original source and the lens and the distance from lens to image, respectively. The diameter of the source at p is demagnified by a factor of $M = q/p$ through the lens (i.e., the beam diameter is reduced by this factor with no loss in intensity). Integrating the Lorentz force through the lens gives a focal length of

$$\frac{1}{f} = \frac{e}{8m_e V} \int_{-\infty}^{\infty} B_z^2 \, dz,\tag{1.8}$$

which depends only on the field strength along the axis.

The electrostatic lenses found in ion beam systems operate in a simpler manner than magnetic lenses, although the underlying principle is analogous. Figure 1.7 depicts a three-electrode design. Positively charged particles enter the lens from the left and encounter an electrical field formed by the large voltage difference between the first two electrodes. The ions follow the field lines, receiving an impulse toward the optical axis and a boost in velocity by the increasingly negative field. As the ions pass the second electrode, they are pulled outward, but since they are now closer to the axis and have more momentum, the change in direction is less than from the first impulse. Upon

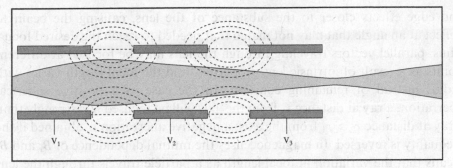

Figure 1.7 Schematic of a three-electrode electrostatic lens in an ion beam column. Field lines generated by the voltages on the electrodes are shown as dotted lines, and the ion trajectories traveling from left to right are in gray.

passing the second field, the beam is once again constricted, but this time the ions are decelerated by the increasingly positive field to near their original velocity [14]. Just as in the magnetic lens, the charged particle beam exits and continues to narrow toward its focal point downstream in the column. It is worth noting that, if the potential difference between the center electrode and end electrodes were reversed, it would create not a converging but a diverging lens system (commonly used in the transmission electron microscope, but not actually used in any SEM or FIB optics).

1.4.2 Aberrations

Most electron columns have more than one lens, usually with apertures along the beam zone axis to reduce aberrations. A small imperfection or a fundamental limit in one lens will be multiplied by the number of lenses and cause a significant loss in resolution. It is known that such optical aberrations in the electrostatic lenses of ion columns are also quite severe. As a result, charged particle lenses are incapable of completely faithful imaging, and to optimize their performance, it is critical to operate them in the paraxial mode. (This means that, for high-resolution images, the angle of the trajectory of particles with respect to the lens axis must be kept extremely small: less than 10 mrad.) The aberrations can be categorized into three basic groups, as follows: spherical aberration, chromatic aberration, and astigmatism.

Spherical aberration is the effect of a nonlinear dependence of beam deflection on radius within a lens. In an ideal lens, the larger the radius from the axis at which a particle is found, the sharper the angle of deflection: a linear relationship that holds true from the center of the lens out to the edges. In a real lens, however, the fields necessarily experience a degree of fringing and edge effects closer to the substance of the lens, causing the beam to deflect at an angle that may not match that needed to reach the desired focus. Thus, parallel vectors traveling through the lens may be focused at different points as a result of intrinsic lens properties, and the focal length varies with radius instead of remaining constant. In the case of a positive spherical aberration, a ray at distance r_1 from the axis will be focused less strongly than a ray at distance $r_2 > r_1$ from the axis; a negative aberration is obtained if the inequality is reversed. In magnetic lenses the mutual dependence of B_r and B_z means that the variation in focal length as a particle travels through the lens further from the center is great enough to cause significant spherical aberration. The phenomenon can be reduced by careful lens design, or with an aperture in place to reduce the aberration; the diameter of the beam resulting from a spherical aberration becomes

$$d_s = 0.5 C_s \alpha^3, \tag{1.9}$$

where C_s is the spherical aberration coefficient, and α is the aperture angle, a function of the focal length and the diameter of the lens field.

Chromatic aberration is the chief limitation on the focusing ability of electron and ion optics. This problem occurs when a spread of energies ΔE is present in the beam. Since the lenses rely on the interaction between fields, charges, and velocities, according to the Lorentz force law, a slight difference in velocity owing to a difference in initial energy will result in a different focal length for a particle. The spread in energies thus translates to a spread in focus. Particles with higher-than-expected energies will be focused beyond the surface of the sample, while particles with lower-than-expected energies will pass through focus above the sample and spread out again, producing a larger spot size than desired. A point object under this condition will be imaged as a disc, the radius of which is proportional to the aperture angle and to the magnitude of the relative energy spread $\Delta E/E$. The chromatic aberration (so called because a similar focal length spread is observed for different energies, and thus different colors, of light) is a fundamental property of the source, and represents the most serious practical limitation on the performance of current ion beam designs. In the approximation implicit in the above equations, it is assumed that all electrons or ions have the same kinetic energy due to having gone through the same accelerating voltage. With an energy spread of ΔE, however, the new focused beam diameter becomes

$$d_c = C_c \frac{\Delta E}{E} \alpha, \tag{1.10}$$

where C_c is the chromatic aberration coefficient of the lens.

The last issue, astigmatism, is an effect of asymmetry in the focusing field, whereby the cross section of the beam in a given plane is not circular, but rather ellipsoid, causing the defocused beam to appear elliptical rather than circular. The astigmatism increases the beam diameter to

$$d_A = \Delta f_A \alpha, \tag{1.11}$$

and the image of a point is mapped onto two perpendicular lines lying in front of and behind the image plane, known as the tangential and sagittal planes, respectively. The distance of these planes from the image plane varies as the square of the distance of the point object from the lens axis and as the aperture angle. This problem can be corrected by using a set of alignment coils near the beam's exit to reshape the beam.

It should be noted that while chromatic and spherical aberrations are properties of the lenses and electron source and can only be corrected by apertures and the addition of other lenses with opposing aberrations, astigmatism can be dynamically corrected by the use of stigmators, a set of magnetic coils with deliberately asymmetric fields that can be used to move the two focal lines until the distortion is corrected. The total beam diameter resulting from the combination of these aberrations is given by

$$d_p^2 = d_0^2 M^2 + d_A^2 + d_s^2 + d_c^2, \tag{1.12}$$

where d_0 is the initial source beam diameter. This calculation is known as the quadrature method.

The first major difference between electron and ion beam columns is in the emission source. As discussed previously, instead of a heated tungsten filament or a field-emission region above a tip, the ion emitter is a liquid surface drawn by electrostatic fields into a sharpened cone about 5 nm wide at the apex. Understanding the balance of forces needed to achieve this shape is somewhat involved, so we will take it as given and assume a beam diameter d_0 as in the electron source with similar parameters for kinetic energy (we will assume that the ions have one electron charge's worth of ionization charge, and are thus given a kinetic energy equivalent to that of the electrons). It is important to note that metal ions used in a focused ion beam system have much higher mass than electrons, meaning that they travel slower at the same kinetic energy and require more force to divert. As discussed above, this means that their lens apparatus needs to use much larger fields than in the electron column. Thus the fact that, in general, electrostatic fields are used for ion lenses, rather than magnetic fields. This comes from the practical problems of generating a large enough magnetic field in the confined space of the lens to focus the ions and the issue of the radius of the spiral path that the ions will follow; the radius of this path will be much larger than that of electrons in a comparable field due to the ions' higher inertia. In order to produce the necessary forces, carefully shaped electrodes with precisely controlled potentials are used, generating electric fields that focus the slower-moving and heavier ions, as seen in Figure 1.7. The ions undergo a slight acceleration in the process, which must be accounted for when considering the impact energy of the beam on the sample.

The high potentials used to generate the electric fields take advantage of the Lorentz force's e-field term; this term is not a cross-product and so the ions do not travel in a spiral path, making them somewhat easier to control. High electrostatic fields are easier to create than magnetic fields, and in general make for a more stable lens. Other than replacing the magnetic term in the focal length equation with the radial component of the electric field (the only component that exists, if edge effects are disregarded, which is a reasonable approximation for such strong fields and high energies), there is functionally no difference between ion optics and electron optics, as far as the physics of operation goes. Since the aberrations depend on the same factors, it should be noted that many ion systems contain stigmator coils even though astigmatism is not as much a concern in ion beams. In fact, as the half-angle α of the beam arriving at the final focal spot diminishes to < 1 mrad, the $d_0 M$ virtual source diameter term dominates in the quadrature equation above. Thus the fundamental limit for present FIB technology, in the minimum beam diameter and the energy spread resulting from mutual repulsion and space charge limitations, overshadows other aberration effects to the point that improvement in any of them will not improve ion beam resolution significantly.

One more point must be made about ion beam systems. In an electron optics column, the apertures serve by and large to correct aberrations. High electron beam intensities and smaller beam size are desirable as they lead to better imaging; the amount of damage an electron beam does to a conductive sample is negligible for most applications. Ion beams, however, must have their beam current carefully controlled, as they constantly damage and change the surface of a sample. As such, apertures serve as current-limiting devices, reducing the ion current to whatever level the user deems appropriate, rather than simply screening out errant ions from the focusing system.

1.5 Detection of electron and ion signals

The differences between ions and electrons can be extended to their respective machines, the FIB system and the scanning electron microscope (SEM).

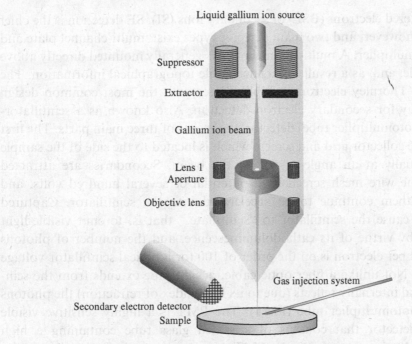

Liquid gallium ion source

Suppressor

Extractor

Gallium ion beam

Lens 1
Aperture

Objective lens

Gas injection system

Secondary electron detector
Sample

Figure 1.8 A schematic diagram illustrates the major components in an FIB system.

Though they are designed and work very similarly, the FIB's use of gallium ions from a liquid-metal ion source rather than electrons provides functionality and applicability different from that of the SEM. It uses the focused beam of gallium ions and rasters the surface of the material of interest. The small amount of material sputtered from the surface during this process may form secondary ions and electrons which are then collected and analyzed as signals to form an image on a screen. This allows high-magnification microscopy with the FIB system.

In both machines a source emits charged particles that are focused into a beam and rastered over small areas of the sample using deflection plates or scan coils. The SEM uses magnetic lenses to focus its beam of electrons; however, since ions are much heavier and, therefore, much slower with a lower corresponding Lorenz force, magnetic lenses are less effective. The FIB system is instead equipped with electrostatic lenses (shown schematically in Figure 1.8), which have proven to be much more effective.

Both the SEM and the FIB form high resolution images by collecting the secondary electrons (SE) that are emitted from the interactions between the beam and the surface atoms, although images may also be formed from backscattered electrons (BSE) or secondary ions (SI). SE detection is the chief method, however, and two main detector types exist: multi-channel plate and electron multiplier. A multi-channel plate is generally mounted directly above the sample, and, as a result, offers negligible topographical information. The Everhart–Thornley electron multiplier detector is the most common design used today for secondary electron detection. Also known as a scintillator-PMT (photomultiplier tube) detector, it consists of three main parts. The first part is the collector grid and screen, which is located to the side of the sample stage, usually at an angle of 45° to the beam. Secondaries are attracted toward the wire mesh screen by a potential of several hundred volts, and most of them continue to be accelerated into the scintillator. Captured electrons cause the scintillator to "scintillate," that is, to emit visible-light photons by virtue of its cathodoluminescence, and the number of photons generated per electron is on the order of 100 for a typical scintillator voltage of 10 kV. Not unlike a fiber optic cable, a light pipe extends from the scintillator and internally reflects (due to its high index of refraction) the photons to the photomultiplier tube (PMT). The PMT is a highly sensitive visible photon detector that consists of a sealed glass tube containing a high vacuum. At the entrance to the PMT, the incoming photons strike a low-work-function material that comprises the photocathode, liberating valence electrons that are subsequently accelerated as photoelectrons toward the first of a series of (usually) eight dynode electrodes. Each dynode is biased positively with respect to the photocathode, and each of them is also biased 100–200 V positively with respect to the preceding one. The photoelectrons generate secondary electrons at the first dynode, and these secondaries are then amplified by a factor of about 10^6 after they have completed striking the remaining dynodes. Recalling that each SE originally produced at the specimen surface generated about 100 photoelectrons, the overall magnification of the scintillator-PMT detector can be as high as 10^8, depending on the applied dynode voltage. Thus, although it may seem unnecessarily complicated to convert SEM secondary electrons into photons, then into photoelectrons, and finally back into secondaries, the high amplification and low electronic noise – versus a simple metal plate to absorb the electrons – in fact fully justifies the system. Backscattered electrons can also be detected by a scintillator-PMT if the bias on the first grid is made negative instead of positive, therefore repelling lower-energy secondaries but not BSE, which retain most of their kinetic energy after impact.

Raised areas of the sample (hills) produce more collectable secondary electrons, while depressed areas (valleys) produce less, thereby creating a contrast that is interpreted by the machine and at the same time intuitively understood by the operator as light and shadow. Also, to increase the secondary electron yield, the whole sample is often tilted away from the horizontal plane and toward the detector in order to increase the SE signal without interfering with the contrast-based topography. A viewing monitor synched to the scan coils controls the beam so that as it scans across the sample surface, the image of the sample is reproduced on the screen, with a magnification inversely proportional to the area of scan.

Images obtained from secondary ions can be below 10 nm resolution and show topographic and materials information complementary to that obtained from an SEM image. Although material contrast arising from differences in specimen chemistry can be significant in FIB secondary electron images, it is most readily observed in secondary ion images, where it is often the dominant effect. While SE images provide uniformly good depth of field, SI images reflect more selective depths that depend on different materials and sample structure. Information about the grain size and crystal orientation can also be obtained using an FIB because of the dependency of the ion–atom interaction upon the crystal grain orientation; this is known as channeling contrast. Thus, from a materials science standpoint, secondary ion imaging is an invaluable capability. SI imaging is also superior for insulating materials when used in conjunction with a charge neutralizer (electron flood gun). Although ions move slower than electrons, they still move faster than the image can be collected. It is important to note that, since the sample is continually sputtering during the FIB imaging process, small beam currents (< 100 pA) are advisable to minimize sample erosion.

Detection methods for secondary ions fall into one of two categories: microprobe mode and microscope (or direct) mode. The microprobe mode is essentially analogous to the process used in SEMs [15]: the primary ion beam is rastered while the SI signal is synchronously detected. The key difference is that the particle detector grid is biased to a highly negative voltage to repel both secondary and backscattered electrons and to attract positive ions, which are subsequently amplified. Typically, SI microprobe images are of the "total detected positive ion" type, in which virtually all positive ions are collected and amplified regardless of mass. A more sophisticated recent development is mass-resolved ion imaging, which is present on FIB systems that are equipped with secondary ion mass spectrometers (SIMS); this allows elemental analysis of the sample combined with imaging [16]. In the alternative microscope or direct mode, ion image formation is nonraster based and relies on electrostatic lenses in the secondary ion column. Several types of position sensitive detectors can currently be used with this process. The most common of these is a microchannel plate connected to a phosphor screen,

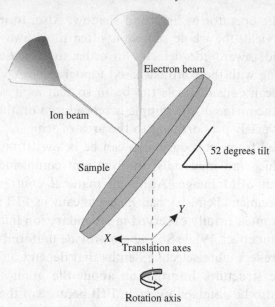

Figure 1.9 Schematic of the two-beam system, in which both electron and ion beams are co-focused at the coincidence point on the sample surface.

and the resulting image is captured by a highly sensitive CCD. Alternatively, a direct ion-imaging detector such as the resistive anode encoder (RAE) can be used to capture the ion images. The resolution in microscope mode imaging is limited by the electric field strengths of the electrostatic lenses to about 1 μm.

1.6 The two-beam system

When we combine both SEM and FIB into one system, called a two-beam system, the ion beam and electron beam are placed in fixed positions, with an angle of 45–52° between two beams for the best performance as illustrated in Figure 1.9. The two beams are co-focused at what is called the "coincidence point," an optimized position for the majority of operations taking place within the machine, with a working distance of typically a few millimeters. The two-beam system allows SEM imaging and FIB sample modification without having to move the sample. In addition, the stage can be tilted, allowing changes in the sample-beam orientation. Similarly to the FIB system, integrated software with a single user interface controls the two-beam.

The two-beam system provides new advantages that simplify as well as improve nanoscale imaging, analysis, and fabrication as detailed in the following chapters. One such advantage is in imaging. Whereas FIB imaging has high contrast abilities but can cause damage to the sample, SEM images have relatively lower contrast, but provide a higher resolution and do not damage the sample. The result is a more complete set of data. A study done at Portland State University showed that using this combination of beams could provide a more comprehensive imaging and characterization of carbon nanotubes [17]. Also the reconstruction of three-dimensional structure and chemistry of a sample can be simplified using the two-beam system to interpolate two-dimensional SEM and FIB images and ionassisted SIMS chemical maps of layers that have been exposed using the milling feature of the ion beam [18].

The charging effects of ion and electron beams are worth consideration. A typical two-beam system setup will have the ion beam working in tandem with a scanning electron microscope, placing both types of beams at the operator's disposal. One of the great challenges in the use of electron beam for imaging is that if the sample is not fairly conductive, the electrons from the beam will build up charge in the material. This charge will then distort both the incoming beam and the outgoing secondary and backscattered electrons, producing electron artefacts and distortions in the data. While the ion beam does to some degree increase local charge at the point of impact, the nature of the ions themselves, generally metals, causes the imbalance to be quickly rectified. This can be exploited to assist in electron-beam imaging of the sample, by using the ion beam at a low setting over a large area to reduce local charging effects and increase surface conductivity [19].

The complementary nature of the negatively charged electrons and the positively charged ions also eliminates the charging problem found in the single beam FIB. The accumulation of charge would hurt the resolution of the image, but because of the availability of both charges, this is no longer a problem with the two-beam system.

Not only does the combined system produce a better and more extensive collection of data, but it also allows for precise monitoring of FIB operation through the SEM. By using the slice-and-view technique to observe the progress of the ion beam cross section, the operator can stop the milling process at a precise point in order to obtain local information. Also, the two-beam system allows for the use of both the ion beam and the electron beam simultaneously without interference, doing away with the necessity to switch back and forth. The sample can be imaged in real time with the SEM while the FIB is in use, providing for higher levels of accuracy in the creation of cross sections [20]. The SEM's damage-free imaging is especially useful in the creation of samples for the TEM, since using an FIB alone causes sample damage while imaging the process. It can also be very useful in the localization of integrated circuit failures, so that more damage is not caused to the sample than is necessary, and the system can combine ion milling, deposition, and SEM imaging to characterize the failures [21].

The two-beam system also improves the deposition of metal or insulating layers. In the case of insulating layers, ion beams often leave the layer with poor insulating properties due to the incorporation of gallium ions in the deposition; however, in the two-beam system, the SEM beam can be used to induce deposition, ensuring a high insulating quality. One such case is that of SiO_x, where the resistance of the layer is two orders of magnitude higher than when an ion beam is used for the deposition [22].

One more advantage of using the two-beam system is the creation of smaller diameter holes. Using the FIB system, a hole can only be accurately milled to a diameter of 10 nm without having material along the sides partially fill in the hole. When the two-beam system is used, however, the nanoscale holes can first be milled using the ion beam and then gently filled using electron beam deposition so that the diameter of these holes can be further reduced by 50%. The ability to have smaller diameter holes has potential applications for single molecule studies, DNA sequencing, and ultra-high-resolution single atom doping.

Yet another important application of the two-beam FIB system is TEM sample preparation. The FIB completes electron transparent samples by thinning out a region of a bulk material, making samples as thin as possible in significantly less time than older methods of sample preparation. When an SEM is used in addition to an FIB the basic techniques of lift-out and micro-pillar sampling are vastly improved.

In addition to machining and TEM sample preparation, we also look into the many imaging capabilities of the FIB in the two-beam system. Whereas electrons in an SEM beam generate far less damage to the material and provide much better resolution, the ion beam of the FIB offers much better sensitivity to details such as crystal orientation and grain structure as well as better contrast. The two-beam system is then an exceptionally practical machine, combining the crisp nondestructive imaging of the scanning electron microscope with the milling capabilities of the focused ion beam. Three-dimensional material information is also available through the two-beam system by shaving off thin layers and imaging them with the SEM, obtaining a series of useful two-dimensional illustrations. Graphs and images provided only by the combined SEM and FIB systems can then be easily interpolated with the two-dimensional data. If secondary ion mass spectrometry (SIMS) is then performed to yield the elemental composition of the material, the set of data can be complete. In summary, the two-beam FIB has immense relevance as one of the most important tools in the study of nanotechnology today.

Acknowledgements

The author is grateful to his students Austin Akey, Alex Epstein, and Franz Sauer, who have partially contributed in preparation of the manuscript, and to the National Science Foundation-MRSEC program and New Jersey Commission of Science and Technology for their support.

References

[1] V. E. Krohn. *Progr. Astronaut. Rocketry*, **5** (1961), 73–80.
[2] V. E. Krohn and G. R. Ringo. *Appl. Phys. Lett.*, **27** (1975), 479–81.
[3] J. Orloff, M. Utlaut and L. Swanson. *High Resolution Focused Ion Beams: FIB and its Applications*, (New York: Kluwer Academic/Plenum Publishers, 2003), pp. 21–77.
[4] J. Zhou. *Handbook of Microscopy for Nanotechnology*, ed. N. Yao and Z. L. Wang, (New York: Springer/Kluwer Academic Publishers, 2005), pp. 287–321.
[5] J. Meingailis. *J. Vac. Sci. Technol.* B, 5:2 (1987), 469–95.
[6] R. Gerlach and M. Utlaut. *Proc. SPIE Int. Soc. Opt. Eng.*, **4510** (2001), 96–105.
[7] L. Reimer. *Scanning Electron Microscopy* (Berlin: Springer-Verlag, 1998), 283–6.
[8] K. Van Doorselaer, M. Van den Reeck, L. Van den Bempt *et al. Proc. 19th Int. Symp. Testing and Failure Analysis* (1993), pp. 405–14.
[9] J. Orloff, M. Utlaut and L. Swanson. *High Resolution Focused Ion Beams: FIB and its Applications*, (New York: Kluwer Academic/Plenum Publishers, 2003), pp. 147–52.
[10] J. M. Lafferty. *J. Appl. Phys.*, **22** (1951), 299–309.
[11] M. Futamoto, M. Nakazawa and U. Kawabe. *Sur. Sci.* **100** (1980), 470–80.
[12] D. B. Williams and C. B. Carter. *Transmission Electron Microscopy: A Textbook for Materials Science*, (New York: Plenum Press, 1996), 76–7.
[13] H. Arimoto, T. Morita, E. Miyauchi and H. Hashimoto. *Jpn. J. Appl. Phys.* **25** (1986), L507–9.
[14] R. P. Feynman, R. B. Leighton and M. Sands. *The Feynman Lectures on Physics*, **2** (1964), 29–42.
[15] W. A. Lamberti. *Handbook of Microscopy for Nanotechnology*, ed. N. Yao and Z. L. Wang (New York: Springer/Kluwer Academic Publishers, 2005), pp. 208–9.
[16] M. W. Phaneuf. *Introduction to Focused Ion Beams: Instrumentation, Theory, Techniques and Practice*, ed. L. A. Giannuzzi and F. A. Stevie (New York: Springer, 2005), pp. 145–72.

[17] L. F. Dong, J. Jiao, D. W. Tuggle and S. Foxley. *Microsc. Microanal.*, **7S2** (2001), 398–9.
[18] R. Hull, D. Dunn and A. Kubis. *Microsc. Microanal.*, **7S2** (2001), 34–5.
[19] P. Gnauck, U. Zeile, W. Rau, G. Benner and A. Orchowski. *Microsc. Microanal.*, **9S2** (2003), 872–3.
[20] L. A. Giannuzzi and F. A. Stevie. *Micron*, **30** (1999), 197–204.
[21] J. M. Soden and R. E. Anderson. *Proc. IEEE*, **81**, 5 (1993), 703–15.
[22] S. Lipp, L. Frey, C. Lehrer *et al. J. Vac. Sci. Technol. B*, **14** (1996), 3920–3.

2

Interaction of ions with matter

NOBUTSUGU IMANISHI
Kyoto University

2.1 Introduction

When a beam of energetic particles enters a solid, several processes are initiated in the area of interaction. A fraction of the particles are back-scattered from the surface layers, whilst the others are slowed down in the solid. The collision induces secondary processes such as recoil and sputtering of constituent atoms, defect formation, electron excitation and emission, and photon emission. Thermal and radiation-induced diffusion contributes to various phenomena of mixing of constituent elements, phase transformation, amorphization, crystallization, track formation, permanent damage, and so on. Ion implantation and sputtering changes the surface morphology; craters, facets, grooves, ridges, and pyramids and/or blistering, exfoliation, and a spongy surface may develop. All those processes are interrelated in a complicated way and several processes have to be included for the understanding of individual phenomena. Therefore, it is necessary to quantitatively understand the experimental observations and to have stringent design abilities for sophisticated applications of these versatile processes in the field of nanotechnology aiming at material modification, deposition, implantation, erosion, nano-fabrication, surface analysis, and so on.

This chapter is composed of basic processes and outline of theoretical models, ion implantation and defect formation, sputtering, and surface morphology. It is focused on the recent experimental findings in the field of interaction of ions with matter and theoretical models including various simulation codes explaining the complicated experimental phenomena.

Focused Ion Beam Systems: Basics and Applications, ed. N. Yao.
Published by Cambridge University Press. © Cambridge University Press 2007.

2.2 Basic processes and outline of theoretical models

2.2.1 General remarks

An energetic ion incident on a solid sequentially collides with constituent atoms. These collisions basically contain a complicated many-body problem because of the atomic composition of a nucleus and many electrons. Fortunately in the interaction of an ion with an atom, the collision between the ion and the nucleus can be treated separately from that of the ion and the electrons because of the large mass difference between the nucleus and the electron. The former collision is called an elastic or nuclear collision and the latter is an inelastic or electronic collision. Kinetic energy and momentum should be conserved during the nuclear collision by which the incident ion recoils the target atom and is scattered at the same time. The electronic collision results in excitation and ionization of the constituent electrons in the atom. When the kinetic energy of the incident ion is not high enough to go deep inside the atom, the nuclear charge is screened by the inner-shell electrons. In this case, the electrons should be included in the nuclear collision for taking into account the screening of the nuclear charge. This screened interaction potential has been one of the fundamental debates for the understanding of the nuclear collision process.

This subsection first treats interatomic potentials, binary scattering process, recoil energy, and scattering cross section concerning the nuclear collision process dominating in a low energy region. Then, nuclear and electronic energy losses are introduced based on several models. Finally, theoretical models and simulation codes which will be presented in this chapter are listed.

2.2.2 Interatomic potentials

In the atomic collision between two particles with charges $Z_1 e$ and $Z_2 e$, an accurate and versatile potential used frequently is the following Coulomb potential multiplied by the screening function Φ_{TF} deduced on the basis of the Thomas–Fermi atomic model [1]:

$$V(r) = \frac{Z_1 Z_2 e^2}{r} \Phi_{TF}(x), \quad (e^2 = 1.44 \text{ eVnm}) \tag{2.1}$$

where r is a distance between the two particles and x is given by $x = r/a$. The screening length a is obtained by

$$a = \left(\frac{3\pi}{4}\right)^{2/3} \left(\frac{\hbar^2}{2m_e e^2 Z^{1/3}}\right) \approx \frac{0.8853 a_B}{Z^{1/3}}, \tag{2.2}$$

where a_B is the Bohr radius and Z is obtained from the Firsov approximation in the region $r < 0.1$ nm [2],

$$Z = \left(Z_1^{1/2} + Z_2^{1/2}\right)^2. \tag{2.3}$$

The Thomas–Fermi screening function Φ_{TF} is calculated from the following equation:

$$\frac{d^2\Phi_{TF}(x)}{dx^2} = \frac{\Phi_{TF}^{3/2}(x)}{x^{1/2}}. \tag{2.4}$$

Examples of approximations of the Thomas–Fermi screened potential are the Moliere potential (applied in the range of $0 < x < 6$) [3]:

$$V(r) = \frac{Z_1 Z_2 e^2}{r} [0.35 \exp(-0.3x) + 0.55 \exp(-1.2x) + 0.10 \exp(-6.0x)], \tag{2.5}$$

and from Lindhard *et al.* [4]:

$$V(r) = \frac{Z_1 Z_2 e^2}{r} \frac{\xi_v}{n} x^{n-1}, \qquad 1 \le n \;\; \infty, \tag{2.6}$$

where $\xi_v = 2/2.7183 \times 0.8853, x = r/a_L$ and $a_L = 0.8853 a_B (Z_1^{2/3} + Z_2^{2/3})^{-1/2}$. Ziegler *et al.* proposed a universal screening function which accurately predicts the interatomic potential between atoms, as given by

$$\begin{aligned} V(r) = \frac{Z_1 Z_2 e^2}{r} &[0.1818 \exp(-3.2x) + 0.5099 \exp(-0.9423x) \\ &+ 0.2802 \exp(-0.4029x) + 0.02817 \exp(-0.2016x)], \end{aligned} \tag{2.7}$$

where $x = r/a_U$, and $a_U = 0.8854 a_B / (Z_1^{0.23} + Z_2^{0.23})$ [5].

2.2.3 Binary scattering and recoil

Collision of the atoms with masses M_1 and M_2 and respective charges of Z_1 and Z_2 can be treated in the center-of-mass system as the classical motion of the particle with the mass of $\mu = M_1 M_2/(M_1 + M_2)$ in the potential of $V(r)$. The following scattering integral can be easily obtained from the equations of motion in the framework of the energy and angular momentum conservations:

$$\zeta = \int \frac{L dr}{r^2 \sqrt{2\mu(E - V(r)) - (L^2/r^2)}} + \text{const}, \tag{2.8}$$

where r is the distance and ζ is the angle displayed in the polar coordinate system, respectively. E and L are the energy of relative motion and the angular momentum respectively and are given by $E = \mu v_0^2/2 = M_2 E_0/(M_1 + M_2)$ and $L = \mu r^2 d\varsigma/dt = p\sqrt{2\mu E}$. E_0 is the kinetic energy of the incident particle. The scattering angle χ is obtained as a function of impact parameter p from the following integral equation:

$$\chi = \pi - 2 \int_{r_{min}}^{\infty} \frac{p \, dr}{r\sqrt{\left(1 - \frac{V(r)}{E}\right) r^2 - p^2}}. \tag{2.9}$$

In the cases of a hard sphere with a radius of r_h, the Coulomb potentials are given by $V(r) = Z_1 Z_2 e^2/r$ and $V(r) = Z_1 Z_2 e^2/r^2$ and the scattering angles are $\chi = \pi - 2\arcsin(p/r_h)$, $\chi = \pi - 2\arctan(2pE/Z_1 Z_2 e^2)$, and $\chi = \pi[1 - (1 + (Z_1 Z_2 e^2/p^2 E))^{-1/2}]$, respectively. The energies E_1 and E_2 of the scattered and recoiled particles are

$$E_1 = \left(\frac{(M_1 - M_2)^2 + 4M_1 M_2 \cos^2(\chi/2)}{(M_1 + M_2)^2} \right) E_0, \tag{2.10}$$

and

$$E_2 = \frac{4M_1 M_2 E_0 \sin^2(\chi/2)}{(M_1 + M_2)^2}, \tag{2.11}$$

respectively.

2.2.4 Cross section

The cross section of the particle scattered to the polar angle of χ is obtained from the following relation:

$$\sigma(\chi) = \frac{p(\chi)}{\sin \chi} \left| \frac{dp}{d\chi} \right|. \tag{2.12}$$

In the cases of the $V = Z_1 Z_2 e^2/r$ and $V = Z_1 Z_2 e^2/r^2$ potentials, the results are

$$\sigma(\chi) = \left(\frac{Z_1 Z_2 e^2}{4E} \right)^2 \sin^{-4}\left(\frac{\chi}{2}\right), \tag{2.13}$$

and

$$\sigma(\chi) = \frac{(\pi^2 Z_1 Z_2 e^2/E)(\pi - \chi)}{\chi^2 (2\pi - \chi)^2 \sin \chi}, \tag{2.14}$$

respectively.

Lindhard *et al.* approximately obtained the following differential scattering cross section using the Thomas–Fermi potential applicable in a low energy region [4]:

$$d\sigma = \pi a^2 \frac{dt}{2t^{3/2}} f(t^{1/2}),$$ (2.15)

where $t = \varepsilon^2 \sin^2(\chi/2)$. The reduced energy ε is the ratio between the screening radius a and the collision diameter b and is given by

$$\varepsilon = \frac{a}{b} = \frac{\mu v_0^2}{Z_1 Z_2 e^2/a} = \frac{M_2 E_0}{(M_1 + M_2) Z_1 Z_2 e^2/a}.$$ (2.16)

The function $f(t^{1/2})$ is approximated as

$$f(t^{1/2}) = \frac{\lambda t^{1/6}}{\left[1 + (2\lambda t^{2/3})^{2/3}\right]^{3/2}},$$ (2.17)

where $\lambda = 1.309$.

The cross sections in the laboratory system can be obtained from those in the center-of-mass system as

$$\sigma(\theta_1) = \frac{\left[1 + \left(\frac{M_1}{M_2}\right)^2 + 2\left(\frac{M_1}{M_2}\right)\cos\chi\right]^{3/2}}{1 + \left(\frac{M_1}{M_2}\right)\cos\chi}\sigma(\chi),$$ (2.18)

and

$$\sigma(\theta_2) = 4\sin\left(\frac{\chi}{2}\right)\sigma(\chi),$$ (2.19)

where θ_1 and θ_2 are the scattering and recoiled angles in the laboratory system, respectively. The energies E_1 and E_2 of the scattered and recoiled particles are respectively given by

$$E_1 = \left(\frac{M_1 \cos\theta_1 \pm \sqrt{M_1^2 - M_2^2 \sin^2\theta_1}}{M_1 + M_2}\right)^2 E_0,$$ (2.20)

and

$$E_2 = \frac{4M_1 M_2 E_0 \cos^2\theta_2}{(M_1 + M_2)^2}.$$ (2.21)

2.2.5 Energy loss

Ions passing through matter collide with constituent atoms and lose their kinetic energies by electronic excitation and ionization and by kinetic collisions. The energy loss by the electron-related collision is called electronic or inelastic energy loss and the kinetic-collision induced energy loss is called nuclear or elastic energy loss. The former and the latter respectively work mainly at a high and a low velocity range. The term of stopping power is also used instead of the energy loss. At the high velocity range the ions lose their energies by the process of photon generation and by inducing nuclear reactions. In the following the nuclear and electronic energy losses will be presented.

The most familiar energy loss formula is the following Bethe–Bloch equation, which is derived on the basis of the plane wave Born approximation applicable at the high velocity range where the screening of the nuclear charge and the nuclear energy loss need not be considered:

$$\frac{\mathrm{d}E}{N\mathrm{d}x} = \frac{4\pi Z_1^2 Z_2 e^4}{mv^2}\left\{\ln\frac{2mv^2}{I} - \left[\ln\left(1 - \left(\frac{v}{c}\right)^2\right) + \left(\frac{v}{c}\right)^2 + \frac{C}{Z_2} + \frac{\delta}{2}\right]\right\} \quad \text{eV-cm}^2,$$

(2.22)

where $\mathrm{d}E/\mathrm{d}x$ is the energy loss per unit length, N the number of atoms in unit volume, m the electron mass, v the ion velocity, I a mean energy of excitation and ionization, c the light velocity, C/Z_2 a shell correction term, and $\delta/2$ a density effect [6,7]. The shell correction term is important in a relatively low velocity range where the ion cannot ionize nor excite inner shell electrons. The terms $(v/c)^2$ and $\delta/2$ contribute in the relativistic velocity range.

In a keV energy range, the velocities of ions, especially of heavy ions are very low and the Bethe–Bloch formula is not applicable any more. The ions capture electrons from the constituent atoms and as a result the screening of the nuclear charge becomes important. The nuclear collision competes with and further predominates over the electronic collision with decreasing ion velocity. Lindhard, Scharff, and Schiott (LSS) derived the universal electronic and nuclear energy loss formulae based on the Thomas–Fermi atomic model at a low velocity range [8]. The electronic energy loss is given by

$$S_e = \frac{\mathrm{d}E}{N\mathrm{d}x} = \eta\frac{8\pi Z_1 Z_2 e^2 a_B v}{NZv_B}, \quad (v < v_B Z^{2/3}),$$

(2.23)

where $\eta \approx Z_1^{1/6}$, and a_B and v_B are the Bohr radius and velocity, respectively. The formula displayed by a reduced energy and length is

$$\left(\frac{d\varepsilon}{d\rho}\right)_e = k\sqrt{\varepsilon}, \tag{2.24}$$

where ε is the reduced energy given in (2.16); the length ρ is given by $\rho = xNM_2 4\pi a_L^2 M_1/(M_1 + M_2)^2$; a_L is the Lindhard screening radius, $a_L = 0.8853 a_B (Z_1^{2/3} + Z_2^{2/3})^{-1/2}$; and the coefficient k depends on Z_1, Z_2, M_1, and M_2 and is given by

$$k = \frac{0.0793 Z_1^{2/3} Z_2^{1/2} (M_1 + M_2)^{3/2}}{\left(Z_1^{2/3} + Z_2^{2/3}\right)^{3/4} M_1^{3/2} M_2^{1/2}}.$$

The nuclear collision is very important for the sputtering and defect creation processes. Lindhard *et al.* derived the differential scattering cross section formula of (2.15) on the basis of the Thomas–Fermi atomic model:

$$d\sigma = \pi a^2 \frac{d\sqrt{t}}{t} f(\sqrt{t}), \tag{2.25}$$

where t is a reduced variable relating to the reduced energy ε and the scattering angle χ: $t = \varepsilon^2 \sin^2(\chi/2)$. Then, the nuclear energy loss is given by

$$\left(\frac{d\varepsilon}{d\rho}\right)_n = \frac{1}{\varepsilon} \int_0^\varepsilon f(\sqrt{t}) d\sqrt{t}. \tag{2.26}$$

As described above, the Bethe–Bloch and LSS equations are underlying energy loss formulae in high- and low-energy ranges, respectively. However, obtained experimental values of energy loss depend on incident ion and energy, and the target atom. Ziegler *et al.* have checked very carefully vast amounts of parameters necessary for obtaining the energy losses covering any incident ion–target atom combinations over an energy range between 1 keV/u and 2 GeV/u [5]. The versatile program named SRIM developed by Ziegler *et al.* is opened for public uses and covers the calculation of energy losses, ranges, range stragglings, damage distributions, sputtering yields, and so on [5].

2.2.6 Outline of theoretical models

Nanotechnology-aimed applications of phenomena induced by energetic ions in matter need more detailed and quantitative theoretical understanding of

Table 2.1 *Monte Carlo (MC) and molecular dynamics (MD) simulation codes discussed in this chapter.*

Section, subsection	Name of code
2.2 Basic processes and outline of theoretical models, 2.2.5 Energy loss	SRIM
2.3 Ion implantation and defect formation,	
2.3.2 Ion range and defect distributions	SRIM, MARLOWE
2.3.3 Ion implantation	DYN-TRIM
2.3.4 Defect formation	MD, MARLOWE DADOS
2.4 Sputtering, 2.4.2 Sputtering yield	
General scope	OKSANA, MD, MD & MC
Dose dependence of yield and preferential sputtering	(1) "Static" Monte Carlo (MC) (a) Crystalline targets MARLOWE, crystal-TRIM, COSIPO, ACOCT (b) Amorphous targets TRIM, ACAT, SASAMAL, (2) "Dynamic" Monte Carlo (MC) TRIDYN, dynamic-SASAMAL ACAT-DIFFUSE, (3) MD, thermal spike model, fluid dynamics
Mass distribution	MD/MC-CEM
Incidental cluster-size dependence of sputtering yield	MD
Sputtering of nonmetals	MD(AIREBO)
2.5 Surface morphology	stochastic nonlinear continuum equation based on Sigmund theory

the interaction of ions with matter. On the basis of the binary-collision approximation (BCA) Sigmund proposed a linear collision cascade model, which has been successfully applied in a low-energy density collision [9]. Sigmund's model is an analytical one. In order to include complex situations depending on individual ion–solid combinations many researchers developed simulation codes written by Monte Carlo method as listed in Table 2.1. Those models were developed in the framework of BCA and resultantly it is difficult to reproduce experimental phenomena appeared in a high-energy density collision. In this situation, various molecular dynamics simulation codes using sophisticated potentials can work well and are listed, too, in

Table 2.1. Results predicted by these models and their comparisons with experimental findings are presented in the respective related subsections.

2.3 Ion implantation and defect formation

2.3.1 General remarks

Ion implantation is a fundamental technique in the fabrication of semi-conductive devices and has been used for materials modification, too. Defect formation following ion implantation plays, in general, unpleasant roles in the above applications but sometimes can be used effectively. This section presents recent experimental and theoretical findings on ion implantation and defect formation.

For example, it was recently found that implanted atoms assemble by themselves into nanoclusters and nanocrystals under thermal annealing and/or radiation induced processes. The formation of these nanostructures is closely related to the nanotechnology. In the high-density energy deposition caused by very heavy ions and cluster ions, nonlinear effects on defect production were found. To understand these experimental facts, a combined code of Monte Carlo (MC), diffusion, and molecular dynamics (MD) simulation codes were developed. In addition, it is necessary to localize defects when one fabricates nanostructures with a focused ion beam. Diffusion lengths were measured and an idea to localize the defects was presented. This chapter includes the above experimental and theoretical results.

2.3.2 Ion range and defect distributions

An ion moving in matter loses its energy by the electronic and nuclear energy losses and finally stops after the travel of a certain path. The total length of the path is named as "range." The atomic collision has a statistical nature and, therefore, the range statistically distributes with a symmetric Gaussian distribution around the mean value with a deviation called "range straggling." The range and the range straggling projected on the incident particle direction are convenient for practical uses and the values are named "projected mean range" and "projected range straggling." The moving ion recoils one of the constituent atoms, which again strikes with and recoils another constituent atom. Then, many atomic defects are generated along the passage of the ion. This defect distribution is also important for the application of energetic ions to material modifications.

For amorphous targets, the range and the range straggling can be calculated from the energy loss formulae given in Section 2.2.5. The more reliable values of the (projected) range, range and defect distributions can be obtained by consulting the SRIM code developed by Ziegler *et al.* [5].

A Monte Carlo simulation code named MARLOWE, which was developed by Robinson *et al.* on the basis of the binary collision approximation, includes the time evolution of cascade collision and can treat the slowing down both in crystalline and amorphous targets [10]. The code is opened for researchers' use and provides the calculation of sizes and shapes of displacement cascades, sputtering, ion ranges, ion reflection, and so on.

In the case of crystalline targets, energy loss and range of an incoming ion depend strongly on an incident angle against the crystal axis and plane. In a channeling condition, that is, when the ion is incident along the crystal axis or plane, the ion passes through the crystal with large impact parameters and as a result the energy loss is very low compared with that for the amorphous target of the same constituents. Then, the ion goes through deep inside the crystal. Therefore, in order to avoid the trouble caused by the channeling phenomena, ion implantation is usually carried out at an incident-angle condition of off-axis or off-plane, or random incidence.

2.3.3 Ion implantation

The ion implantation technique has been a key element in the currently established Si technologies and in the fields of materials modification because of its well-known advantages, such as low-temperature processing, control of distribution and concentration of dopants, overcoming solubility limits which cannot be realized via conventional techniques. Ion implantation induces mixing of constituent elements, alloy formation, and clusterization and crystallization in nanometer size in matrix. These processes are very complex and depend not only on ion-related parameters of species, dose, and energy but on a high-temperature chemical reactivity between the implanted and matrix elements.

Related to the nanotechnology, formation processes of nanoclusters and nanocrystals are one of the most exciting and promising topics. These nanostructures are evidently produced by thermal annealing and/or radiation irradiation after ion implantation as schematically shown in Figure 2.1. In special conditions implanted atoms assemble at interfaces of different materials and form nanostructures by themselves. A comprehensive review was made by F. Gonella *et al.* on several evidences of clusterization of metal atoms implanted in glass matrices [11,12]. Figure 2.2 shows that the chemical bond of element implanted into silica glass changes depending on its concentration. The introducing criterion of cluster formation of implants in SiO_2

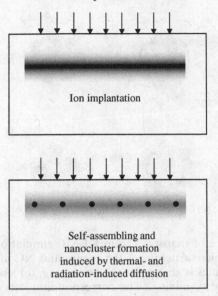

Figure 2.1 Schematic view graphs of nanocluster/nanocrystal formation.

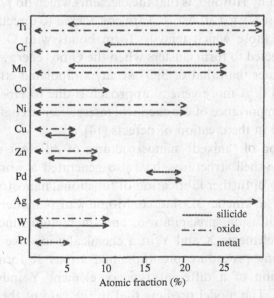

Figure 2.2 Chemical bonds of the element implanted in silica glass as a function of its concentration. (Reprinted with permission from *Nucl. Instr. Meth. B*, **166–167** (2000), 831–9. Copyright 2000 Elsevier Science B.V.)

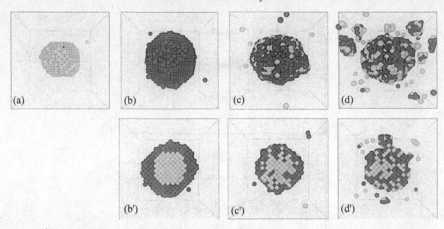

Figure 2.3 Sequence of snapshots from KLMC simulations. (a) The X-type "core" NC. The microstructure of the compound NC after the deposition of $N^Y = 2N^X$ impurities is shown for (b) no mixing, (c) low intensity mixing, and (d) high intensity mixing. The corresponding cuts through the NCs (perpendicular to the plane of view) are shown in (b'), (c'), and (d'). (Reprinted with permission from *Nucl. Instr. Meth. B*, **148** (1999), 104–9. Copyright 1999 Elsevier Science B.V.)

glasses, presented by Hosono, is that the elements which do not react with the matrix (for example, Ag, Cu, Au, ...) are considered to directly form metallic precipitates and those which tend to form bonds with O from the silica network are expected to form clusters when the Gibbs energy for metal oxide formation is greater than that of SiO_2 at an extrapolated effective temperature of 3000 K [13]. A more general approach is due to Hosono *et al.*, who pointed out the importance of the chemical interaction strength among atoms coming into play in the creation of defects [14].

The preparation of "mixed" nanostructures of different metals, or nanometer-scale core-shell structures, has also generated a worldwide concern for the possibility of further fabrication of functional nanostructures. Strobel *et al.* developed a kinetic 3D lattice Monte Carlo model applicable for simulating the diffusion, precipitation, and interaction kinetics of sequentially implanted elements X and Y in a chemically inactive matrix [15]. By applying annealing procedure pre-implanted atoms X form nanocrystals. Then, implantation of a different type of element Y induces collisional mixing of atoms. The model predicts that in the case of the following conditions, the elements X and Y form a core/shell structure, shown in Figure 2.3 without making a random distribution of X and Y: if (i) the solubility of Y exceeds the solubility of X in the matrix, if (ii) the atomic mass of Y is small, i.e. low collisional mixing, and (iii) for a rather low ion dose rate of Y and/or

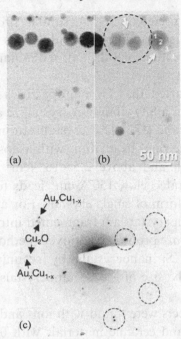

Figure 2.4 Bright-field TEM cross-sectional micrograph of the sample Au₃Cu₃H (annealed in $H_2(4\%)$-N_2 atmosphere at 1173 K for 1 h) (a) before and (b) after a thermal annealing in air at 1173 K for 15 min. The arrows indicate some of the two-fold clusters made of a Au-enriched alloy and Cu_2O: (c) SAED (selected area electron diffraction) diffraction pattern from the circular region of part (b). (Reprinted with permission from *Phys. Rev. Lett.*, **90** (2002), 085502. Copyright 2003 The American Physical Society.)

a high second implantation temperature, which favors a compensation of displacements by thermodynamically driven atomic movements.

Mattei *et al.* sequentially implanted Au and Cu ions into silica glass and annealed the sample in either oxygen or hydrogen atmosphere [16]. They found that the sequential alloying directly forms alloyed Au-Cu colloids caused by an enhanced diffusion of copper in small gold clusters during implantation. Subsequent annealing in hydrogen drives copper atoms dispersed in the matrix toward the clusters, increases the Au/Cu concentration toward the normal 1:1 ratio, and consequently induces the structural ordering of the alloy. Annealing of the formed Au-Cu alloy nanoclusters in oxygen atmosphere leads to separation of gold and copper by the possible reason that Cu is extracted from the nanoclusters by chemical interaction with the incoming oxygen (Figure 2.4). Another dealloying is caused by irradiation with light ions, which forms vacancies in the nanocrystal and resultantly Au diffuses preferentially. A similar procedure of subsequent implantation of two elements and post annealing was successfully applied by White *et al.* for the formation of CdS and CdSe nanocrystals in SiO_2 and Al_2O_3 [17].

Klimenkov *et al.* reported that Ge nanoparticles fabricated by the ion implantation technique in a SiO_2 thin film crystallize after irradiation with a high-energy electron beam [18]. The crystallization process depends on irradiation dose and dose rate. Irradiation with a dose above $3.7 \times 10^{22}/cm^2$ results in cluster growth and above $2.5 \times 10^{23}/cm^2$ in crystallization. An irradiation with a dose rate below $150\,A/cm^2$ leads to the crystallization of Ge nanoparticles in the form of single crystals. For an irradiation dose rate above this value, the nanocrystals are fragmented into twinned and multiple particles. Takeguchi *et al.* proposed a novel method of fabricating size- and position-controlled Si nanocrystals by scanning an electron beam intermittently, because the size of the Si nanocrystals depends on the beam diameter [19].

Similar irradiation effects were found with ions and neutrons. Wang *et al.* irradiated intermetallic and ceramic materials with 0.5 to 1.5 MeV Kr^+ or Xe^+ ions [20]. As an example, when a U_3Si target composed of large grains tens of micrometers in size and large martensite plates is irradiated above a critical temperature of 570 K for amorphization, martensite plates disappear and small grains of $< 20\,nm$ in size appear below a dose of $2.7 \times 10^{15}\,Kr^+/cm^2$ and increase in number with increasing dose keeping their sizes. At a dose of $1 \times 10^{20}\,Kr^+/cm^2$, the sample completely changes to polycrystalline materials with grains of $\sim 20\,nm$ in size. The mechanism of the radiation-induced nanocrystallization is very complicated and needs a farseeing theoretical understanding for its versatile application.

Formation of shallow bands of nanoclusters was observed for various implanted species [e.g., Ge [21], Sn [22], Sb [23]] in the vicinity of a Si/SiO_2 interface. Heinig *et al.* explained the evolution of the near-interface nanocluster band by a model taking into account collisional ion-beam mixing and chemical reactions near the Si/SiO_2 interface [24]. That is, atomic displacements during ion implantation and the action of the Si interface as a sink for free oxygen result in a Si excess in the first few nanometers of SiO_2 parallel to the interface. Then the excess Si acts as nucleation centers for diffusing metal, which results in the formation of the near-interface nanocluster band. The model predicts two prerequisites of a damage rate of about one displacement per atom at the Si/SiO_2 interface and the diffusion capability of the implanted impurities for the near-interface nanocluster band formation. On the other hand, a study of Ge implanted into a SiO_2 film

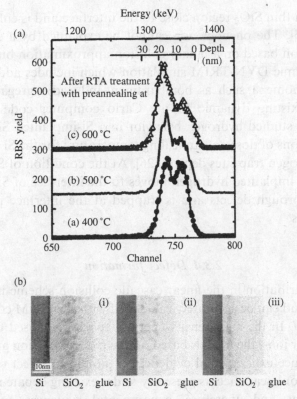

Figure 2.5 (a) Ge depth distribution in SiO$_2$ films with a thickness of 20 nm after rapid thermal annealing (RTA) treatment and preannealing in 7%H$_2$-Ar atmosphere. (b) The Ge nanocluster formation after three different RTA steps at 1223 K in N$_2$ atmosphere is shown: for 30 s with the formation of a bulk cluster band (i), for 4 min with the additional occurrence of an interface-near cluster band (ii), and for 12 min with the single δ-like cluster band near the interface (iii). (Reprinted with permission from *J. Appl. Phys.*, **91** (2002), 10062–67. Copyright 2002 American Institute of Physics.)

revealed that annealing in hydrogen-containing atmosphere (7% H$_2$ + 93% Ar) significantly accelerates the Ge mobility at elevated annealing temperatures ($T > 1173$ K, $t = 60$ min). This results in the formation of very large Ge clusters (typical size of 30–40 nm) together with a remarkable Ge loss in the oxide. Klimenkov *et al.* concluded that by appropriate choice of the post-implantation treatment conditions, especially the application of two-step annealing procedure involving low-temperature hydrogen preannealing, the nanoclusters in the bulk SiO$_2$ layer can be suppressed (Figure 2.5) [21].

Ignatova *et al.* studied the sputtering of SiO$_2$/Si interfaces with Ga ions by using secondary neutral spectrometry and found an oscillation of ion-implanted Ga concentration at SiO$_2$/Si interface, that is, the Ga concentration is depleted in a thin SiO_2 region close to the interface and is enhanced in a Si thin region [25]. The phenomenon cannot be explained by a simple Monte Carlo simulation based on a binary collision approximation but is explained well by a dynamic DYN-TRIM simulation which includes additional defect transport phenomena such as bombardment-induced segregation incorporated in an existing dynamic Monte Carlo computer code DYNTRIM. Imanishi *et al.* studied hydrogen behavior in a Si-implanted SiO_2 sample at several conditions of dose and annealing temperature for the Si implantation, on which hydrogen trap sites depend [26]. At the condition of Si nanocrystal formation, the implanted hydrogen moves to the interface of Si nanocrystal/ SiO_2 matrix through defects and is trapped at the interface passivating Si dangling bonds.

2.3.4 Defect formation

The defect distribution in the linear cascade collision scheme is discussed in Section 2.3.2 and can be accurately calculated with the SRIM code developed by Ziegler *et al.* In the high-density energy deposition caused by very heavy ions and cluster ions, the models based on the binary collision approximation cannot reproduce experimental evidences of nonlinear effect on defect production. The nonlinear effect is presently under exciting debate relating to the nanotechnology, and, therefore, experimental evidences and theoretical understanding of them are mainly treated in this subsection.

Canut *et al.* measured damage cross sections for silicon single crystals irradiated with $200 \, keV/atom$ Au_n^+ clusters and found that the cross section varies roughly as n^2 [27]. Shen *et al.* found a similar dependence in the case of Si irradiated with a fullerene ion [28]. These experimental results provide evidence for the existence of nonlinear spike effects in high-density collision cascades. The spike effect cannot be treated by the models based on BCA but can be done by MD simulation. Molecular-dynamics computer simulation study was made by Koster *et al.* for processes occurring in amorphous Si after irradiation with a projectile with energy $E < 500 \, eV$ focusing on the effects of high and low deposited energy density by Au and Si impact [29]. They found that damage production (creation of over- and under-coordinated Si atoms) and recoil implantation are strongly enhanced under Au impact. The Au impact deposits high-energy density at a local point and creates a strong collision spike. That is, its center is strongly under-dense – a "hole" appears – and consequently suffers strong tensile pressure, which induces a compressive shock wave radiating outwards. The process occurs on a time scale of $0.4 \, ps$ and produces abundant defects and relocation of atoms.

Relaxation processes follow it, reducing the number of defects formed (of under-coodinated atoms, in particular) and transporting relocated atoms back into the vicinity of their original sites. The Si impact does not create such a collision spike and leads to a low damage production.

A series of MD simulation studies done by Nordlund *et al.* predicts the following general trends [30]: Low-mass ions (like Ne in Pt) tend to mostly produce linear cascades with no large heat spikes. Likewise, cascades in materials with a low mass or open crystal structure like silicon also do not exhibit a large surface enhancement of damage production. For heavy ions in dense, close-packed materials like Ni, Cu, W, Pt, and Au, on the other hand, a liquid-like state caused by a large heat spike is produced, and the thermodynamic and mechanical aspects of materials become important. In this case, the presence of a surface can strongly affect the behavior of cascades by hot liquid flow to the surface, microexplosion of the hot liquid zone moved out to the surface by the inside pressure wave, and coherent displacement. In addition, a comparison of cascades in pairs of materials (Pt and Au, Ni and Cu) expected to behave similarly in a ballistic collision revealed that melting point and elastic hardness of a material are important parameters in determining the behavior of cascades. Because in all these materials the liquid has a lower density than the solid, the heat spike formed by local high-energy deposition leads to the subsequent liquid flow and microexplosion processes. Then, for similar irradiations, Pt and Ni with relatively high melting points suffer smaller defects than Au and Cu with low melting points. On the other hand, semiconductors have open crystal structures, which lead to medium-energy recoils traveling further than in face centered cubic (fcc) lattices and thus diminishing the energy density in the heat spike. In addition, since the density of liquid phase is high compared with that of solid in these materials and the heat spike tends to relax inwards rather than outwards, making the "thermodynamic" surface mechanisms unlikely to occur. Thus, heavy ions produce a liquid-like state in dense, close-packed materials and its flow onto the surface results in a nonlinear sputtering regime and crater formation on the surface.

Accumulation of defects in a crystal leads to the formation of a local amorphous zone. Pelaz *et al.* studied the effects of dose, dose rate, and implant and annealing temperatures on amorphization resulting from ion implantation in Si by simulating with a combined code of MARLOWE and DADOS [31]. They first obtained coordinates of the displaced atoms in the lattice with MARLOWE [10] which uses the binary collision approximation. Then, the coordinates were continued in DADOS [32], which is the three-dimensional kinetic Monte Carlo diffusion code and was improved to include

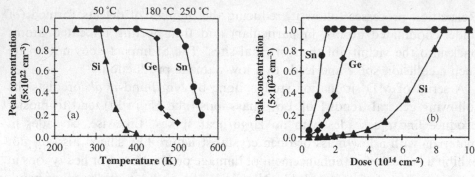

Figure 2.6 Peak concentration of the I–V complexes for 80 keV, $10^{12}\,cm^{-2}s^{-1}$ implants for Si, Ge, Sn ions. (a) Amorphization is not possible above the transition temperature at a $10^{15}\,cm^{-2}$ dose. (b) A superlinear dependence is observed near the amorphizing dose. (Reprinted with permission from *Comp Mater. Sci.*, **27** (2003), 1–5. Copyright 2002 Elsevier Science B.V.)

damage ranging from individual defects to continuous amorphous pockets and layers. Calculated results are shown in Figure 2.6 for a critical transition temperature where the rate of defect production and annihilation are comparable. Here, peak concentration of interstitial–vacancy (I–V) complex is plotted as a function of implantation temperature and implantation dose. It reproduces and explains features of the crystalline-to-amorphous transition. When implantation cascades are annealed independently, many interstitials escape recombination for a long enough period to contribute significantly to transient enhanced diffusion (TED). This situation happens at low doses, high implant temperatures, and low dose rates. At high doses, low temperatures, and high dose rates, the damage is accumulated during ion implantation. During a high-temperature post-implant anneal, recombination of Frenkel pairs is efficient and only the extra interstitial generated by the implanted ion controls TED [33].

It is necessary to localize defects especially for a nanostructure fabrication with a focused ion beam. Vieu *et al.* measured the diffusion length of defects in GaAs/$Ga_{0.67}Al_{0.33}$As multi quantum-well layers implanted with Ga^+ ions to a dose of 1×10^{15} ions/cm^2 by means of a photoluminescence (PL) technique [34]. Obtained values are 65 and 270 nm for the depth and lateral directions, respectively, at room-temperature implantation. That is, defect diffusion in the III–V heterostructures occurs much more efficiently in the lateral direction than in depth. Nonradiative recombination of electron–hole pairs formed in the GaAlAs barrier layers by the irradiation possibly enhances the defect diffusion. The fast diffusion of nonequilibrium defects is drastically reduced, for a liquid nitrogen temperature (LNT) irradiation, to values of 36 and 70 nm for the depth and lateral directions, respectively. The low-temperature implantation is inevitable for localizing the implantation-caused damages in the advanced nanostructure fabrication.

2.4 Sputtering

2.4.1 General remarks

In a low energy region where the nuclear collision process is dominant over the electronic one, an incident ion makes a kinetic collision with an atom in a solid and transfers a part of kinetic energy to the atom. The recoiled atom successively collides with another atom and induces a cascade collision process. During the process some of the recoiled atoms are ejected from the solid surface as mostly neutral particles and partly positively or negatively charged ions. Sigmund developed an analytical formula called the linear collision cascade model which successfully describes the sputtering process [9]. In a high-deposit energy condition, the recoiled atoms collide with not only stationary atoms but moving atoms, and as a result the colliding local region attains a thermal equilibrium. A thermal spike and a shock wave model were proposed to describe the process in the condition that the atoms behave collectively and the linear cascade collision process cannot be applied anymore. Recently, several simulation codes have been developed using Monte Carlo treatment, dynamic Monte Carlo treatment, molecular dynamics, combined treatment of Monte Carlo and molecular dynamics, rate equation, and so on. These models successfully reproduce the experimental results on yields, emission energies, and angular distributions sputtered from metal targets. Some of them cover mass spectra and yields of sputtered neutral and ionized clusters.

On the other hand, in the cases of semiconductive and insulating materials, even in the nuclear-collision-dominant low energy region, the electronic collision process contributes to the sputtering process in a complex way and the sputtering mechanism is material dependent and different from that applied for the metals. In spite of the importance of such materials as semiconductor, insulator, and organic- and bio-molecules, reliable models are, therefore, very few.

Many excellent publications reviewed the experimental results and the mechanisms of the sputtering process in the case of metal targets [35–38]. However, descriptions have been scarcely devoted space on dose and dose rate dependences, an incident cluster effect, chemical and deposition effects, and incident particle broadening (dispersion) that will become important for the focused ion beam process. This section is focusing on the experimental and theoretical details of the above mentioned various effects on sputtering.

2.4.2 Sputtering yield

General scope

The sputtering yield, Y, is defined as the average number of atoms removed from a solid surface per incident particle, and is given by $Y = \sum Y_i$ for multi-elemental materials, where Y_i is a partial sputtering yield for the element i. Energy and angular distributions of sputtered particles are given by $\partial^3 Y / \partial K \partial \Omega^2$ with an emission energy K and a solid angle Ω. The sputtering yield depends on the following various parameters: incident ion (mass, energy, dose rate, angle of incidence, and clustering), target materials (masses and fractions of atoms, crystallinity, crystal orientation, surface binding energies, conductivity, surface curvature), and so on. A threshold energy is about 20–40 eV for normally incident ions and the yield generally increases with incident energy having a broad maximum in the energy region of 5–50 keV. Sigmund developed an analytical theory describing the linear collision cascade [9]. According to the theory, the sputtering yield is given by the simple formula

$$Y(E, \theta, d) = \frac{0.042 F_D(E, \theta, d)}{N U_0},\qquad(2.27)$$

where F_D is the deposited-energy depth distribution, d the depth of the sputtering surface from the entrance point of the projectile, E the projectile energy, θ the angle of incidence to the surface, N the density of the target material, and U_0 the average surface-binding energy. For backward sputtering at perpendicular incidence, the formula is simplified into

$$Y = \frac{0.042 \alpha (M_2/M_1) S_n(E, Z_1, Z_2)}{N U_0},\qquad(2.28)$$

where S_n is the nuclear stopping power and α is a function depending on the mass ratio between the target (M_2) and projectile (M_1) atoms.

The linear collision cascade theory predicts the yields very well for light targets, but for heavier targets, in particular in combination with heavier projectiles, the yield increases substantially faster with Z_1 than that predicted by the theory. The density and binding effects in sputtering by ions of Ne, Ar, and Xe were studied by Shulga using the code OKSANA which is based on the binary collision approximation and takes into account weak simultaneous collisions at larger distances [39]. The ion energy and target cover ranges from 100 eV to 100 keV and from Mg to U, respectively. The linear cascade theory describes correctly the binding and mass effects at an incident energy higher than 1 keV but fails to include the density effect. To study the binding and density effects the following fitting functions are applied:

$$Y = A \frac{N^p}{U_0^q}(S_L U_0),\tag{2.29}$$

and

$$Y = A \frac{N^p}{U_0^q}(S_H U_0),\tag{2.30}$$

where $S_L = (3/4\pi^2)(\alpha(M_2/M_1)K_m/U_0)$, (applied at the incident energy less than 1 keV) and $S_H = 0.042\alpha(M_2/M_1)S_n(E, Z_1, Z_2)/NU_0$; A, p, and q are fitting parameters; and K_m is the maximum transferred energy in the binary collision. The sputtering yield is shown in Figure 2.7 as a function of target atomic number. The calculated data are obtained by applying the effects to the Sigmund's linear cascade theory.

For ion bombardment at an oblique angle of incidence the sputtering yield increases monotonically with increasing angle of incidence up to a maximum near 60–80°. A dependence on the angle of incidence is expected to be given by $Y(\psi)/Y(0) = 1/\cos^\beta \psi$, where ψ is measured from the surface normal. But surface morphology influences the reproducibility of the experimental data and the dependence is not clear. For single crystals the sputtering yields depend also on the direction of particle incidence relative to the crystal orientation.

Particles sputtered from a solid surface during particle bombardment are mostly neutral atoms emitted with a cosine like angular distribution and with a broad energy distribution that peaks at a few eV, about half the surface binding energy. A small fraction of the sputtered particles are clusters containing up to some 100 atoms. A very good review was presented by Betz and Wien on the energy and angular distributions of atomic, molecular, and cluster particles sputtered from solid surfaces under ion bombardment [38]. Experimental data are often compared with the Sigmund–Thompson distribution as described by [9]

$$\frac{d^3 Y}{dK d^2 \Omega} \propto \frac{K}{(K + U_0)^{3-2m}} \cos \Theta,\tag{2.31}$$

where Θ is the polar angle of the emitted particles with respect to the surface normal and m is an adjustable parameter close to 0.

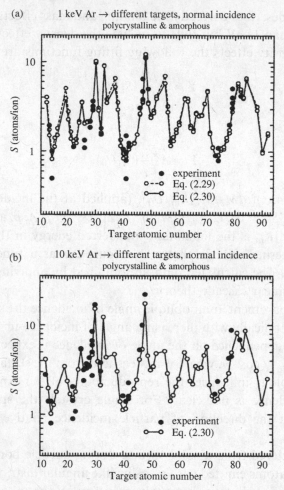

(a) 1 keV Ar → different targets, normal incidence
polycrystalline & amorphous

(b) 10 keV Ar → different targets, normal incidence
polycrystalline & amorphous

Figure 2.7 The experimental yield data for (a) 1 keV and (b) 10 keV Ar and their fits by (2.29) and (2.30). (Reprinted with permission from *Nucl. Instr. Meth. B*, **195** (2002), 291–301. Copyright 2002 Elsevier Science B.V.)

In the case of single crystalline targets, the angular distribution is strongly affected by the surface structure. Rosencrance *et al.* measured energy-resolved angular distributions for Ni and Rh atoms desorbed from Ni and Rh fcc{001} single crystals bombarded by keV Ar^+ ion and found that the experimental spectra for the two elements are almost identical, implying that the crystal and surface structures dominate the ejection process [40]. From the comparison between the experimental results and the MD calculations which are based on the molecular-dynamics/Monte Carlo corrected effective-medium interaction potential, it was found that about 90% of the ejected atoms originate from the first top layer and at least three major ejection mechanisms exist for the surface depending on the sequential collisions between atoms on (i) the first top layer, (ii) on the first and second layers, and (iii) on the first and third layers. The dominant mechanism of ejection is the above second case.

As described thus far, the main parts of the sputtering phenomena can be physically understood by the analytical linear cascade models and by the associated computer simulation schemes of binary-collision approximation and Monte Carlo treatments. That is, these models can represent well the experimental results as far as the energy deposit to the target is small enough to be treated in the linear cascade regime and the contribution of the electronic excitation is small. In the case of high-energy deposition, various phenomena have been observed as spikes, large cluster emission, chemical effects, and so on, which cannot be explained in the linear cascade regime. Therefore, the further understanding of these sputtering phenomena needs the molecular dynamic treatment, because these phenomena contain many-body interactions, electron–phonon coupling, and internal excitation of the limited region. In addition, some of the MD calculations were applied to the problems of bulk-binding energy and single-crystal sputtering, and further to the understanding of preferential sputtering and electronic sputtering. These will be discussed at each subsection described below.

Dose dependence of yield and preferential sputtering

Various BCA codes have been developed on the basis of the linear collision cascade theory for crystalline targets (named MARLOWE [10], COSIPO [41], ACOCT [42], crystal-TRIM [43], etc.) and for amorphous targets (named TRIM [5], ACAT [44], SASAMAL [45,46], etc.). These codes do not take into account compositional changes caused by the collision cascades and are called "static" Monte Carlo (MC) code. The "dynamic" codes (named TRIDYN [47], ACAT-DIFFUSE [44], dynamic-SASAMAL [45,46], etc.) are improved to treat the compositional changes during the ion bombardment and can reproduce the dose dependence of depth profiles of implanted ion and preferential sputtering. An example of calculations by the dynamic-SASAMAL code for Ar ion incidence, as shown in Figure 2.8, is the effect of mass ratio of two components on the preferential sputtering [45,46]. The TRIDYN code predicts that implantation of heavy ions into light materials (low atomic number) leads to oscillations in partial sputtering yields as a function of incident dose [47], as shown in Figure 2.9. In the BCA regime, the kinetic energy is effectively transferred to atoms with similar mass as the incoming particle. In addition, lower-mass atoms have larger ranges in the solid than

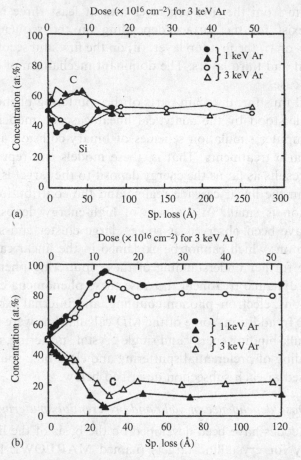

Figure 2.8 Change of surface atomic concentration in SiC (a) and WC (b) during 1 and 3 keV Ar sputtering. (Reprinted with permission from *Nucl. Instr. Meth. B*, **190** (2002), 256–60. Copyright 2002 Elsevier Science B.V.)

heavier atoms, so that lighter atoms are depleted in the solid at the depth of maximum energy deposition. As a result the lighter atoms are in general preferentially sputtered at the surface.

Besides those reasons based on the BCA regime, factors influencing preferential sputtering are different binding energies of different atoms at the surface, segregation of one element at the surface, and diffusion in the implanted layers of the solid. These preferential sputtering have been studied in more detail with various simulation codes such as molecular dynamic treatment, thermal spike model, and fluid dynamics calculation [48–52]. An example of sputter preferentiality and partial sputter yields calculated by molecular dynamics simulation [49] is presented in Table 2.2. The results

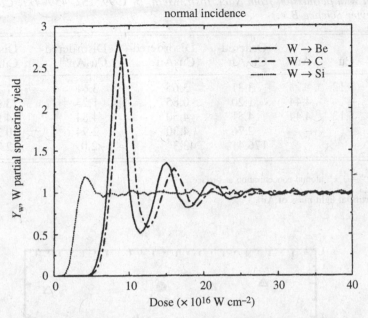

10 keV W → Be, C, Si

normal incidence

Figure 2.9 Dose dependence of partial sputtering yield of W. Be, C, and Si are bombarded with 10 keV W at normal incidence. (Reprinted with permission from *Nucl. Instr. Meth. B*, **171** (2002), 435–42. Copyright 2000 Elsevier Science B.V.)

show that the sputter preferentiality in the Cu-Au alloy is mainly caused by a mass difference rather than a binding energy and furthermore an enhanced sputter preferentiality of Cu atoms for an ordered Cu_3Au alloy compared to a disordered alloy demonstrates a strong influence of local ordering on sputtering emission.

Mass distribution

Mass distributions of emitted neutral large clusters have been systematically measured by Wucher group using a post ionization method with a laser induced single ionization technique [53,54]. Obtained cluster yields are very large and have power law dependences n^δ on the cluster size n. The exponent δ depends on wavelength of ultraviolet, and the respective values for 193 nm and 157 nm of ArF and F_2 lasers are Al (−8.6, −7.7), Ag (−6.0, −5.3), Nb (—, −8.1), and Ta (—, −8.5). Further, the exponent depends on the total sputtering, as shown in Figure 2.10. Start and Wucher carefully corrected the apparent cluster yields on size dependent detection efficiency, and found that a power-law yield distribution for In clusters by 15-keV Xe incidence has two

Table 2.2 *Sputter preferentiality* δ *calculated by a molecular dynamics simulation. (Reprinted with permission from Nucl. Instr. Meth. B, 1999, 152, 459–471. Copyright 1999 Elsevier Science B.V.)*

	Cu	Au	Ordered-Cu_3Au	Disordered-Cu_3Au	Disordered-Cu_3Au'[b]	Disordered-$CuAu_3$
Y_{Cu}	5.12	–	3.31	3.65	3.61	1.24
Y_{Au}	–	4.44	1.20	0.85	1.23	3.31
Y_{tot}	5.12	4.44	4.51	4.50	4.84	4.55
Y_{Cu}/Y_{Au}	–	–	2.76	4.30	2.94	0.38
$\delta(\%)$[a]	–	–	176.4	43.3	−2.0	12.7

[a] $\delta = \dfrac{Y_{Cu}C_{Au}^s}{Y_{Au}C_{Au}^s} - 1$, c_i^s: atomic concentration at surface.

[b] Au′: an artificial light mass of Au.

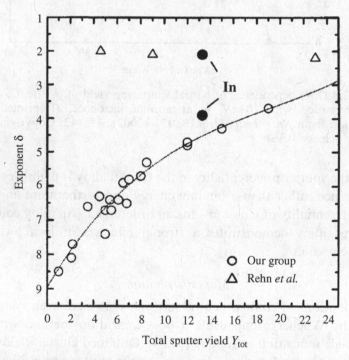

Figure 2.10 Compilation of power-law exponents extracted from cluster yield distributions versus total sputtering yield Y_{tot}. The In data (solid symbols) have been determined by correcting the cluster-size-dependent detection probability: the remaining exponents (open symbols) are non-corrected values for Ag, Cu, Al, Ge, Nb, and Ta. For comparison, the Au data of Rehn *et al.* [55] have been included. (Reprinted with permission from *Phys. Rev. B*, **66** (2002), 075419. Copyright 2002 The American Physical Society.)

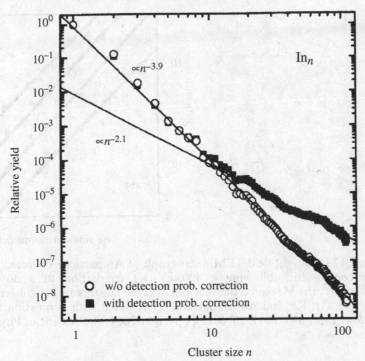

Figure 2.11 Relative yields of In_n clusters sputtered from a pure, polycrystalline indium surface under bombardment with 15 keV Xe^+ ions. The data have been normalized to the partial yield of In atoms. (Reprinted with permission from *Phys. Rev. B*, **66** (2002), 075419. Copyright 2002 The American Physical Society.)

decay components of -3.9 for small ($n \leq 20$) and -2.1 for large ($20 \leq n \leq 100$) clusters (Figure 2.11) [54]. On the other hand, Rehn *et al.* measured size distributions of very large clusters ($n \geq 500$) by a transmission electron microscope (TEM) and obtained a power law dependence of cluster yield with a rather low value of exponent of -2, which is independent of total sputtering yield as shown in Figures 2.10 and 2.12 [55]. These results are explained in the following way: the inverse-square dependence reveals that the very large clusters are produced when shock waves, generated by sub-surface displacement cascades, impact and then they ablate the surfaces [56,57]. Many smaller clusters detected with the post ionization method can result from the break up of larger ones, which provides a plausible explanation for the large negative exponents. This idea was simulated by Wucher *et al.* with so-called "molecular dynamics and Monte Carlo-corrected effective medium potential (MD/MC-CEM)." That is, first, the formation of so-called very large highly excited nascent clusters, is calculated by the MD

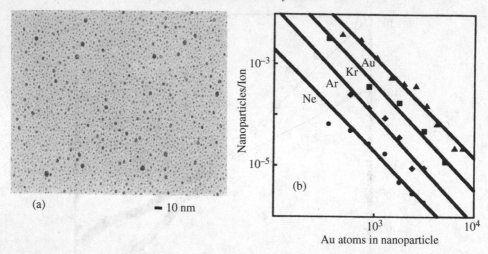

Figure 2.12 (a) Bright-field TEM micrograph of Au particles collected on a carbon-coated grid following a 400 keV Kr irradiation to a dose of 1×10^{14} cm^{-2}. (b) Measured number of collected clusters versus cluster size (n) for Ne, Ar, Kr, and Au irradiations. (Reprinted with permission from *Phys. Rev. Lett.*, **87** (2001), 207601. Copyright 2001 The American Physical Society.)

part [53]. Then, the decomposition of the produced clusters on their flight away from the surface is followed by either the MD or by the MC scheme. The fragmentation produces smaller clusters. Though the absolute values of the calculated exponent are significantly too low compared to the experimental data, the power-law-like yield distribution with varying exponent depending on the yield is reproduced very well.

Incidental cluster-size dependence of sputtering yield

Experimental and theoretical works have been done on total sputtering yields and secondary ion emission for cluster-ion incidence. Brunelle *et al.* found that the yield shows a Y/n^2 dependence (incident Au clusters Au$_n$; $n = 3$–13) with the same maximum at about 150 keV and 200 keV for Ag and Au sputtering yields, respectively, as shown in Figure 2.13 and the data show no simple relation between the yield and the nuclear stopping power [58]. A sophisticated calculation was done by Colla and Urbassek with molecular dynamic simulations for sputtering of a Au(111) surface bombarded by Au$_n$ cluster in the equi-velocity ($E/n = 16$ keV/atom, $n = 1, 2, 4$) and equi-energy ($E = 16$ keV, $n = 1, 2, 4, 8, 12$) conditions [59,60]. The temporal evolution of the sputtering yield shows that the yield increases with increasing time after the cluster impact and the duration of the evolution increases with cluster

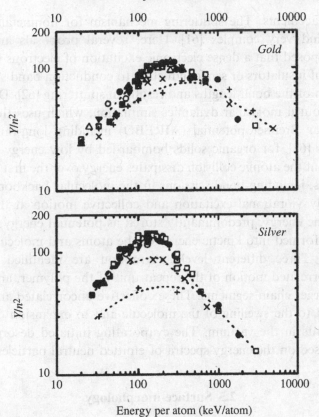

Figure 2.13 Gold and silver sputtering yields divided by n^2, as a function of the energy per atom of the Au_n ($n = 1$–13) cluster projectiles. The dashed lines are guides for the eye. Symbols used correspond to following values of n: $+ 1$, $\times 2$, $\square$ 3, $\triangle$ 4, $\bullet$ 5, $\circ$ 7, $\blacktriangledown$ 9, $\blacksquare$ 11, $\blacktriangle$ 13. (Reprinted with permission from *Phys. Rev. B*, 2002, **65**, 144106. Copyright 2002 The American Physical Society.)

size. As a result, for an equi-velocity impact condition, the sputter yield has a nonlinear dependence on n and, for example, the yield for a Au_2 impact is five times as high as that for the two Au-atom incidence. On the other hand, for equi-energy impact, a saturation or even a slight decline of the yield for cluster sizes beyond $n = 4$ is predicted.

Sputtering of nonmetals

In sputtering experiments on nonmetals the measured yields are generally different than expected from collisional theory. Here besides the energy transferred to target atoms by elastic collisions, also the energy transferred to electrons producing electronic excitation and ionization can contribute to atomic displacements. The sputtering mechanism for nonmetals is material dependent and very complex [61]. Here, several proposals are presented. Stampfli proposed that a dense electronic excitation of electrons from valence band states of insulators or semiconductors to conduction band states induces an expansion of the bond lengths and results in sputtering [62]. Delcorte *et al.* [64] carried out a molecular dynamics simulation, which uses the adaptative intermolecular Brenner potential (AIREBO) including long-range Van der Waals forces [63], for organic solids bombarded by low energy Ar ions and concluded that the atomic collision dissipates energy over the first few hundred femtoseconds, including bond-scissions in the molecular backbone and creating ultimately vibrational excitation and collective motion at the molecular scale [64]. The energy, predominantly stored as potential energy in the molecule, is transformed into kinetic energy of the atoms and molecular segments after 500 fs. Three different levels of motion are identified: interatomic vibration, correlated motion of the repeat unit of the polymer, and correlated motion of larger chain segments. These collective uncorrelated and correlated motions lead to the swelling of the molecule and to expansion of kilodalton chain segments in the vacuum. The evaporating initiated desorption is also proposed based on the energy spectra of emitted neutral particles.

2.5 Surface morphology

The sputtering process has been widely used to erode and to smooth out the solid surface. However, it sometimes roughens at random the solid surface and in rare cases forms periodically structured ripples with a minimum wavelength of a few nm. The ripple formation by sputtering was discovered by Barber *et al.* for amorphous target materials [65]. The phenomena depend on many factors, such as incident ion energy, mass, angle of incidence, sputtered substrate temperature, and material composition. Ripples appear only for a limited range of incident angles, depending on materials and ions involved [66]. At high temperatures, the ripple wavelength depends on temperature, while the wavelength is constant at low temperatures. MacLaren *et al.* interpreted this constant behavior as the predominance of radiation enhanced diffusion over thermal diffusion at low temperatures, because the former gives a temperature independent contribution to the diffusion constant [67]. The roughness, which is proportional to the ripple amplitude, increases linearly with dose and with the square of dose rate. The ripple wavelength is proportional to incident energy.

The theoretical works describing the process of ripple formation were done by Brandy *et al.* [68] and in a more extended way by Makeev *et al.* [69] for

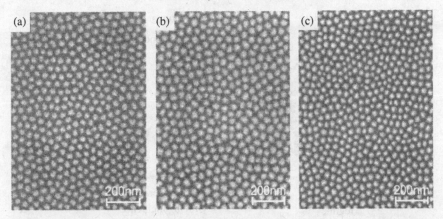

Figure 2.14 Scanning electron micrographs of dot patterns on GaSb films with initial layers of: (a) c-GaSb(100), (b) c-GaSb(111), and (c) amorphous GaSb deposited on Si(111). The patterns are created at an Ar^+-ion energy of 500 eV, an ion dose rate of $1 \times 10^{16}\,cm^{-2}\,s^{-1}$, and an erosion time of 200 s. (Reprinted with permission from *Appl. Phys. Lett.*, **80** (2002), 130–2. Copyright 2002 American Institute of Physics.)

amorphous solids. Makeev *et al.* derived a sophisticated stochastic nonlinear continuum equation based on the Sigmund theory of sputter erosion to understand the effect of the sputtering process on the surface morphology. The nonlinear theory predicts the development of a periodic ripple structure at short time scales, while at large time scales the surface morphology may be either rough or dominated by new ripples. The transitions are predicted to take place between various surface morphologies depending on the experimental parameters (e.g. angle of incidence, energy deposition depth). The theoretical predictions were compared with abundant experimental results of ripple amplitudes, ripple wavelengths, ripple orientation, and random roughness and reproduced most of the results [69].

Self-organizing process of crystalline surface bombarded by an ion beam was recently discovered by Valbusa *et al.* in Ag (110) and in Cu (110) [70,71], and lately by Facsko *et al.* in GaSb(100) and GaSb(111) (Figure 2.14) [72,73]. These surface structures with dimensions of some tens of nanometers can form quantum dots with a high aspect ratio and are therefore particularly attractive for quantum electronic applications. As shown in Figure 2.15 as an example, Ag (110) and Ag (100) surfaces after Ne^+ or Ar^+ bombardment have different morphologies of the ripples along crystallographic direction symmetry and a regular checkerboard patterns, respectively. In the case of a grazing angle incidence, both (100) and (110) present a well-defined ripple morphology, which depends on beam orientation and not on crystallographic

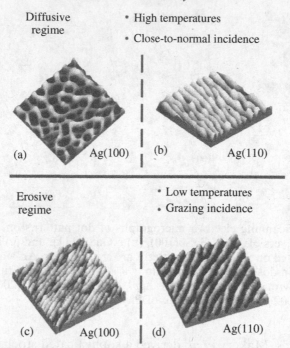

Diffusive regime
• High temperatures
• Close-to-normal incidence

(a)   Ag(100)   (b)   Ag(110)

Erosive regime
• Low temperatures
• Grazing incidence

(c)   Ag(100)   (d)   Ag(110)

Figure 2.15 Three-dimensional view of the surface morphology resulting after 20 min Ne^+ bombardment of Ag(100) (a and c) and Ag(110) (b and d) at an ion dose rate of $\Phi = 5\,\mu A/cm^2$. Images (a) and (b) are taken after normal incidence sputtering at $T = 400\,K$ and $320\,K$, respectively (diffusive regime). Images (c) and (d) are taken after grazing incidence sputtering $\theta_i = 70°$ at $T = 200\,K$ and $T = 180\,K$ respectively (erosive regime). (Reprinted with permission from *Materials Sci. Eng. C*, **23** (2003), 201–9. Copyright 2002 Elsevier Science B.V.)

directions. Valbusa *et al.* proposed two regimes of diffusion dominant "diffusive" and sputtering dominant "erosive" [70]. The former is responsible for the formation of crystal-orientation dependent patterns and the latter do not reflect the crystal orientation. Then, the substrate temperature and the ion dose rate are competing parameters to select and enhance only certain diffusion processes and consequently to tune the final surface morphology. Valbusa *et al.* interpreted the obtained results for Ag (110) and Ag (001) in the following way: on an anisotropic surface, such as Ag (110), they select which of the two crystallographic directions the inter-layer mobility is inhibited, leading to the growth of two different and perpendicular ripple structures, while, on an isotropic surface, such as Ag (001), temperature and ion dose rate only determine whether or not the diffusion instability is active. They pointed out that in the case of the normal incidence, the

Erlich–Schwoebel barrier plays an important role in the diffusion process. Then, the tuning of incidence angle or surface temperature makes it possible to shift from the "erosive" regime characterized by ripples oriented along the direction of the ion beam, to a checkerboard pattern, which is aligned along the crystal axis and occurs in the "diffusive" regime. This method combined with focused ion beam will be cultivated extensively as a new technique in the fabrication of quantum dots.

2.6 Summary and future perspectives

Ion beam techniques using especially ion implantation and sputtering have been used widely in manufacturing electronics devices and partly in materials modification because of their very versatile and controllable properties. In addition, the application has been promoted by the reason that the basic process of energetic particles in solid was simple and understandable in the framework of the binary collision approximation, which is briefly introduced in Section 2.2. However, recent sophisticated experiments have revealed that the collision process induces fertile and nanometer-size effects on solid in complicated ways. In Section 2.3, after the general guidance of the implantation process, the embedded nanostructure formation caused by ion implantation is reviewed from experimental and theoretical points of view. In Section 2.5, the topographical change of surface caused by the simultaneous operation of sputtering and diffusion is guided focusing on the self-organizing process of crystalline surface to nanometer-size quantum dots with a high aspect ratio. These nanostructure effects are particularly attractive and the methods combined with focused ion beam will be cultivated extensively as new techniques in the fabrication of quantum dots.

Other experimental findings are the nonlinear effects on the damage production and sputtering in the high-density energy deposition caused by very heavy ions and cluster ions. In Section 2.3, the nonlinear effects of defect formation are discussed by combining experimental and theoretical results. In addition, it is necessary to localize defects when one fabricates nanostructures with a focused ion beam. Measurement of diffusion lengths and an idea to localize the defects are presented. In Section 2.4, after the description of the linear collision cascade model, the experimentally observed nonlinear effects in the high-energy deposition are reviewed focusing on yields, emission energy and angular distributions, mass spectra, and yields of clusters along with the comparison with the results calculated by recent simulation codes based on Monte Carlo treatment, dynamic Monte Carlo treatment, molecular dynamics, combined treatment of Monte Carlo and molecular dynamics. More detailed and systematical experiments of dose and dose rate dependences, an incident cluster effect, chemical and deposition effects, and incident particle broadening (dispersion) are necessary for the focused ion beam process.

Thus, the recent experimental findings and theoretical understanding are tightly connected to the application of focused ion beams in the field of nanotechnology. However, experiments done with focused ion beams are scarce. Therefore, it is an urgent request to take accurate and systematic data related to ion implantation, defect formation, sputtering, and surface morphology with focused ion beams, and to refine theoretical models in a more quantitative way.

References

[1] See for an example, M. Nastasi, J. W. Mayer and J. K. Hirvonen. *Ion–Solid Interactions: Fundamentals and Applications* (Cambridge: Cambridge University Press, 1996).
[2] O. B. Firsov. *Sov. Phys. JETP*, **36** (1959), 1076–80.
[3] G. Molière. *Z. Naturforsch., A*, **2** (1947), 133–45.
[4] J. Lindhard, V. Nielsen and M. Scharff. *Mat. Fys. Medd. Dan Vid. Selsk.*, **36**: 10 (1968).
[5] J. F. Ziegler, J. P. Biersack and U. Littmark. *The Stopping and Range of Ions in Solids* (New York: Pergamon Press, 2003).
[6] H. Bethe. *Ann. de Phys.*, **5** (1930), 325–400.
[7] F. Bloch. *Zeit. F. Phys.*, **81** (1933), 363–76.
[8] J. Lindhard, M. Scharff and H. E. Schiott. *Mat. Fys. Medd. Dan Vid. Selsk.*, **33**: 14 (1963).
[9] P. Sigmund. *Phys. Rev.*, **184** (1969), 383–416.
[10] M. T. Robinson. *Phys. Rev. B*, **40** (1989), 10717–26.
[11] F. Gonella and P. Mazzoldi. *Handbook of Nanostructured Materials and Nanotechnology*, Vol. 4, ed. S. Nalwa (San Diego, CA: Academic Press, 1999), pp. 81–158.
[12] F. Gonella. *Nucl. Instr. Meth. B*, **166–7** (2000), 831–9.
[13] H. Hosono. *Jpn. J. Appl. Phys.*, **32** (1993), 3892–4.
[14] H. Hosono and N. Matsunami. *Phys. Rev. B*, **48** (1993), 13469–73.
[15] M. Strobel, K.-H. Heinig and W. Möller. *Nucl. Instr. Meth. B*, **148** (1999), 104–9.
[16] G. Mattei, G. De Marchi, C. Maurizio *et al. Phys. Rev. Lett.*, **90** (2003), 085502.
[17] C. W. White, J. D. Budai, S. P. Withrow *et al. Nucl. Instr. Meth. B*, **141** (1998), 228–40.
[18] M. Klimenkov, W. Matz, S. A. Nepijko and M. Lehmann. *Nucl. Instr. Meth. B*, **179** (2001), 209–14.
[19] M. Takeguchi, M. Tanaka and K. Furuya. *Appl. Surf. Sci.*, **146** (1999), 257–61.
[20] L. M. Wang, S. X. Wang, R. C. Ewing *et al. Mater. Sci. Eng. A*, **286** (2000), 72–80.
[21] M. Klimenkov, J. von Borany, W. Matz, R. Grötzschel and F. Herrmann. *J. Appl. Phys.*, **91** (2002), 10062–7.

[22] A. Markwitz, R. Grötzschel, K.-H. Heinig, L. Rebohle and W. Skorupa. *Nucl. Instr. Meth. B*, **152** (1999), 319–24.

[23] A. Nakajima, H. Nakao, H. Ueno, T. Futatsugi and N. Yokoyama. *Appl. Phys. Lett.*, **73** (1998), 1071–3.

[24] K. H. Heinig, B. Schmidt, M. Strobel and H. Bernas. *Mater. Res. Soc. Symp. Proc.*, **647** (2001), O14.6.1.

[25] V. Ignatova, I. Chakarov, A. Torrisi and A. Licciardello. *Appl. Surf. Sci.*, **187** (2002), 145–53.

[26] M. Ikeda, R. Mitsusue, N. Imanishi *et al. Nucl. Instr. Meth. B*, **209** (2003), 154–8.

[27] B. Canut, M. Fallavier, O. Marty and S. M. M. Ramos. *Nucl. Instr. Meth. B*, **164–5** (2000), 396–400.

[28] H. Shen, C. Brink, P. Hevelplund *et al. Nucl. Instr. Meth. B*, **129** (1997), 203–6.

[29] M. Koster and H. M. Urbassek. *Nucl. Instr. Meth. B*, **202** (2003), 125–31.

[30] K. Nordlund, J. Keinonen, M. Ghaly and R. S. Averback. *Nucl. Instr. Meth. B*, **148** (1999), 74–82.

[31] L. Pelaz, L. A. Marqués, M. Aboy *et al. Comp. Meter. Sci.*, **27** (2003), 1–5.

[32] M. Jaraiz, L. Pelaz, E. Rubio *et al. Mater. Res. Soc. Symp. Proc.*, **54** (1998), 532.

[33] L. Pelaz, G. H. Gilmer, M. Jaraiz *et al. Appl. Phys. Lett.*, **73** (1998), 1421–3.

[34] C. Vieu, J. Gierak, M. Schneider, G. B. Assayag and J. Y. Marzin. *J. Vac. Sci. Technol. B*, **16** (1998), 1919–27.

[35] R. Behrisch. (ed.) *Sputtering by Particle Bombardment I*, (Berlin: Springer-Verlag, 1981).

[36] R. Behrisch. (ed.) *Sputtering by Particle Bombardment II*, (Berlin: Springer-Verlag, 1983).

[37] R. Behrisch and K. Wittmaack. (eds.) *Sputtering by Particle Bombardment III*, (Berlin: Springer-Verlag, 1991).

[38] G. Betz and K. Wien. *Int. J. Mass Spec. Ion Process.*, **140** (1994), 1–110.

[39] V. I. Shulga. *Nucl. Instr. Meth. B*, **195** (2002), 291–301.

[40] S. W. Rosencrance, J. S. Burnham, D. E. Sanders *et al. Phys. Rev. B*, **52** (1995), 6006–14.

[41] M. Hautala. *Phys. Rev. B*, **30** (1984), 5010–18.

[42] Y. Yamamura and W. Takeuchi. *Nucl. Instr. Meth. B*, **29** (1987), 461–70.

[43] M. Posselt. *Radiat. Eff. Def. Solids*, **130–1** (1994), 87–119.

[44] Y. Yamamura. *Nucl. Instr. Meth. B*, **28** (1987), 17–26.

[45] Y. Miyagawa, H. Nakadate, F. Djurabekova and S. Miyagawa. *Surf. Coatings Tech.*, **158–9** (2002), 87–93.

[46] Y. Miyagawa and S. Miyagawa. *Nucl. Instr. Meth. B*, **190** (2002), 256–60.

[47] W. Eckstein. *Nucl. Instr. Meth. B*, **171** (2000), 435–42.

[48] H. Gades and H. M. Urbassek. *Nucl. Instr. Meth. B*, **102** (1995), 261–71.

[49] T. J. Colla and H. M. Urbassek. *Nucl. Instr. Meth. B*, **152** (1999), 459–71.

[50] M. M. Jakas and E. M. Bringa. *Phys. Rev. B*, **62** (2000), 824–30.

[51] E. M. Bringa, M. Jakas and R. E. Johnson. *Nucl. Instr. Meth. B*, **164–5** (2000), 762–71.

[52] M. M. Jakas. *Nucl. Instr. Meth. B*, **193** (2002), 727–33.

[53] A. Wucher and M. Wahl. *Nucl. Instr. Meth. B*, **115** (1996), 581–9.

[54] C. Staudt and A. Wucher. *Phys. Rev. B*, **66** (2002), 075419.

[55] L. E. Rehn, R. C. Birtcher, S. E. Donnelly, P. M. Baldo and L. Funk. *Phys. Rev. Lett.*, **87** (2001), 207601.

[56] H. M. Urbassek. *Nucl. Instr. Meth. B*, **31** (1988), 541–50.

[57] I. S. Bitensky and E. S. Parilis. *Nucl. Instr. Meth. B*, **21**, (1987), 26–36.

[58] S. Bouneau, A. Brunelle, S. Della-Negra *et al. Phys. Rev. B*, **65** (2002), 144106.

[59] T. J. Colla and H. M. Urbassek. *Nucl. Instr. Meth. B*, **164–5** (2000), 687–96.

[60] T. J. Colla, R. Aderjan, R. Kissel and H. M. Urbassek. *Phys. Rev. B*, **62** (2000), 8487–93.

[61] S. Ninomiya, S. Gomi, N. Imanishi *et al. Nucl. Instr. Meth. B*, **209** (2003), 233–8.

[62] P. Stampfli. *Nucl. Instr. Meth. B*, **107** (1996), 138–45.

[63] S. J. Stuart, A. B. Tutein and J. A Harrison. *J. Chem. Phys.*, **112** (2000), 6472–86.

[64] A. Delcorte, B. Arezki, P. Bertrand and B. J. Garrison. *Nucl. Instr. Meth. B*, **193** (2002), 768–74.

[65] D. J. Barber, F. C. Frank, M. Moss, J. W. Steeds and I. S. Tsong. *J. Mater. Sci.*, **8** (1973), 1030–40.

[66] A. Karen, Y. Nakagawa, M. Hatada *et al. Surf. Interf. Anal.*, **23** (1995), 506–13.

[67] S. W. MacLaren, J. E. Baker, N. L. Finnegan and C. M. Loxton. *J. Vac. Sci. Technol. A*, **10** (1992), 468–76.

[68] R. M. Brandy and J. M. E. Harper. *J. Vac. Sci. Technol. A*, **6** (1988), 2390–5.

[69] M. A. Makeev, R. Cuerno and A. -L. Barabási. *Nucl. Instr. Meth. B*, **197** (2002), 185–227.

[70] U. Valbusa, C. Boragno and F. Buatier de Mongeo. *Materials Sci. Eng. C*, **23** (2003), 201–9.

[71] S. Rusponi, G. Costantini, C. Boragno and U. Valbusa. *Phys. Rev. Lett.*, **81** (1998), 4184–7.

[72] S. Facsko, T. Dekorsy, C. Koerdt *et al. Science*, **285** (1999), 1551–3.

[73] S. Facsko, T. Bobek, H. Kurz *et al. Appl. Phys. Lett.*, **80** (2002), 130–2.

3

Gas assisted ion beam etching and deposition

HYOUNG HO (CHRIS) KANG

IBM Microelectronics, East Fishkill, NY

CLIVE CHANDLER AND MATTHEW WESCHLER

FEI Company, Hillsboro, OR

3.1 Introduction

Fundamental to the area of nanotechnology is the ability to modify surfaces. There are many methods available to researchers to achieve surface modification over large areas but there are limited choices for small samples or sections of samples. The main method for small sample manipulation is via focused ion beams (FIB). These devices enable the user to define complex patterns covering hundreds of square micrometers all the way down to submicrometer feature formation.

FIB enables various materials to be deposited such as conductors, insulators and carbon based materials. FIB also enables users to etch materials selectively. FIB induced deposition and etching has been widely used in the field of mask repair, circuit modification, formation of contacts in semiconductors, atomic force microscope (AFM) tip fabrication, maskless lithography, and TEM sample preparation.

In this chapter, the materials of deposition, the basic concepts of focused ion beam induced deposition and etching, their parameters, and application examples will be introduced. The material presented will mostly draw from work with Ga^+ ion beams but other ion sources will also exhibit similar behavior with the addition of gas.

3.2 Gas assisted focused ion beam etching

The wide use of FIB systems as micro-machining tools stems from their ability to precisely mill away material from a localized area. This may be done to expose buried structures for failure analysis, as in the semiconductor

Focused Ion Beam Systems: Basics and Applications, ed. N. Yao.
Published by Cambridge University Press. © Cambridge University Press 2007.

field, or to create free standing structures for nanotechnology. To assist with these processes, a reactive neutral gas is introduced to the sample through a fine capillary tube at the milling site. The gas is able to enhance the etching rate compared to straight sputtering of the material and also reduce the amount of redeposited material from the sputtering of the sample. Other benefits from the use of gas are the different etch rate enhancements obtained for different materials that can allow for preferential etching of one material in the presence of another. The use of gas can also result in structures that cannot be formed by the use of the ion beam alone, such as high aspect ratio vias.

3.2.1 Gas assisted focused ion beam etching – principles

Gas assistance of the FIB induced etching process relies on gases that react with the sample to produce a volatile compound that is removed by the vacuum system. Overall, there are many similarities in the steps used to describe the deposition processes and those used to describe etching. For example, deposition and etching gases can adsorb and desorb without reacting (chemisorption) with the surface they are on. However, some etching gases behave differently in that instead of just physisorbing on the surface, the etching gas is able to chemisorb. For example, the use of chlorine on a clean silicon surface results in a chlorosilyl layer being formed on the silicon.

The steps in the gas assisted focused ion beam etching process are as follows:

(1) A chemically neutral reactive gas is introduced through a fine gas nozzle and is adsorbed (or chemisorbed) on the sample surface.
(2) The gas reacts with the sample either spontaneously, as in silicon and fluorine, or in the presence of the ion beam the gas decomposes and reacts with the sample, as with water and carbon based polymers.
(3) The volatile products leave the surface (desorption).

Winters and Coburn [1] have investigated the spontaneous etching and effect of ion enhanced spontaneous etching of silicon with fluorine-based gases. These studies have shown that in the silicon-XeF_2 system the incoming XeF_2 gas is able to spontaneously react with the silicon surface without the interaction of the ion beam according to the following reaction:

$$XeF_2(g) + Si(s) \rightarrow \text{``}SiF_2\text{''}(s) + Xe(g), \tag{3.1}$$

$$SiF_2(s) + XeF_2(g) \rightarrow SiF_4(g) + Xe(g). \tag{3.2}$$

Table 3.1 *Selection of halide gases and their etch rate enhancements for semiconductor materials using a 30 kV Ga$^+$ ion beam.*

Gas/sample	Si	SiO$_2$	Al	W	GaAs	InP
XeF$_2$ [13]	7–12	7–10	1	7–10	NA	NA
Cl$_2$ [13,14]	11	1	5–10	1	50	4
I$_2$ [15,16,17]	5–10	1–1.5	5–15	NA	10–15	11–13

The symbols (g) and (s) indicate gas phase and solid phase respectively. The SiF$_2$, while stoichiometrically correct, is probably a mixture of SiF, SiF$_2$, and SiF$_3$. The means by which the incoming ion from the FIB is able to enhance the etch rate above the rate for the spontaneous process is by removing the "SiF$_2$" layer through recombination to form SiF$_4$(g):

$$2SiF_2(s) + ion \rightarrow SiF_4(g) + Si(s), \tag{3.3}$$

$$SiF(s) + SiF_3(s) + ion \rightarrow SiF_4(g) + Si(s), \tag{3.4}$$

through sputtering of SiF$_2$:

$$SiF_2(s) + ion \rightarrow SiF_2(g), \tag{3.5}$$

or by increasing the desorption rate of SiF$_4$. The main gaseous species detected from the ion enhanced etching of silicon is SiF$_4$ and some SiF$_2$. It should be kept in mind that most of the material sputtered during ion beam exposure originates from the first atomic layer and that the yield is inversely proportional to the binding energy. Equations (3.1)–(3.5) result in the formation of a volatile species with a low binding energy and hence can increase the etching rate.

Equations (3.3), (3.4), and (3.5) will result in fresh silicon surfaces being exposed to the reactive XeF$_2$ gas to form a fresh layer of SiF$_2$ as in (3.1). This way more XeF$_2$ is able to react with silicon in the presence of the ion beam than without, increasing the rate of removal of silicon.

The interaction of molecular chlorine on silicon is known to behave in a similar fashion as the XeF$_2$ case with the exception of (3.2) which does not occur with chlorine at room temperature. However, chlorine is still able to give a significant boost to the etch rate of silicon in the presence of the ion beam as shown in Table 3.1. The lack of spontaneous etching is more the preferred situation with gas enhanced ion etching. Here the addition of the gas causes no etching, even though in some cases the gas will chemisorb to the sample and the ion beam is required to drive the etching process. This provides greater control of the etching process than is possible with the presence of spontaneous etching.

Gas assisted etching of metals

The use of halide etching gases on metals to increase the etching rate is widely used in the semiconductor industry. In this case the halide helps to remove the metal lines as a step to rewiring the circuit for failure analysis or parameter testing. Here the halide not only aids in the etching rate but is also able to create an electrically isolated cut since any sputtered metal is oxidized by the halide and rendered nonconducting. For applications requiring the removal of aluminum metal the use of either chlorine or iodine is preferred due to their ease of use, formation of volatile products, and excellent enhancement rates. XeF_2 is not used here since the reaction product, AlF_3, is not volatile and there is no improvement in the etching rate.

The case with tungsten is the opposite of aluminum. In this case XeF_2 is the preferred gas since the product of the reaction WF_6 is a gas and is thermodynamically stable. Chlorine does not provide any enhancement over sputtering since the WCl_6 product is a solid with a low vapor pressure.

Gas assisted etching of SiO_2

Gas assisted etching of SiO_2 is achieved by the use of fluorine based gases. XeF_2 is widely used because of the ease of handling this fluorine source. There is no spontaneous component in the etching of SiO_2. XeF_2 does not chemisorb to the SiO_2 surface but only physisorbs in a low concentration initially. This is due to the paucity of adsorption sites on the oxide surface prior to milling with the ion beam. However, after a few passes of the ion beam the damaged surface will readily adsorb up to ten times more fluorine, which is about the same coverage as one monolayer [2,3]. Even when damaged, the oxide surface does not spontaneously etch which is advantageous and provides excellent control of the etching based solely on the ion beam current. Studies using argon ions and XeF_2 have shown that the majority of the desorbed product from the etching of SiO_2 is SiF_4. This has been shown to be independent of the energy and type of ion used to excite the etching [4,5].

Etching of carbon based materials

The use of halide materials as etch enhancing agents provides solutions to many materials issues. However, they are by no means universal etching agents and are known to have limits with organic, carbon based materials. Water has been found to be a very effective etch enhancing material for

Table 3.2 *Yield and etch rate enhancement for various gases on PMMA with a 30 kV Ga$^+$ ion beam [7].*

Gas	Yield (μm^3/nC)	Enhancement
No gas	0.4	–
H$_2$	0.4	1
Iodine	0.8	2
XeF$_2$	1.7	4.2
Oxygen	1.8	4.5
CH$_3$OH	4.2	10
Water	7.5	18

polymers and biological samples. The effect of water has been noted in the SEM community for many years in papers and books [6]. This work was then applied by Stark *et al.* [7] to focused ion beam systems with great success. Table 3.2 shows a comparison of the etching enhancement of various gases on poly(methyl methacrylate) (PMMA) and clearly shows the efficiency of water over other gases including XeF$_2$.

The products from the reaction of water on polymers have not been determined but Stark *et al.* did speculate that CO$_x$ is formed and pumped away. However, this would seem to be unlikely in light of the results reported in the Si/XeF$_2$ system where the XeF$_2$ is able to chemisorb and attack the Si surface prior to the ion beam arriving. In this system, the rate limiting step is desorption since the gas reacts with the Si to form a saturated layer of fluorosilanes. The ion interaction with the fluoridated surface results in the maximum removal rate of Si of ∼25 atoms/ion [8]. These studies have also shown the products of the etching reaction to be SiF$_4$ and SiF$_2$. Comparing this to the etching data of water, where no reaction occurs prior to the arrival of the ion beam and the water is simply adsorbed to the surface, the authors observed an etching rate of ∼100 atoms/ion striking the surface. The rate is four times higher than that observed in the XeF$_2$/Si system. It would seem likely that the role of the water in the PMMA/water reaction is to cleave the polymer chains and the volatile products are chunks of monomer or methyl methacrylate fragments.

3.2.2 Inhibition of the etching process

The addition of gas is also able to inhibit ion beam sputtering of the sample. In the previous discussion the focus was on gases that produced acceleration in the etching rate of materials by the production of volatile products.

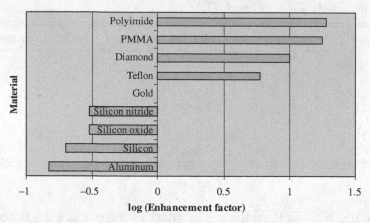

Figure 3.1 Water etch enhancement and inhibition for various materials [7].

In some cases there is no acceleration or even inhibition of the etching process. The situations where the gas does not produce any enhancement can be due to the gas not reacting with the substrate or the sputter rate of the product is the same as for the substrate. For a reduction of the etching rate to occur the gas and the sample must react to form nonvolatile products or products with a high binding energy.

Water is a good example of a gas that is able to produce all three results; enhancement, no change, and inhibition of etching. The chart below shows the log of the enhancement factor versus different materials and clearly shows enhancement, no affect, and inhibition. The enhancement factor is calculated by dividing the yield ($\mu m^3/nC$) with gas by the yield without gas:

$$\text{enhancement factor} = \frac{\text{yield with gas}}{\text{yield without gas}}. \tag{3.6}$$

It can be seen from Figure 3.1 that the etching rate of gold is unaffected by the addition of water because it is inert and does not react. The etching rate of aluminum or silicon is greatly reduced by the addition of water and is one-sixth the normal sputtering rate of the material under a 30 kV Ga^+ ion beam. This is very advantageous and allows the user to preferentially etch polymers from around aluminum lines in integrated circuits. The result of this etching is seen in Figure 3.2, where the surface layer of resist with contact holes has been removed to expose the profile of the contact. Note that the silicon exposed to the ion beam at the bottom of the hole does not show any apparent damage from the ion beam.

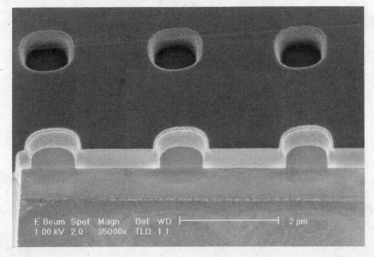

Figure 3.2 Water enhanced etching of Photoresist. (Courtesy of FEI Company.)

The presence of a chemically active gas is able to influence the rate of physical sputtering of a material by three factors:

(1) A modified surface is formed by the reaction of the gas with the material. For silicon and XeF_2 this will be a saturated SiF_x layer where $x = 1\text{-}3$ and for water the formation of SiO_{2-y}, $y < 1$.
(2) The binding energy of the surface atoms may increase, as in the case of silicon and water leading to inhibition of etching, or decrease as with silicon and XeF_2 leading to enhancement of the etch rate.
(3) The number of atoms of the original element or material in the top surface layers decreases.

For example, using water in the presence of aluminum and silicon creates a surface where most of the atoms in the top layer are oxygen which originate from the gas passed over the surface. As mentioned above since most of the sputtered material originates from the surface layer the ion beam is mostly removing the oxygen from the top oxide layer. These oxygen vacancies are rapidly replaced by the gas leading to a net decrease in the rate of sputtering of the aluminum or silicon in the sample.

The variation in the etching rate with different materials is very useful if many materials are present in a cross section and the user needs to see where they are. This is a common practice in the semiconductor industry where engineers need to see the layers of a device during manufacturing to monitor the FAB (fabrication plant for semiconductor materials) processes. This is called delineation etching and utilizes the varying etch rates of materials

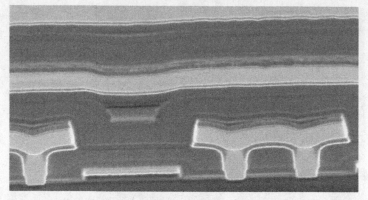

Figure 3.3 Results of delineation etching of a semiconductor device. (Courtesy of FEI Company.)

when exposed to a constant ion beam current and gas flow. Some gases are designed to have a great variation in etching rate and an example is delineation etch™ from FEI which is used to delineate the layers in semiconductor devices. The use of these gases can change a cross section from a blank and uninformative view into a series of topographic features based on the materials present. Figure 3.3 is an example of the end result of a delineated semiconductor device.

3.2.3 Etching with other ion sources

Other ion sources apart from gallium will also produce similar results. Noble gas ion sources have been used successfully to etch samples by sputtering and with the addition of reactive gases. There are differences in the etching rates achieved with the use of other ion sources due to the size and mass of the ion produced. The interaction of an ion passing through a solid is explained in depth in the book by Utlaut *et al.* [9]. The energy delivered to the sample is via the nuclear losses of the collision cascade as the ion passes into the substrate. The smaller and lighter the ion is the less energy it is able to deliver to the sample and hence a lower sputter yield results.

Coburn *et al.* [8] were able to demonstrate this clearly with various 1 kV noble gas ions in a set of experiments with silicon and chlorine etching gas. The heaviest gas used was argon and when used in conjunction with chlorine a maximum yield of four silicon atoms ejected per argon ion was observed. When helium was used instead of argon the yield dropped to less than one silicon atom per helium ion. Figure 3.4 is a summary of these results.

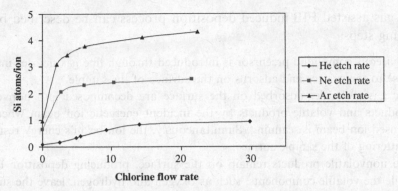

Figure 3.4 Yield of silicon atoms sputtered per 1 kV He^+, Ne^+, and Ar^+. The chlorine flow rate is 10^{15} molecules/s [8].

3.2.4 Sample damage

As with any physical process the ion beam will change the characteristics of the top layers of a sample. With a 30 kV gallium ion beam the top 30 nm of material is implanted with gallium and any atomic structure that was present will be altered or destroyed depending on the dose the sample received. This damage layer is very apparent in TEM samples obtained by FIB unless the operator is careful to follow known damage minimizing procedures, which can greatly reduced the thickness of the damaged layers.

Another effect of ion beam exposure is the movement of species within the top most layers of the sample. Material from the surface can move into the bulk and bulk material can move to the surface. This can be caused by recoil implantation, cascade mixing, or radiation enhanced diffusion. Thus, etching gases can be found implanted into the sample and not only on the top surface of the material [1].

3.3 Gas assisted focused ion beam deposition

In terms of ion beam assisted deposition, FIB induced deposition and conventional ion beam assisted deposition (IBAD) are similar. However, FIB induced deposition with gas assistance significantly differs from the IBAD. In general, ion beam assisted deposition is used in conjunction with other sources such as evaporation, sputtering, pulsed laser, and molecular beam epitaxy. The gas assisted FIB deposition is a direct deposition by using only the focused ion beam itself along with a gas. In addition, FIB induced deposition is also a deposition in a very local area only where the controlled ion beam is scanning.

The gas assisted FIB induced deposition process can be described by the following steps:

(1) A gaseous compound precursor is introduced through fine gas nozzles inserted close to the surface and adsorbs on the surface of the sample.
(2) The gas molecules adsorbed on the surface are decomposed into nonvolatile products and volatile products by the incident energetic ion beam where the focused ion beam is scanning. Simultaneously, the ion beam's energy results in sputtering of the sample surface.
(3) The nonvolatile products remain on the surface, producing deposition layers, while the volatile components, such as oxygen and hydrogen, leave the surface.

Deposition conditions can be optimized by controllable system parameters such as the properties of precursor gases, gas flux (needle location, the heating temperature of gas), ion beam current, dwell time, beam overlap, beam scanning area, and raster loop time.

Once the gas adsorbs to the surface of the sample, the precursor is required to have sufficient sticking probability to remain on the surface long enough before it is decomposed. Furthermore, the precursor on the surface must be easily decomposed by the incident energetic ion beam for any deposition to occur.

The incident ion beam decomposes the precursor gas on the surface as well as sputters or mills the materials on the surface. So the total deposition yield (Y_{dep}), the number of atoms deposited per incident ion, can be expressed as a function of the decomposition yield (Y_{decomp}) of the precursor gas and the sputtering yield (Y_{sputt}) on the surface:

$$Y_{dep} = Y_{decomp} - Y_{sputt}. \tag{3.7}$$

The decomposition yield is proportional to the number of adsorbed molecules on the surface and the cross section for molecular dissociation. The number of adsorbed molecules on the surface is known as a function of the gas flux onto the substrate, the substrate temperature, and the gas/substrate interaction on the deposited area [10].

For the success of deposition many beam parameters need to be adjusted properly. In general the deposition rate is a combination of local gas flux or local pressure, ion beam current, pattern size, dwell time per pixel, beam overlap, and raster refresh time. The following sections introduce the type of deposition precursor gases, and parameters to obtain good deposition.

3.3.1 Deposition precursor gases

In 1984, the first FIB induced deposition of Al was reported by Gamo *et al.* from metal organic material, trimethyl aluminum (TMA) $Al_2(CH_3)_3$ precursor [11].

Table 3.3 *Precursor gases for FIB induced deposition.*

Deposition Material	Precursor gas	Reference
Aluminum	Trimethyl aluminum (TMA) $Al_2(CH_3)_3$	[11]
Aluminum	Trimethylamine alane (TMAA)	[18]
Aluminum	Triethylamine alane (TEAA)	[18]
Aluminum	Tri-isobutyl aluminum (TIBA), $Al(C_4H_9)_3$	[19]
Tungsten	Tungsten hexafluoride WF_6	[20]
Tungsten	Tungsten hexacarbonyl, $W(CO)_6$	[21]
Tungsten	Tungsten hexafluoride, WF_6	[22]
Gold	Dimethyl gold hexafluoro acetylacetonate $C_7H_7O_2F_6Au$	[23, 24]
Platinum	(methylcyclopentadienyl) trimethyl platinum $C_9H_{16}Pt$	[25, 26]
Copper	Cu(hfac)TMVS	[27]
Tantalum	PMTA, $Ta(OC_2H_5)_5$	[20]
Tantalum	Pentaetoxy tantalum, $Ta(OC_2H_5)_5$	[20]
Iron	Iron pentacarbonyl, $Fe(CO)_5$	[19]
Palladium	Palladium acetate $[Pd(O_2CCH_3)_2]_3$	[28]
Insulator (TEOS):	$(C_2H_5)_4Si$	[29]
Silicon dioxide	Si(OCH3)4	[30]
Silicon dioxide	A combination of siloxane and oxygen gases	[31]
Carbon	–	[32]
Carbon	Naphthalene ($C_{10}H_8$)	[33]
PMMA	–	[34]

So far there are dozens of different precursors that have been successfully reported for FIB induced deposition of metal (aluminum, tungsten, tantalum, platinum, gold, copper, palladium, and iron), insulator (silicon dioxide), and carbon. The detail of precursor gases for the successful deposition is shown in Table 3.3.

As one condition of a successful deposition in the focused ion beam system the proper amount of gas precursor in a local area should be introduced by the focused ion beam system. If the gas flux is too high the deposition rate becomes slow at the same beam conditions. If the gas flux is too low all the gases quickly get consumed resulting in the dominant sputtering of the sample surface by the incident ion beam. The gas flux can be adjusted by the nozzle location, e.g. the height from the surface and the distance from the scanning ion beam, and the temperature of gas crucible. Figure 3.5 shows the schematic diagram of the gas needle location in a focused ion beam system. The gas flux or local pressure decreases as the height of the nozzle in the system increases. However, the nozzle must be high enough to be able to

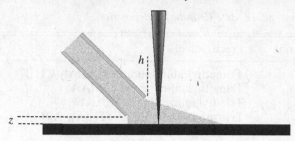

Figure 3.5 Schematic diagram of a gas needle.

work without it causing damage to the sample surface. When the nozzle location is too close to the beam location the incident ion beam can be perturbed by the nozzle due to extra field effects near the nozzle. Typically, most focused ion beam units use a mechanical system to insert and retract the gas nozzle. The location of the inserted nozzle must be consistent so that the gas flux does not change over time. The ideal nozzle height is generally a few hundred micrometers above the sample surface, although the optimum location varies with different systems.

The temperature of the gas crucible also affects the gas flux in the nozzle. A higher temperature causes a higher gas flux. However, if the gas crucible temperature is too high the gas chemistry will be changed.

3.3.2 *Definition of beam overlap*

Now we need to understand how the focused ion beam scans the sample surface. The operator sets the ion beam current acceleration voltage, focus, and stigmatism to define the beam diameter (D). The ion beam spot with diameter D scans within a pattern that is drawn by the operator. The pattern contains the digital pixels that indicate the location of the beam spot. At each pixel the ion beam stays for a certain period of time, called the dwell time, and decomposes the gas precursor on the surface to produce a deposition. Then the ion beam is blanked and moves with step size (S) to the next pixel. Depending on the step size of the beam we can also define the beam overlap (OL). The overlap of the scanning beam can be expressed as the following equation:

$$OL = \frac{D - S}{D}. \tag{3.8}$$

The beam overlap is one of the critical parameters in order to achieve successful deposition in FIB. If the step size is half the beam diameter, the first and second beam spots overlap by 50%. A negative overlap happens when the raster beam moves from one location to another with a larger step

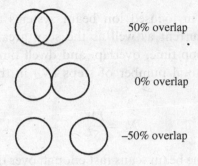

Figure 3.6 Examples of beam overlap.

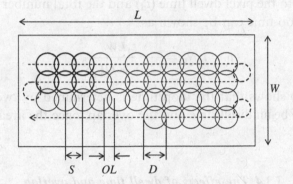

Figure 3.7 Schematic diagram of a digital pattern showing the step size (S), overlap (OL) and diameter (D) of the beam spot.

size than its spot size. Figure 3.6 shows an example of beam overlap schematically. When we set up the beam overlap in the system we need to take into account the Gaussian beam profile. Even though the beam is controlled digitally, the Gaussian beam profile of the beam spot contributes an extra ion beam current in the space between beam spots. In general, positive overlap is used for milling, zero overlap is used for etching, and negative overlap is used for deposition.

3.3.3 Beam control for deposition pattern

The beam scans over a patterned area through all individual pixels in one loop time. Once the ion beam has reached the last pixel of the pattern it moves back to the first pixel of the pattern. This ion beam scanning cycle repeats a number of times until it reaches the input setup time or achieves the correct deposition thickness. Figure 3.7 shows a digital pattern of an ion beam scan with an area of $L \times W$.

The deposition rate in focused ion beam systems also depends on the amount of ion beam current as well as the scan area and loop time. The relationship between loop time, overlap, and dwell time can be understood from Figure 3.7. The total number of steps (N_s) in the area $L \times W$ can be expressed as follows:

$$N_s = \frac{LW}{S^2}.$$

(3.9)

In practical situations the beam scans fast enough over the pattern so that the beam blanking time between pixels can be ignored. The loop time (t_l) is directly related to the pixel dwell time (t_d) and the total number of step in the pattern. The loop time can be shown as

$$t_l = t_d N_p = \frac{t_d LW}{(D(1 - OL))^2}.$$

(3.10)

Equation (3.10) shows that the loop time is determined by two operational parameters, the beam dwell time and the overlap, once the area of pattern is defined.

3.3.4 The effects of dwell time and overlap

Now we will take account of the effects of beam dwell time and overlap on deposition rate and quality. Figure 3.8 shows a series of SEM images of tungsten deposition on Si substrate with various overlaps and dwell times. The size of the drawn pattern was $6\,\mu m \times 1.5\,\mu m$ and the target thickness was $1\,\mu m$. For all depositions, a beam current of $1000\,pA$ was used. Figures 3.8(a) to (d) show tungsten deposition for a beam overlap of 50%, 0%, -150%, and -500%, respectively. For the depositions with positive and zero overlap, the thickness of the deposition appears uneven and does not reach to the target thickness. For positive beam overlap the high beam current density between beam spots can provide a strong sputtering yield during the deposition process. Figure 3.8(c) shows a good deposition with -150% overlap with target and even thickness. Figure 3.8(d) shows that the target deposition thickness can obtained with a deposition of -500% overlap. However, the deposition surface is not smooth. The space between the ion beam spot diameters is too large so that neither gas decomposition nor extra sputtering occurs in the space. Figures 3.8(e) and (f) show the deposition with an overlap of -150% and dwell times of $2\,\mu s$ and $20\,\mu s$, respectively. All gas precursors are consumed in the first few tenths of a microsecond of exposure to the beam. In addition, extra long dwell time provides a sputtering effect on

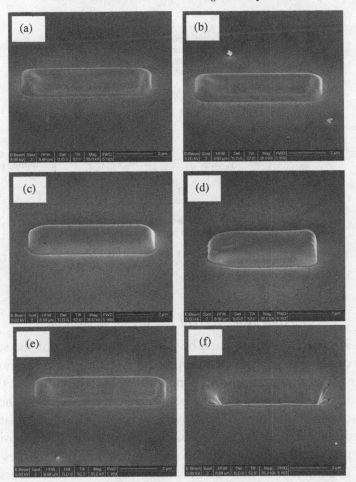

Figure 3.8 Images of a series of different overlaps and dwell times, with a viewing angle of 52°. (a) 0.2 μs dwell time, 50% OL, (b) 0.2 μs dwell time, 0% OL, (c) 0.2 μs dwell time, −150% OL, (d) 0.2 μs dwell time, −500% OL, (e) 2 μs dwell time, −150% OL, (f) 20 μs dwell time, −150% OL. (Courtesy of FEI Company.)

the deposited surface. If the dwell time is too long the deposited materials are sputtered away leaving a milling hole on the surface of the sample, as shown in Figure 3.8(f). The examples show that the proper setting of dwell time and overlap are essential to achieve a successful deposition in FIB systems. The optimum overlap and dwell time for good depositions can be achieved by series of experiments. A sophisticated system can store the optimum overlaps and dwell times for different deposition conditions so that the operator can easily load the optimum parameters when required.

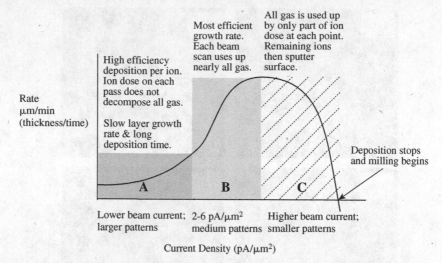

Rate
μm/min
(thickness/time)

High efficiency
deposition per ion.
Ion dose on each
pass does not
decompose all gas.

Slow layer growth
rate & long
deposition time.

Most efficient
growth rate.
Each beam
scan uses up
nearly all gas.

All gas is used up
by only part of ion
dose at each point.
Remaining ions
then sputter
surface.

Deposition stops
and milling begins

A   B   C

Lower beam current;   2-6 pA/μm²   Higher beam current;
larger patterns   medium patterns smaller patterns

Current Density (pA/μm²)

Figure 3.9 Deposition efficiency of Pt [12].

3.3.5 Ion beam current density on deposition

The ion beam current density must be taken into account for the success of deposition of material in the focused ion beam system. Figure 3.9 shows three distinct ranges in the graph of the deposition rate (μm/min) and the beam current density (A/μm²). At the low beam current density both the decomposition rate and sputtering rate are low, and result in a low deposition rate. At intermediate beam current density values, both the gas decomposition and sputtering rates increase. However, the decomposition rate reaches its maximum and is dominant so that the maximum deposition rate occurs. With a high ion beam current density the decomposition rate does not increase but the sputtering rate keeps increasing.

The range of intermediate beam current density is most efficient for the deposition rate, with minimum sputtering. As an example, it has been accepted that the most efficient deposition rates are within the current density range of 2–6 pA/μm² for $C_9H_{16}Pt$ (Pt deposition) and 100–150 pA/μm² for $W(CO)_6$ (W deposition) [12].

In practical situations it can be easier to use the deposition time rather than the beam current density. Once we know the time of maximum beam current density we can avoid unwanted sputtering as well as obtain the maximum deposition rate during deposition process. In this calculation the time of the intermediate current density range (2–6 pA/μm²) for Pt deposition is introduced. This calculation can be based on a Pt volume deposition yield of 0.5 μm³/nC at −50% overlap and 0.2 μs dwell time. For a simple calculation let's assume that we would like to deposit $100\,\mu m^3 (= 10\,\mu m \times 10\,\mu m$ and $1\,\mu m$ thick) of platinum. In this case, we need $0.6\,nA$ ($= 6\,pA/\mu m^2 \times 100\,\mu m^2$) beam current to have the beam current density of $6\,pA/\mu m^2$ over $100\,\mu m^2$. Based on the definition of deposition time the calculation will be:

$$\text{Deposition time} = \frac{\text{volume } (\mu m^3)}{\text{beam current (nA)} \times \text{volume deposition yield } (\mu m^3/nC)},$$

(3.11)

and the calculated deposition time for $6\,pA/\mu m^2$ is

$$\text{mill time} = \frac{100 \text{ cubicmicrometers}}{0.6 \text{ nanocoulombs per second} \times 0.5} = 333\,s = 5\,min\,30\,s.$$

Similarly, the deposition time for the boundary of $2\,pA/\mu m^2$ is about 16 min 37 s. So, an approximate corresponding time for a current density between $2\,pA/\mu m^2$ and $6\,pA/\mu m^2$ will be between 16 min 37 s and 5 min 30 s, respectively. So to minimize the unwanted sputtering during Pt deposition the deposition time must be set longer than 5 min 30 s for a thickness of $1\,\mu m$.

3.4 Filling metal in high aspect ratio vias

As the semiconductor industry moves toward smaller dimensions and greater numbers of charge carriers, the need for high aspect ratio vias increases. Aspect ratio can be defined as the ratio of the depth of a via to the width of the via at some given point. In today's integrated circuits, it may be necessary to make contact to a metal line that is more than $10\,\mu m$ deep, while at the same time avoiding making electrical contact to other metal traces that are in close proximity. This could mean that one would have to create a platinum filled via that is $10\,\mu m$ deep and only $0.50\,\mu m$ wide (an aspect ratio of 20:1). Certain challenges exist when attempting to make high aspect ratio vias:

• reaching the required depth;
• keeping critical dimensions small to avoid shorting with other traces;
• getting uniform, solid, metal deposition in the via;
• achieving low resistivity in FIB deposited metal within a high aspect ratio via;
• knowing when the desired track is contacted (end-point detection).

As you attempt to mill a deep via, it becomes difficult to reach the desired depth. It can be thought of as digging a trench while throwing dirt over your shoulder. As you get deeper and deeper, it gets more difficult to continue making progress. Another major concern, of course, is avoiding hitting other metal traces. This becomes challenging as the number of metallization layers increases and device planarization increases. As more time is spent milling a via, more edge rounding and widening of the via will occur. It can be challenging to avoid contact with other metal lines. Milling the via at low beam current decreases the effects of edge rounding and increases the aspect ratio of the via. Low beam current milling, however, decreases the quality of the end-point signal.

If the via cannot be filled solidly with metal, then it will have such a high resistivity that it will be of no use. To increase the chances of a solid platinum fill, the deposition is slowed by lowering the stage relative to the needle. Another option would be to decrease the gas flow from the injector. An organic carrier typically transports the metal that is deposited. When depositing platinum on the surface of a sample, there is a much greater solid angle for escape of the organic carrier than within a high aspect ratio via. As more of the organic carrier gets trapped in the platinum, the resistivity increases (due to the hydrocarbon-based carrier compound). By slowing down the deposition rate, we also increase the amount of time for the carbon-based material to escape from the hole, thus decreasing the resistivity.

Selecting the area for the via to be milled can be done by using a CAD layout, optical images, or by other means of referencing a feature visible in the FIB. Once the area is selected, the procedure outlined below can be used as a guideline. It is often difficult to obtain a sharp and definitive end-point trace when milling a high aspect ratio via. End-point detection done by measuring the stage current, which is essentially a measure of the amount of secondary electrons that are coming off the sample. When milling a very narrow and deep via, it is often difficult to get many secondary electrons out of this hole, thus making end-point detection more challenging. However, the end-point may not be as pronounced as an end-point for a via of lower aspect ratio. If possible, it is often helpful to characterize the amount of time it takes to reach a given metal line, and use this time as a gauge in conjunction with the end-point detection trace. Gas chemistry is essential for the via milling process. Through chemical reaction of the gas with the milled material, the material is removed and not redeposited. If gas is not used then only relatively low aspect ratio holes are possible due to the sputtered material not being able the escape the milled hole once the depth is approximately four times the width of the hole. Once this condition is reached further milling does not increase the depth and just results in sputtered material being pushed around inside the milling area by the ion beam. It has been found that, on most samples, the best results are obtained using insulator enhanced

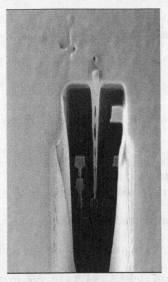

Figure 3.10 High aspect ratio hole milled in SiO_2 to an aluminum contact. Image has been stretched to reflect edge-on viewing. (Courtesy of FEI Company.)

etch gas such as XeF_2. It has also been found that the use of a stovepipe needle is a necessity when milling a high aspect ratio via. The beam must be paused periodically on the deposition process. During the beam pausing time the byproduct gases in the via come out and fresh deposition gas precursors go in. Without beam pausing the deposition of the large void will be left inside the void. The desired beam on/off time is about 10/90. Figure 3.10 shows the results of milling a high aspect ratio via in an integrated circuit and filling platinum in the via.

References

[1] H. F. Winters and J. W. Coburn. *Surface Sci. Rep.*, **14** (1992), 161–270.
[2] S. Joyce, J. G. Langan and J. I. Steinfeld. *Surf. Sci.*, **195** (1988), 270–82.
[3] M. A. Loudiana and J. T. Dickinson. *J. Vac. Sci. Technol. B*, **3** (1985), 1393–6.
[4] Y. Y. Tu, T. J. Chuang and H. F. Winters. *Phys. Rev. B*, **23** (1981), 823–35.
[5] H. F. Winters. *J. Vac. Sci. Technol. B*, **1** (1983), 927–31.
[6] J. J. Hren. *Introduction to Analytical Electron Microscopy*, ed. J. J. Hren, J. I. Goldstein and D. C. Joy (New York: Plenum, 1979), p. 481.
[7] T. J. Stark, G. M. Shedd, J. Vitarelli, D. P. Griffis and P. E. Russell. *J. Vac. Sci. Technol. B*, **13** (1995), 2565–9.
[8] U. Gerlach-Meyer, J. W. Coburn and E. Kay. *Surf. Sci.*, **103** (1981), 177–88.
[9] J. Orloff, M. Utlaut and L. Swanson. *High Resolution Focused Ion Beams* (New York: Plenum, 2003).

[10] A. D. Dubner and A. Wagner. *J. Appl. Phys.*, **66** (1989), 870–4.

[11] K. Gamo, N. Takakura, N. Samoto, R. Shimizu and S. Namba. *Jpn. J. Appl. Phys.*, **23** (1984), L293–5.

[12] *Platinum Deposition Technical Note* (Hillsboro, OR: FEI Company, 2003), PN 4035 272 21851-B.

[13] R. J. Young, J. R. A. Cleaver and H. Ahmed. *J. Vac. Sci. Technol. B*, **11** (1993), 234–41.

[14] J. D. Casey Jr, A. F. Doyle, R. G. Lee and D. K. Stewart. *Microelectron. Eng.*, **24** (1994), 43–50.

[15] M. L. Thayer. Tutorial. *IEEE Int. Reliability Phys. Symp.* (1996).

[16] A. Yamaguchi and T. Nishikawa. *J. Vac. Sci. Technol. B*, **13** (1995), 962–6.

[17] R. J. Young. *Vacuum*, **44** (1993), 353–6.

[18] M. E. Gross, L. R. Harriott and R. L. Opila, Jr. *J. Appl. Phys.*, **68** (1990), 4820–4.

[19] R. L. Kubena, F. P. Stratton and T. M. Mayer. *J. Vac. Sci. Technol. B*, **6** (1988), 1865–8.

[20] K. Gamo, N. Takehara, Y. Hamaura, M. Tomita and S. Namba. *Microelectron. Eng.*, **5** (1986), 163–70.

[21] D. K. Stewart, L. A. Stern and J. C. Morgan. *Electron Beam X-ray and Ion Beam Technologies: Submicrometer Lithographies VIII, Proc. SPIE*, **1089** (1989), 18–25.

[22] Z. Xu, T. Kosugi, K. Gamo and S. Namba. *J. Vac. Sci. Technol. B*, **7** (1989), 1959–62.

[23] G. M. Shedd, A. D. Dubner, C. V. Thompson and J. Melngailis. *J., Appl. Phys. Lett.*, **49** (1989), 1584–6.

[24] P. G. Blauner, J. S. Ro, Y. Butt and J. Melngailis. *J. Vac. Sci. Technol. B*, **7** (1989), 609–17.

[25] T. Tao, J. S. Ro, J. Melngailis, Z. Xue and H. Kaesz. *J. Vac. Sci. Technol. B*, **8** (1990), 1826–9.

[26] J. Puretz and L. W. Swanson. *J. Vac. Sci. Technol. B*, **10** (1992), 2695–8.

[27] A. D. Della Ratta, J. Melngailis and C. V. Thompson. *J. Vac. Sci. Technol. B*, **11** (1993), 2195–9.

[28] L. R. Harriott, K. D. Cummings, M. E. Gross and W. L. Brown. *Appl. Phys. Lett.*, **49** (1986), 1661–2.

[29] R. J. Young and J. Puretz. *J. Vac. Sci. Technol. B*, **13** (1995), 2576–9.

[30] H. Komano, Y. Ogawa and T. Takigawa. *Jpn. J. App. Phys.*, **28** (1989), 2372–5.

[31] D. K. Stewart, A. F. Doyle and J. D. Casey, Jr. *Electron-Beam, X-Ray, EUV, and Ion-Beam Submicrometer Lithographies for Manufacturing V, Proc. SPIE*, **2437** (1995), 276–307.

[32] P. J. Heard and P. D. Prewett. *Microelectron. Eng.*, **11** (1990), 421–5.

[33] *Carbon Deposition Technical Note* (Hillsboro, OR: FEI Company, 2003), PN 4035 272 27241-A.

[34] T. Shiokawa, Y. Aoyagi, P. H. Kim, K. Toyoda and S. Namba. *Jpn. J. Appl. Phys.*, **23** (1984), L232–3.

4

Imaging using electrons and ion beams

KAORU OHYA

The University of Tokushima

TOHRU ISHITANI

Hitachi High-technologies Corporation

4.1 Overview of imaging using FIB

Scanning electron microscopes (SEM), transmission electron microscopes (TEM), and scanning TEM (STEM) have been used for structural observation of microdevices, advanced materials, and biological specimens. In recent years, a gallium (Ga) focused ion beam (FIB) has been used for preparing their cross-sectional samples [1,2,3, and Chapter 9 in this book]. Here, FIB works both as a milling beam and as a probe for a scanning ion microscope (SIM). SIM images are used during the whole milling processes, that is, drawing the milling area, milling monitoring, confirmation of the final milling, and observation of the FIB milled sections of interest. As SIM image resolution has been improved and about 5 nm is achievable at present, SIM observation has been increasingly used in place of SEM observation when there is no especial need for high image resolution.

However, we have found that the properties of SIM images are somewhat different to SEM images. For example, SIM and SEM images of identical FIB milled cross sections are shown in Figures 4.1(a)–(d): (a) and (c) show a solder (Pb-Sn) on copper (Cu) and (b) and (d) show a Si device (static random access memory) [4]. Black-and-white contrast among materials is opposite between the SIM and SEM images. In addition, grain contrast (or channeling contrast) can be observed more clearly in the SIM image and its contrast is sometimes stronger than the material contrast. Figure 4.2 shows

Focused Ion Beam Systems: Basics and Applications, ed. N. Yao.
Published by Cambridge University Press. © Cambridge University Press 2007.

Figure 4.1 Comparison of image contrast between 30 keV Ga SIM and 5 or 10 keV SEM images of the FIB cross-sectioned samples (by the SEM/ FIB dual beam system of Hitachi S-3000FB): (a,c) solder (Pb-Sn) on Cu base and (b,d) Si device (static random access memory) [4]. The solder is covered with a carbon (C) layer. In the SIM image (c), the Cu grains show various contrasts depending on their crystal orientations, that is, grain contrast or channeling contrast. Main elements are shown on the images with their atomic numbers in parentheses.

SIM and SEM images of FIB cross-sectioned human hair [5]. The SIM image shows cuticles more clearly than the SEM image, suggesting that SIM images are more sensitive to the surface than SEM images. The opposite material-contrast between SIM and SEM images has been reported by others [1,6,7], but the origins of it have been discussed in only a few papers [7,8]. Greater utilization of SIM images depends on understanding the secondary electron (SE) properties under ion impact. In this chapter we review the SE properties under ion impact using Monte Carlo (MC) simulation, compare it with those under electron impact, and clarify the origins of the essential differences in characteristics between the SIM and SEM images.

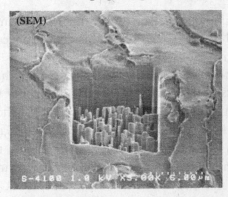

Figure 4.2 Comparison of image contrast between 30 keV Ga SIM and 1 keV SEM images of the FIB cross-sectioned human hair [5].

4.2 Background physics for imaging formation in SIM

4.2.1 Comparison in penetration behaviors between energetic ions and electrons in solids

Before focusing on the SE properties, we will briefly describe the penetration behaviors of energetic ions [9,10], compared with energetic electrons. However, only the basics of the electron penetration are described (refer to the texts [11,12] for more discussion on the equivalence). Here, the beam characteristics of the Ga-FIBs for SIM imaging and/or FIB milling of interest are as follows: an energy of 10–40 keV, a current of 1 pA–30 nA (i.e., a difference of over four orders of magnitude), a diameter of about 5 nm to 1 μm (i.e., a difference of over two orders of magnitude), and a current density of 0.1 to several tens A/cm^2. For the SEM imaging, on the other hand, the primary electron energy is mostly in the region of 1 to 30 keV for conventional SEM systems and 100 to 200 keV for high-voltage (HV) SEM imaging in TEM/STEM.

Interaction of energetic ions or electrons with solids

Interaction of energetic ions or electrons with solids is slightly complex. Before ultimately losing their energies or escaping the sample, each incident electron or ion may undergo a large number of scattering events, distributed between elastic and inelastic processes. While elastic collisions result from collisions of energetic electrons or ions with nuclei of the target atoms and alter their undergoing directions, inelastic collisions result in transfer of their energies to the target, leading to generation of SE, Auger electrons, photons, electron hole pairs, and so forth. The MC method using a stepwise simulation of various scattering events is useful to microscopically understand beam interactions (refer to the special issue on electron beam/specimen interaction modeling in *Scanning* **17**, 4 and 5, 1995). MC programs also allow estimating the various signals, yields, distributions, and so forth.

Electron and Ga ion trajectories are MC simulated so that their beam interactions are visually understood. Figures 4.3(a) and (b) show their trajectories in silicon (Si) and tungsten (W) targets, respectively, where beam energies E_0 are 30 keV and 1.5 keV for electrons and 30 keV for Ga ions [9]. Here, Si and W are chosen as typical lighter and heavier atomic mass elements relative to Ga (i.e., atomic mass $M = 28.1$, 69.7, and 183.5 amu for Si, Ga, and W, respectively). A mass ratio of the strike to struck particles governs the elastic scattering, as will be discussed later. The MC programs for electrons and ions employed here are based on single scattering models proposed by Joy [13] and Ishitani [14], respectively.

Range

The electron interaction volume indicates little dependency on E_0 and atomic number of the target atoms (Z_2) except for its scale magnification. The 30 keV ranges R (defined as a maximum projected range) are about 6 and 0.6 µm for Si and W, respectively, and the 1.5 keV R are about 0.05 and 0.01 µm for Si and W, respectively. Although a factor of 10 exists in 30 keV R between Si and W targets, there is only a slight difference between their reduced mass ranges in µg/m^2. It is known that an approximately Z_2-independent mass range is valid at $E_0 = 10$–100 keV [15]. On the other hand, the ranges R for 30 keV ions are as short as only about 0.04 and 0.02 µm for Si and W, respectively, and are roughly equal to R for 1.5 keV electrons. In addition, a Z_2-independent mass range is no longer valid for ion R.

Backscattering

The larger the sample Z_2, the higher the backscattering yield η for the electron beam. On the other hand, η for an ion is rather low compared with that for an electron. This is expected as Ga ions are never deflected backwards in Ga–Si elastic two-body collisions because $M_{Si}/M_{Ga} < 1$ (see *Elastic collision*, p. 92).

Total path length

As to total path length (defined as an accumulated path length along each zigzag trajectory of the electrons or ions stopped in the sample), there is little difference among the electrons. The reason is that electrons suffer negligible energy loss in elastic two-body collisions because $M_{target-atom}/M_{electron} \gg 1$. On the other hand, the larger the ion deflection angle, the larger the ion energy loss (see the next subsection). Thus, the ion range after the large-angle

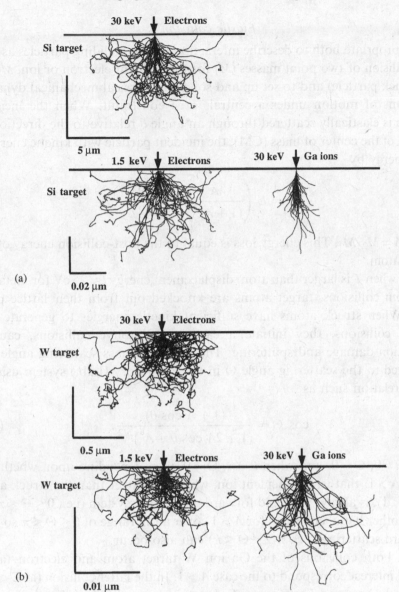

Figure 4.3 Monte Carlo simulations of 30 keV and 1.5 keV electron and 30 keV Ga$^+$ ion trajectories in the targets: (a) Si target and (b) W target. The number of trajectories drawn for each plot is 50 [9].

scattering becomes short on average. If the energy transferred from the ion to the target atom exceeds the sharp threshold energy, the target atom is displaced and is added to the cascade with kinetic energy. This cascade brings about lattice damage and sputtering.

Elastic collision

It is appropriate both to describe interaction of two colliding particles as that of a collision of two point masses (M_1 for the incident electron or ion, M_2 for the struck particle) and to set up and solve the classical mechanical dynamic equations of motion under a central force constraint. When the incident particle is elastically scattered through an angle θ relative to the direction of motion of the center of mass (CM), the incident particle with kinetic energy E loses energy by

$$T = E \left(\frac{4A}{(1+A)^2} \right) \sin^2 \left(\frac{\theta}{2} \right), \qquad (4.1)$$

where $A = M_2/M_1$. This energy loss is equal to the post-collision energy of the struck atom.

Only when T is larger than atom displacement energy ($\approx 25\,\text{eV}$ for metal) in ion–atom collisions, target atoms are knocked out from their lattice positions. When struck atoms have sufficient kinetic energies to generate secondary collisions, they initiate a cascade of atomic collisions, causing irradiation damage and sputtering. The center of mass scattering angle θ is converted to the scattering angle Θ in the laboratory (Lab.) system using a simple relation such as

$$\cos \Theta = \frac{(1 + A \cos \theta)}{(1 + 2A \cos \theta + A^2)^{1/2}}. \qquad (4.2)$$

When $\theta = 0$, $\Theta = 0$, and when $\theta = \pi$, $\Theta = 0$ or π, depending upon whether $A < 1$ or $A > 1$, that is, the incident ion, which is heavier than the struck atom (i.e., $A < 1$), is always scattered forward in the Lab. system (i.e., $0 \leq \Theta \leq \pi/2$). On the other hand, collisions for $A > 1$ cover a full range of $0 \leq \Theta \leq \pi$ so that backward scattering (i.e., $\pi/2 < \Theta \leq \pi$) can also occur.

Here, both collisions of the Ga ion–W target atom and electron–target atom of interest correspond to the case $A > 1$. In the latter collision (at $E < 100$ keV), both $\Theta \approx \theta$ and $T \approx 0$ are satisfied at any θ because $A \gg 1$. In other words, primary electrons can change their directions without losing their kinetic energies through elastic collisions. No target atoms are knocked out from their lattice positions by the electrons of inorganic substances such as metals, semiconductors, etc. On the other hand, organic substances (such as amino acids, higher hydrocarbons, and aromatic compounds), which are little taken up in the present study, are much more sensitive to radiation damage than inorganic specimens.

Inelastic collision

Regarding the approach to ion inelastic scattering, the continuous slow-down approximation has been widely used. A schematic representation of the stopping power combines two energy regimes. The stopping power increases from zero, passes through a maximum when the incident velocity v is of the order of the orbital velocities of lattice electrons ($\approx Z_2^{2/3} v_B$), and finally falls off inversely as the first power of energy. Here, $v_B = 2.2 \times 10^6$ [m/s] is the velocity of the Bohr electron in the hydrogen atom. In the lower velocity region (i.e., $v < Z_1^{2/3} v_B, Z_2^{2/3} v_B$), a v-proportional stopping power is derived. The velocity of $v = Z_1^{2/3} v_B$ is converted to energy of E [keV] $\approx 25 Z_1^{4/3} M_1$ [amu], which corresponds to 25 keV for the H ion and 170 MeV for the Ga ion. Using dimensionless energy ε and distance ρ, v-proportional electronic stopping power is given as (e.g., by Dearnaley *et al.* [16], Ryssel and Ruge [17], and Orloff *et al.* [10])

$$-\left(\frac{d\varepsilon}{d\rho}\right)_{electronic} = \kappa \varepsilon^{1/2}. \tag{4.3}$$

The incident energy E_0 of several tens keV has too low a velocity for the Ga ions to excite X-rays, in contrast to the case for electrons. The X-ray excitation requires the incident's velocity to be in the same order as that of the target-atom shell-electrons in classical mechanics. The SE excitation of interest is in the inelastic processes.

Stopping power

The stopping power of an energetic particle in a solid results from a sum of two components (i.e., the nuclear and electronic stopping powers), which may be taken in good approximation as independent of each other, and the total stopping power given by

$$-\left(\frac{dE}{dR}\right)_{total} = -\left(\frac{dE}{dR}\right)_{nuclear} -\left(\frac{dE}{dR}\right)_{electronic}. \tag{4.4}$$

The nuclear stopping depends on the cumulative effect of statistically independent elastic scattering of the incident particle and the target atom. The stopping power is given by

$$-\left(\frac{dE}{dR}\right)_i = -\left(\frac{d\varepsilon}{d\rho}\right)_i \left(\frac{(\rho/R)}{(\varepsilon/E)}\right), \tag{4.5}$$

where i = nuclear, electronic, or total. Here, $-(d\varepsilon/d\rho)_i$ is the i-component universal stopping power per atom. Values of ε/E, ρ/R, and κ for the

Table 4.1. *Various parameters in the beam–specimen interactions.*

Target elements	Atomic number Z	Atomic mass M [amu]	Density ρ [kg/m^3]	M/M_{Ga}	LSS parameters		
					e/E [keV]	ρ/R [μm]	κ
Si	14	28.086	2.42×10^3	0.4	5.44×10^{-3}	18.7	0.117
W	74	183.85	1.93×10^4	2.64	1.96×10^{-3}	11.8	0.287

Ga: $Z_{Ga} = 31$ $M_{Ga} = 69.72$

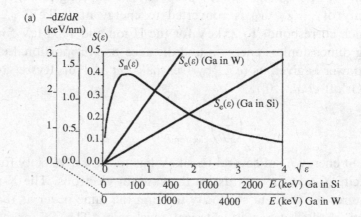

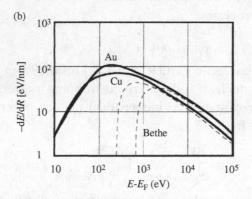

Figure 4.4 Comparison of stopping power between ion and electron; (a) ion stopping powers for Si and W targets [9], and (b) electron stopping powers for Au and Cu targets [18].

combination of Ga ions with Si and W targets are given in Table 4.1. The curves of $S_n(\varepsilon)$ [$\equiv -(d\varepsilon/d\rho)_{nuclear}$] and $S_e(\varepsilon)$ [$\equiv (d\varepsilon/d\rho)_{electronic}$] versus ε are shown in Figure 4.4(a). The E and $-(dE/dR)$ axes, which are reduced for the combinations of Ga ion with Si and W targets, are also given in the figure.

It is found that $S_n(\varepsilon)$ is dominant in energy loss for the 30 keV Ga ion penetration in Si and W targets. As to the electron stopping power, the nuclear term in (4.5) is negligible as discussed before. The electron inelastic stopping powers for Au and Cu targets [18] are shown in Figure 4.4(b). The stopping power has a maximum plateau of 60–100 eV/nm in a range of $0.1 \, keV < E < 1 \, keV$. The Bethe equation is valid only at sufficiently high electron energies, i.e., $E > 3 \, keV$ [15]:

$$-\left(\frac{dE}{dR}\right) = \left(\frac{2\pi e^4 Z_2 N}{E}\right) \ln\left(\frac{1.166E}{I_{av}}\right). \tag{4.6}$$

It was found that the electron stopping powers at $E_0 = 1$–30 keV are smaller by one to two orders of magnitude than those for Ga ions. This predicts that Ga ion ranges are shorter by one to two orders of magnitude than electron ranges under the same incident energies.

4.2.2 Ion-induced secondary electron emission

Bombardment of a solid surface by an ion causes secondary electron (SE) emission through many kinds of ion–solid interactions. Within the past few decades considerable progress has been achieved in the research field of ion induced SE emission from solid surfaces both experimental and theoretical respects, as reviewed in [19–25].

Ion-induced SE emission may proceed via two independent effects which differ by the mechanism of the energy transfer from the incoming ion to the electrons of the solid. If the potential energy of the ion is twice (or more) the work function of the solid, SE emission may proceed via resonance neutralization and subsequent Auger de-excitation or Auger neutralization. This process, which proceeds in front of the solid surface and operates at low (several keV) energies, is called potential emission. Its basic principles were developed by Hagstrum [26,27]. As long as singly charged ions are concerned, an empirical formula on the yield of SEs, γ_p, was obtained by Baragiola *et al.* [28] from a least-squares fit to experimental data on potential emission:

$$\gamma_P = 0.032(0.78E_i - 2\Phi), \tag{4.7}$$

where E_i and Φ are the ionization potential of the projectile ion and the work function of the solid surface in eV. However, gallium (Ga) ion bombardment will not lead to potential emission because of its low ionization potential ($\sim$6 eV). The Φ for normal metals is approximately 4–5 eV.

The second effect which covers all other sources of SE emission and which shall be discussed here is called kinetic emission. In this case SEs are excited within the solid by direct transfer of kinetic energy from the impinging ion. The SEs that are excited near the surface may eventually be emitted if their energy is high enough to overcome the surface barrier. The kinetic emission is the major or only source of SEs at medium and high impact energies.

The kinetic emission is strikingly similar to electron-induced SE emission which occurs for primary electron bombardment. The emission of SEs is described by a three-stage process:

(1) production of SEs within a solid;
(2) migration of some of these SEs to the surface;
(3) escape through the surface.

The last two processes in kinetic emission are almost identical to those of electron-induced SE emission. The important properties, i.e., characteristic of the migration and escape, the energy-loss rate, the mean free path for elastic scattering, and the magnitude and shape of the surface barrier, are common for both primary particles. The difference between the SE emission induced by the two types of primary particle is primarily in the production stage.

For electron bombardment the emitted electrons are conventionally divided into two groups: the "true" low-energy SEs with energies below 50 eV, and the backscattered (or reflected) electrons (BSE) and high-energy SEs with energies from 50 eV up to the primary energy. This division is formal, since SEs from ionization events may also occur above 50 eV, and primary electrons may have slowed down to be a few eV before ejection. The SEs induced by ion bombardment are not expressed in a similar manner, because there is no backscattered electron and only a minor fraction of the SEs have energies above 50 eV.

There is a clear difference in the energy dependence of the SE yield between electron and ion bombardments in the energies of tens of keV, because of a large difference between electron and ion masses. At a given impact energy, the initial velocity of a primary electron is two or three orders of magnitude larger than that of an ion. This causes a large difference between the mean free paths for excitation of an electron by primary electrons and by ions. In general, therefore, the SE yield has a maximum for primary electron energies below 1 keV, whereas the corresponding maximum is at the ion energies of hundreds of keV. As a result, at the energies of tens of keV, the SE yield for electron bombardment becomes small as the impact energy is increased, whereas an opposite trend is found for ion bombardment. The general energy dependence of the SE yield closely resembles that of the electronic stopping power of the impinging fast, light ions [29,30]. It must be admitted that the same simple relation does not hold for impact by slow, heavy ions.

Excitation of secondary electrons in solids

Different excitation mechanisms are responsible for the SE emission phenomena. The interaction of the impinging particle (v in velocity) with the conduction electrons results, first, in the ionization of single conduction electrons (commonly also called "excitation" of electrons from the Fermi sea). Second, for impact energies above a projectile- and target-dependent threshold energy the impinging particle can create plasmons in real metals. The decay of these plasmons leads to an excitation of conduction electrons. Third, the interaction of the impinging protons with the core electrons results in the direct excitation of these electrons. The inner-shell vacancies produced in this way are filled immediately by electrons from occupied higher states. At the same time another electron would be excited.

The maximum energy transfer in such a binary encounter to a conduction electron with the Fermi velocity, v_F, is

$$\triangle E_{max} = 2m_e \left[v + \frac{v_F}{2} \right]^2, \tag{4.8}$$

where m_e is the electron mass [31]. This equation applies equally well for bound electrons with a kinetic energy determined from the orbit velocity. This process is important mainly for incidence of light ions. This includes electron production induced by the projectile ion or by secondary processes (e.g., ionization by recoil target atoms or by energetic SEs); for proton impact on metals the excitation of conduction electrons of the target is generally assumed to be the major source of SEs.

The SE production processes at low-velocity ion impact ($v < v_B$, where v_B is the Bohr velocity) are characterized by the fact that the target electrons are much faster than the projectile ion; for 30 keV Ga ions, $v = 0.13v_B$. Instead of regarding a fast ion that hits an electron at rest, one has to consider a slow ion which is hit by a fast orbital or conduction electron. By such a collision the energy transfer from the ion to the electrons may be adequate to eject it from the atom excite it considerably above Fermi energy.

For low-velocity heavy ions electron promotion is the dominant process [19]. If the collision between the ion and a target atom proceeds slowly, a temporary molecule can be formed. One or more electrons may be lifted up to higher levels than before the collision and ejected from the atoms immediately after the collision. Electron promotion can be efficient in producing inner-shell vacancies that may lead to electron emission via Auger transitions [21].

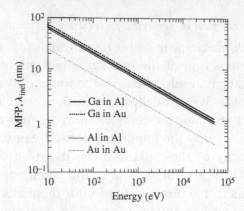

Figure 4.5 Energy dependence of MFPs for electron excitation by a Ga ion, and recoiled Al and Au atoms in Al and Au [34]. The MFPs are calculated using the nonlinear response theory developed by Echenique *et al.* [32].

A slow ion (with velocity v lower than the velocity of the conduction electrons of the target) represents a strong perturbation to the collision system. A linear response theory for fast ions is not valid to treat the slow ion–target interaction [32]. The recent nonlinear response theory of the electronic stopping power expresses the electron excitation in a free electron gas, in terms of elastic scattering of the electrons at the Fermi energy, E_F, with the intruding ion [33]. According to the nonlinear response theory, the inverse mean free path (MFP) of projectile ion or recoiled target atoms to excite the electron of interest is calculated as:

$$\frac{1}{\lambda_{inel}} = \frac{3\pi n v}{4\sqrt{2}v_F^2}\sum_{l=0}^{\infty}\sum_{m=0}^{\infty}(2l+1)(2m+1)$$
$$\times\{1-\cos 2\delta_l(E_F)-\cos 2\delta_m(E_F)+\cos[2(\delta_l(E_F)-\delta_m(E_F))]\}$$
$$\times\int_{-1}^{1}(1-x)^{1/2}P_l(x)P_m(x)\mathrm{d}x, \tag{4.9}$$

where, $\delta_{l,m}$ are the phase shifts for the scattering of a conduction electron at E_F by the potential of the intruding ion or the recoiled target atom at rest. The quantities n and $P_{l,m}$ are the density of conduction electrons and the Legendre polynomials, respectively. Figure 4.5 shows the energy dependence of the calculated MFPs of a Ga ion, and recoiled target atoms in aluminum (Al) and gold (Au). The inverse MFP is directly proportional to v (or the square root of the energy) and n, so that the number of excited electrons increases with increasing v and depends upon the electron shell structure of the target atoms, and the atomic number, Z_2.

Transport of secondary electrons to solid surfaces

The second stage in kinetic emission is the transport of excited SEs to the surface. Energetic electrons may lose their original energy in a sequence of inelastic collisions to other target electrons, including an electron multiplication by the cascade process.

For a metal with conduction electrons the electron excitations are very efficient in the slowing down processes. The stopping power for electrons in metals has a maximum between 100 eV and 1 keV (see Figure 4.4), and depends strongly on the density of conduction electrons. For electron excitation by a primary electron, the differential inverse MFP, $\tau(E, \omega)$, of an electron with an energy E is described by the complex dielectric response function, $\text{Im}[-1/\varepsilon(\omega, k)]$, i.e.,

$$\tau(E, \omega) = \frac{1}{\pi E} \int_{k_-}^{k_+} \frac{dk}{k} \text{Im}\left[-\frac{1}{\varepsilon(\omega, k)} \right], \qquad (4.10)$$

and the integration of $\tau(E, \omega)$ over the allowed values of k yields the inverse MFP. Here ω and k are the energy and momentum transfers to conduction electrons, and $k\pm = 2^{1/2}[E^{1/2} - (E - \omega)^{1/2}]$ in atomic units. Optical data of the target material for the $k = 0$ limit, $\varepsilon(\omega, 0)$, can be connected to $\varepsilon(\omega, k)$ according to an "optical-data" model developed by Penn [35], and by Ashley [36]. Since the optical data are based on the experimental results, it includes complicated processes of the inter-band, intra-band, and some other transition mechanisms automatically, in addition to the single and bulk-plasmon excitation of conduction electrons, which can be treated by the Lindhard dielectric function. Figure 4.6 shows the energy dependence of the calculated MFPs of electrons in Al and Au.

The electron energy loss is substantially influenced by the possible existence of an energy gap E_g, i.e., whether or not the material is an insulator. Electrons, which possess less energy than E_g, cannot lose energy by electronic excitation. Therefore, the energy loss of SEs in the insulator takes place via inefficient collisions with core atoms though their density fluctuation, i.e., phonons.

The elastic scattering is caused by the core atoms, but electron–electron collisions may lead to directional changes as well. The frequent collisions lead to a completely isotropic distribution of SEs in the solid. The MFP for elastic scattering can be calculated using the partial wave expansion cross section with appropriate solid-state potential [37] or using screened Rutherford formula where the energy-dependent screening parameters are used [38].

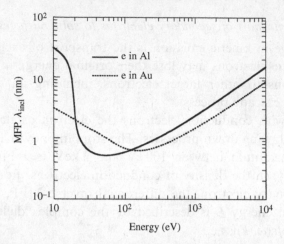

Figure 4.6 Energy dependence of inelastic MFP of electrons in Al and Au [34]. The MFPs are calculated using the optical-data model by Ashley [36].

Escape of secondary electrons in vacuum

The third stage, the ejection over the surface barrier, accounts for the selection of the SEs that are emitted. Since the majority of the SEs have an energy in the ranges of eV and tens of eV, the SE emission is strongly affected by the surface barrier, the magnitude of which ranges from a few eV up to 20 eV. The commonly accepted treatment implies that the SEs have to pass a barrier of height, $U = \Phi + E_F$, which is the distance from the bottom of the conduction band to the vacuum level, in such a manner that the energy component parallel to the surface is conserved. The component perpendicular to the surface is reduced with U, so that the internal threshold energy for escape is $E = U$.

The kinetic emission occurs if the projectile ion gives a conduction electron sufficient kinetic energy to overcome the surface barrier U. This leads to a threshold energy for kinetic emission, $E_{th} = (1/2)M_1 v_{th}^2$, where $v_{th} = (1/2) v_F[(1 + \Phi E_F)^{1/2} - 1]$ and M_1 is the ion mass. The threshold energy for Ga ions is 10.7, 18.8, and 26.6 keV for Al, Cu, and Au, respectively, which is inconsistent with experiments on the kinetic electron emission observed at only a few keV. Such inconsistencies have already been observed for other heavy ions [39,40]. This has been considered to be due to the electron promotion mechanism modeled recently for electron excitation within kinetic emission processes [41], including the inner shell electrons, as well as conduction electrons. The mechanism promotes the electrons to higher and vacuum levels during close collisions with material atoms. For simplicity, the kinetic emission below an energy E_{th} may be also treated as a reduction in the surface barrier energy U; e.g., $0.5E_F + \Phi$, which has been arbitrarily chosen. With decreasing surface barrier energy, the SE yield increases and the threshold energy is significantly reduced. For electron-induced SE emission, however, the reduction of the surface barrier energy causes a much smaller increase in the SE yield [42].

For insulators the surface barrier U is determined by the electron affinity E_A. Usually, the value of E_A from 0 to 1 eV. The average electronic stopping power of insulators for ions is similar for metals of comparable atomic numbers, but the stopping power for the migration of low-energy SEs and the magnitude of surface barrier are much smaller for the insulating material. As a result, the SE yield from insulators is much larger than the yield for metals of comparable atomic numbers.

4.2.3 Monte Carlo simulation of ion-induced kinetic secondary electron emission

As mentioned above, the kinetic emission is modeled in SE excitation in a solid, transport of the SEs to the surface, and their escape through a surface barrier. The SE is excited by three collision processes; one due to collisions between projectile ions and target electrons, one due to collisions between recoiled target atoms and target electrons, and one due to collisions between excited SEs and other target electrons.

The basic concept for a Monte Carlo model is to simulate trajectories of a projectile ion penetrating into the target bulk and of recoiled target atoms and excited SEs traveling towards the surface in the basis of the binary collision approximation [43] with given mean free paths (MFPs) for elastic and inelastic collisions. The motion of projectile ions and recoiled target atoms are treated in the same way: the straight free flight path is determined from the total MFP, λ_{tot}, defined as $1/\lambda_{el} + 1/\lambda_{inel}$, using a random number. The elastic MFP, λ_{el}, is fixed at $N^{-1/3}$ where N is the atomic density of the target. Depending on each of the inverse MFPs, λ_{el} and λ_{inel}, either elastic collision or inelastic collision (i.e., SE excitation) is chosen using another random number.

If elastic collision is chosen, the scattering angle is determined using a so-called scattering integral with appropriate interatomic potential, or any asymptotic procedure [44]. The elastic energy loss is calculated within a classical collision scheme, and in each elastic collision, a new recoiled target atom is generated. If inelastic collision is chosen, the particle loses its energy and excites an SE. The energy of the excited SE is equal to the energy loss of the particle, e.g., (4.8), which is calculated from a head-on collision of the particle with a conduction electron. The initial directional angle of the excited electron is calculated using the energy and momentum conservation law.

The excited SEs interact with the target through elastic collisions with the target core atoms, and through excitations of other target electrons. The trajectory of each SE is chosen as a series of random numbers to determine the path length between collision events, the type of collision that has taken place and the energy loss or the scattering angle. In each inelastic process, SEs are excited, so that an electron cascade is generated. The electron cascade model can also be applied for the simulation of SE emission by electron bombardment [45–51].

4.3 Comparison between secondary electron images in SIM and SEM

Contrast mechanisms in SIM images have been discussed [9,52,53] and they fall largely into the categories of material contrast, topographic contrast, and channeling contrast. Although the contrast mechanism present in the SIM images is similar to that in SEM, there are some differences. In this section SIM and SEM images are discussed from a viewpoint of interaction between ion or electron beams and samples. While SIM advantages include high contrast sensitivity to surface topography and crystal grain (channeling), drawbacks include sample surface damage and ion implantation. Therefore, beam charging for insulator samples and image deterioration due to surface sputtering and contamination will be presented in this section.

4.3.1 Material contrast

Material contrast arises from differences in the yield of SEs as a function of atomic number Z_2 of samples. In experiments, the SE emission in SIM shows a decrease with increasing Z_2 [7,8,54]. Figure 4.7 shows the relative SE intensities as a function of Z_2 for SIM and SEM images. The SE intensities were measured for Al, Si, Cr, Fe, Ni, Cu, Ag, and Pt bombarded both by 30 keV Ga ions in the FIB (Hitachi FB-2000A) system and by 5 keV electrons in the SEM (Hitachi S-4200). The SE intensities for the SIM broadly descend with increasing Z_2, being superimposed with a fine structure. The SE intensities for the SEM, on the other hand, show Z_2-dependency increasing with Z_2, commonly known as material contrast.

Recently, Monte Carlo (MC) simulations on the SE emission have revealed the origin of the difference in the material contrast between SIM and SEM images [4,34,42,54,55]. SEs are produced along the entirety of the trajectory of a projectile particle penetrating into the sample material. Due to their low

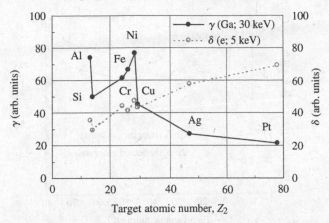

Figure 4.7 Relative SE intensities as a function of Z_2 for the Ga-SIM and SEM images [54].

excitation energy and energy loss in inelastic collisions before reaching the surface, only a small fraction of the SEs can overcome the surface barrier to escape into the vacuum. The SE yield increases due to the existing back-scattered electrons (BSEs), which scatter backwards inside the material and approach the surface to excite additional SEs that can escape from the surface barrier [11;34]. Since the electron backscattering coefficient increases with increasing Z_2, the SE yields of high-Z_2 metals under electron bombardment are dominated by the additional electron excitation by the BSEs on their way out, as shown in Figure 4.8(a). Furthermore, it shows that the general increase in the SE yield with increasing Z_2, which has been observed in experiments [7,8,54], is mainly caused by the increase in the number of SEs excited by the BSEs.

For bombardment with light ions, such as hydrogen (H) ions, the projectile ion backscattering is effective in the kinetic emission as well [56], whereas for relatively heavy, Ga ions, it is much less important, due to the small back-scattering coefficient. A more important factor in the kinetic emission due to the bombardment of Ga ions is the electron excitation by recoiled material atoms, which receive a large fraction of the total kinetic energy of the projectile ion in frequent elastic collision events. Therefore, the electron excitation can be split into three components, as shown in Figure 4.8(b): one due to excitation by a projectile ion, one due to excitation by the recoiled material atoms and one due to excitation by the excited SEs, i.e., the electron cascade.

With increasing Z_2, the component from the projectile ion, which dominates the total SE yield, generally decreases while showing a fine structure. This general decrease is because a large energy transfer from the Ga ion to

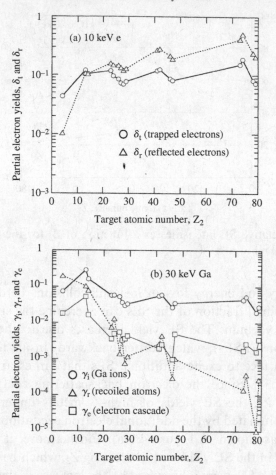

Figure 4.8 Partial SE yields due to electron excitations (a) by primary electrons backscattered from the surface and finally trapped inside the material for 10 keV electron bombardment, and (b) by projectile ions, recoiled material atoms, and electron cascades for 30 keV Ga ion bombardment, as a function of the atomic number, Z_2, of material atoms [34]. The lines are only intended as a guide for the eye.

the heavier material atoms causes the ion velocity to decrease steeply just after incidence. The lowering of the ion velocity leads to a lowering in the electron excitation probability or the inverse MFP for electron excitation (see (4.9)). Due to the decreasing excitation energy with increasing mass of the recoiled atoms, the component from the electron excitation by the recoiled material atoms decreases with increasing Z_2. As a result, a strong decrease in the total SE yield with increasing Z_2 is obtained for Ga ion bombardment. The component from the electron cascade is less important for such heavy ions. For light ions, however, most of the SEs are excited by the electron cascades in the same manner as electron bombardment [57].

The Z_2 dependence is characterized by a fine structure relating to the periods of the periodic system. Furthermore, a different structure from the case for electron bombardment was observed in Z_2 dependence for heavy ion bombardment [58]. Such fine structures are calculated for both Ga ion and electron bombardments. For Ga ion bombardment, the components from the electron excitation by the recoiled material atoms and the electron cascade change with Z_2 in a different manner from the SE yield for electron bombardment. The SE yields for electron bombardment correlate with the density of conduction electrons, so that they are proportional to the stopping power. This indicates the dominant contribution of bulk properties of the target materials. However, due to large elastic energy loss of projectile ions and surface effects (e.g., the work function), the SE yields for slow, heavy ion bombardment are not directly proportional to the electronic stopping power [59].

There is much difference in the potential of electron multiplication between the SEs excited by Ga ions and by primary electrons [42,54,57]. The mean energies of SEs excited by 10 keV Ga ions are as low as approximately 7–15 eV, decreasing with an increasing Z_2. The mean velocity of recoiled target atoms also decreases with an increasing mass under identical energy. The excited SEs with low energies have poor potential in the SE multiplication and also have less chance than the SEs to escape from the surface. Therefore, the SE yield generally decreases with an increasing Z_2. For electron impact in SEM, on the other hand, the mean energy of electrons excited by 10 keV electrons is as high as approximately 40–60 eV, increasing with an increasing Z_2 (contrary to the case of Ga ion impact). Then, the SE yield increases with an increasing Z_2, as it is contrary to the SE yield under Ga ion bombardment.

4.3.2 Topographic contrast

Topographic contrast is familiar to anyone who has seen an SEM image of an irregular object. In general, topographic contrast in SIM has been explained in a similar manner to topographic contrast in SEM, in terms of differences in the SE yield as a function of the angle of incidence of primary beam relative to sample surface normal as the local inclination of the sample surface varies.

At high energy bombardment, a simple geometric model [29], which assumes constant electronic stopping along the ion trajectory in the SE

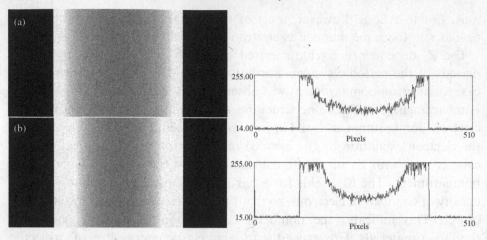

Figure 4.9 Reconstruction of pseudo SE images from the incident angle dependence of SE yield, $\gamma(\theta)$, assuming (a) $f=1$ and (b) $f=2$ where $\gamma(\theta)/\gamma(0) = (\cos \theta)^{-f}$ [60]. Random noise at $S/N = 10$ is artificially added to each pixel intensity. The images intensities saturate artificially at the intensity of 255.

escape depth and disregards scattering of the ion, leads us to a simple law in the form

$$\frac{\gamma(\theta)}{\gamma(0)} = (\cos \theta)^{-1}, \qquad (4.11)$$

which is called the "inverse cosine law"; $\gamma(\theta)$ and $\gamma(0)$ are the SE yields at an incident angle, θ measured from the surface normal and at normal incidence, respectively. Figure 4.9 shows pseudo SE images calculated for a rod with a circular cross section, assuming different incident-angle-dependences, including the inverse cosine law [60]. Since the "inverse cosine" approaches to infinity at grazing angles, the boundary between the rod and the background is enhanced in the pseudo image. However, at the energy ranges of tens of keV or less, due to the breakdown of the assumptions in the model, the incident angle dependence of the SE yield largely deviates from the inverse cosine law and varies complicatedly, depending on the ion energy and solid materials.

A systematic investigation on the incident angle dependence of the SE yield for various ion species and materials has been performed by means of Monte Carlo calculations at wide energy range of $100\,\text{eV}$–$1\,\text{MeV}$ [61,62,63]. The deviation from the inverse cosine law was represented by a fitting parameter f to a modified form of (4.11) where the exponent of (-1) is replaced to $(-f)$. At low energies, the ion energy loss within the SE escape depth contributes to

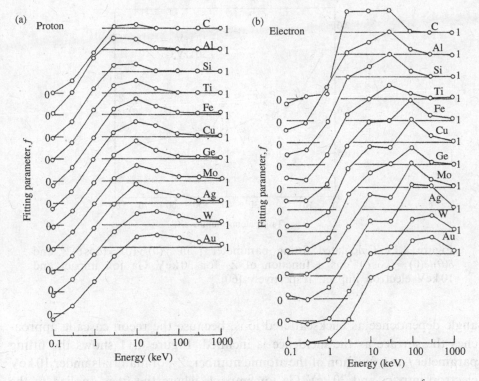

Figure 4.10 Calculated fitting parameter f in $\gamma(\theta)/\gamma(0) = (\cos \theta)^{-f}$ and $\delta(\theta)/\delta(0) = (\cos \theta)^{-f}$ as a function of energy (a) for proton impact and (b) electron impact, respectively, on different materials [61].

the incident angle dependence negatively ($f < 1$). The negative ($f < 1$) and positive ($f > 1$) effects of the backscattered ions on the incident angle dependence can be apparently seen at low and intermediate energies, respectively. The positive effect is due to the additional excitation of SEs near the surface when the ion was backscattered, while the negative effect is due to the appearance of the ions backscattered from the surface just after incidence without excitation of SEs. When the incidence is inclined, the positive (negative) effect is enhanced at intermediate (low) impact energy. At high energy, however, the $\gamma(\theta)/\gamma(0)$ approaches the inverse cosine ($f = 1$), because the backscattering coefficient becomes small as the impact energy increases. For light ion impacts, as well as electron impact in the SEM, the backscattering effects are a major reason for the deviation from the inverse cosine law (Figures 4.10(a) and b).

For Ga ion impact, although the contribution of backscattered ions to kinetic emission is negligible, a large number of fast recoiled atoms are generated. The fast recoiled atoms impose the same effects on the incident

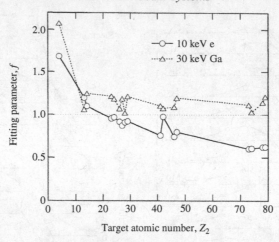

Figure 4.11 Calculated fitting parameter f in $\gamma(\theta)/\gamma(0) = (\cos\theta)^{-f}$ and $\delta(\theta)/\delta(0) = (\cos\theta)^{-f}$ as a function of Z_2 for 30 keV Ga ion impact and 10 keV electron impact, respectively [60].

angle dependence as backscattered ions, because the recoil cascade approaches the surface as the incidence is inclined. Figure 4.11 shows the fitting parameter f as a function of the atomic number, Z_2, of materials under 10 keV electron impact and 30 keV Ga ion impact, where the fit is applied to the calculated results for the angles of less than 60° [60]. For electron impact, the backscattering effect changes the f value from above 1 to below 1 as the backscattering coefficient increases with increasing Z_2. On the other hand, due to the small contribution of the backscattered ions and the recoiled target atoms to kinetic emission, the general dependence approximately obeys the inverse cosine law ($f = 1$), except for low-Z_2 materials ($f > 1$) where the contribution of recoiled target atoms largely increases.

Further study of topographic contrast is required to resolve differences in topographic contrast of SE imaging in the FIB and SEM systems. The topographic contrast is convolved with detector geometry and configuration to produce an image with many features familiar to the casual observer. It should be noted that topographic contrast is frequently overwhelmed by channeling contrast in certain material systems and geometries, pointed out by Prewett and Mair [64].

4.3.3 *Channeling contrast*

Channeling (orientation or grain) contrast is perhaps the most striking feature of FIB microscopy of metallic samples, and arises from channeling of the

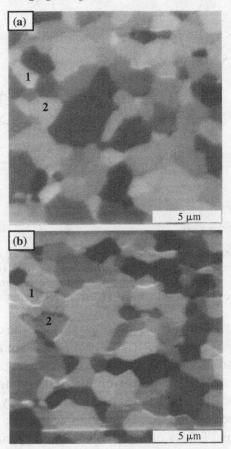

Figure 4.12 Channeling contrast SIM images of Al layer on Si substrate. (a) Image observed at an incident angle of $\alpha = 15°$ and (b) observed at $\alpha = 30°$ after a few minutes bombardment at $\alpha = 15°$. Note: the bright grains in (a) are more etched than the dark ones.

incident ions between lattice planes of the sample. Depth of penetration of the channeled incident ion varies with the relative angle between the ion beam and the lattice plane and the interplanar spacing of the lattice. A typical SIM image showing the channeling contrast from the aluminum (Al) surface on a silicon (Si) specimen is shown in Figure 4.12. The dark grains with low SE yield correspond to the channeled grains.

Through channeling contrast of the SEM images, grains in polycrystalline samples are seen with BSE (backscattered electron) contrast of the SEM images that depends sensitively on the sample tilt [15]. This sensitivity results from the dependence of the BSE yield on the orientation of the primary beam relative to the lattice planes and is caused by the primary Bloch wave field.

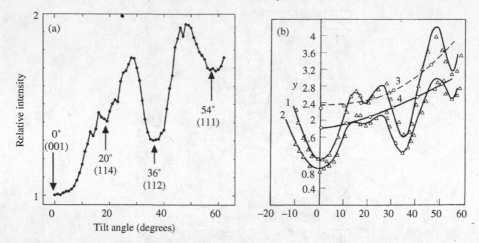

Figure 4.13 (a) Relative SE intensity for the FIB-SIM as a function of the Ga ion beam angle tilted on a Cu crystalline surface [66]. (b) Incident angle dependence of SE yield γ of a Cu crystalline due to the impacts of Ar ions with the energies of 30 keV (1) and 20 keV (2) [67]. (Adapted from [66].)

Due to the sensitivity of the variation on the surface, the channeling contrast can decrease with an increasing electron-probe focusing angle which should not exceed 1–10 mrad. The depth of average channeling varies with Z_2, but for most elements, the additional contribution to the contrast is much smaller from depths > 50 nm. The channeling contrast is also very sensitive, either on an amorphous surface oxide layer or multiple scattering in the crystal.

As to the directional effects of kinetic emission, experimental and theoretical work has been reviewed by Brusilovsky [65]. Fast particles entering crystalline materials in low-index lattice directions are steered into the interior of the target by many small-angle scattering collisions. This causes a drastic reduction in the SE yield for two reasons: first, kinetic emission depends on inelastic energy loss and channeled ion transfers relatively small energy in each collision along its path; second, SEs generated deep in the material have difficulty in escaping to the surface. Very recently, an experiment [66] revealed that the fine changes in SE intensities in SIM with the tilt angle was in reasonable agreement with the incident angle dependence of the SE yield observed using an argon ion beam by Mashkova *et al.* [67] (Figures 4.13(a) and (b)).

According to the continuum model of Lindhard [68], the critical angle for ion channeling is approximately estimated to be

$$\Psi_c \propto (Z_1 Z_2 E_0)^{1/2}. \tag{4.12}$$

The experimental Ψ_c values mostly range from 5 to about 20°, which are sufficiently larger than the FIB focusing angles of a few mrad. A clear contrast, as shown in Figure 4.12, has been obtained after removing the native oxide surface layer with a high SE yield. On SIM imaging, channeling contrast has been mostly observed for metals such as Al, Cu, Fe, Ni, and Au, but not for semiconductors of Si, GaAs, etc. At first sight one might expect severe disruption of the crystalline caused by the ion beam to amorphize the surface. A plausible explanation by R. Shimizu (Osaka Institute of Technology, Pers. Comm.) is that many metals remain crystalline as a result of fast recrystallization under ion bombardment at room temperature.

4.3.4 Spatial resolution and information depth

The SEs have information along their own trajectories in addition to the trajectories of the projectiles that excite the SEs of interest. Most of the SEs are produced in the surface layer, as they are independent of the incident probe. Under electron impacts, the SEs consist of two components which correspond to the SEs excited by PEs (primary electrons) and BSEs (backscattered electrons), respectively. The SEs excited by PEs have a small spatial spread of < 10 nm, but the SEs excited by BSEs have a large one of several tens to several hundreds of nanometers, decreasing PE energy. Under the Ga ion impacts, the spatial spread of SE information is roughly as small as 10 nm, decreasing with an increasing Z_2.

Figures 4.14(a) and (b) show birthplaces of the SEs that have escaped from the surface at various beam incident positions near the Al–Au boundary [4,55]. Incident probes for SIM and SEM imaging are assumed to be the point beams of 30 keV Ga ions and 10 keV electrons, respectively. The SEs for the Ga ion impact are classified into three types because of the collision partners: projectile ion (black), recoiled target atom (white), and cascade electron (gray). It can be observed that the three components contribute equally to the SE yield for the low-Z_2 sample (Al). For the high-Z_2 sample (Au), however, the electron excitation by the projectile ions dominates those by both the recoiled target atom and the cascade electrons. A large number of SE birthplaces for both ion and electron impacts are < 5 nm in depth. For the ion impacts, a laterally narrow spread of the SE birthplace shown in Figure 4.14(a) is caused by the short ranges of the projectile ions, recoiled target atoms, and cascade electrons. Under electron impacts, on the other hand, the SEs excited by PEs have a small lateral spread of < 10 nm, but the SEs excited by BSEs have a large one of several hundreds of nm, as mentioned above [34].

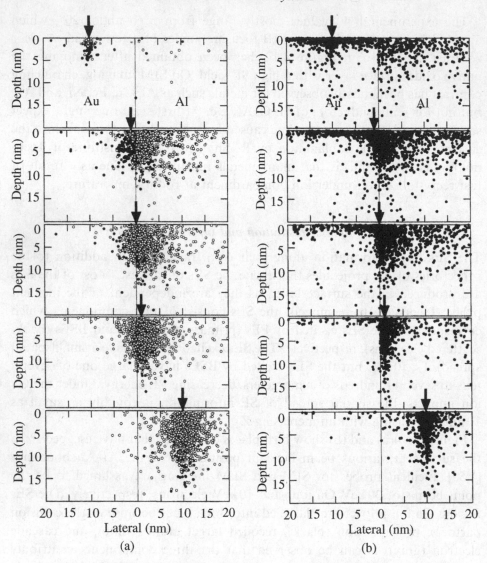

Figure 4.14 Birthplaces of SEs that have escaped from the surface (made of Au and Al by halves) due to the beam impacts of (a) 30 keV Ga ions and (b) 10 keV electrons [4,55]. The arrows indicate the impact points on the surface. The number of projectiles is 10^4. In (a), the SEs are excited by projectile ion (black), recoiled target atom (white), and cascade electron (gray).

Assuming the SIM and SEM observations of an FIB milled cross section, we consider a layered sample to have an Al–Au–Al structure, the cross section of which is tilted by $\pi/4$ radians with regard to their probe incident directions, as shown in an inset in Figure 4.15 [4]. All the yields of SEs for

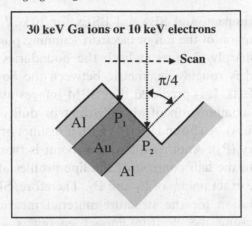

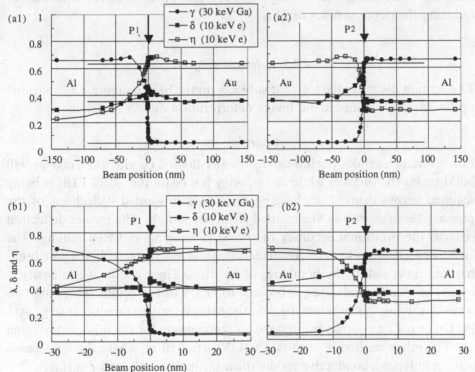

Figure 4.15 Line profiles of SE yield γ for 30 keV Ga ions and SE yield δ and BSE yield η for 10 keV electrons as a function of the beam scanning position for $\pi/4$-tilted sample revealing a cross-sectional Al-Au-Al structure [4]; (a1) near the boundary (P_1) between Al and Au and (a2) near the boundary (P_2) between Au and Al; (b1) and (b2) magnified profiles at the boundaries P_1 and P_2 corresponding to the profiles of (a1) and (a2), respectively.

30 keV Ga ion impact, and SEs and BSEs for 10 keV electron impact are plotted as a function of the ion or electron scanning position in Figure 4.15. The γ profile sharply changes at both the boundaries of Al–Au (P_1) and Au–Al (P_2) and is roughly symmetric between the boundaries due to the substructure effects. It is predicted that SEM images using the SE and BSE signals show a combination of an indistinct or dull change of the image brightness at the Al–Au boundary (P_1) with a distinct or sharp change at the Au–Al boundary (P_2). Another interesting point is that the beam positions corresponding to the half contrast of the line profiles at the boundaries are shifted from the exact points of P_1 and P_2. Therefore, SIM imaging is better in spatial resolution for the structure/material measurements than SEM imaging. Decreasing the electron impact energy is a direct method for reducing the depth profiles of SEs.

4.3.5 Relevant subjects

This section discusses the relevant subjects on the SIM imaging, i.e., charging up for insulator samples and image deterioration due to sputtering.

Charging up

For insulator samples (including electrically floated conductors), charge will build up on the surfaces while the primary ion beam (i.e., Ga$^+$ FIB) is being scanned across them. The charging up causes unwanted deflections of the primary beam as well as SEs emitted from the sample. The former deflection reduces the positional accuracy of the SIM imaging and FIB milling. The latter deflection makes the SE detection yield fluctuate with the ion bombarding time and/or the bombarding position. Then, the image intensity is locally or wholly modulated to become too dark or too bright. There are two ways of solving this problem: (i) coat the samples with a conductor (e.g., Au, Pt, C), or (ii) an electron beam flood to compensate for the positive charging up. The other method is ultra-violet (UV) irradiation, where the light penetration induces a conductive modulation in the semiconductor surfaces.

Coating is usually done using a thin film coater (or evaporator). As a variation of the external coating, an FIB assisted deposition (FIB-AD) of conductor films is used to do the selective coating and to connect the charging site to a known ground site. In the case of the use of electron flooding of low energies to neutralize the incident positive charge on the sample, SEs are no longer used as the SIM imaging signal because no discrimination can be made between the SEs and the flood electrons. Positive secondary ions (SIs) are now used instead of the SEs, but the SI intensity is rather smaller than the SE one to deteriorate the SIM image quality. In the FIB cross-sectioning of nonconductive materials, for example, the rough milling using a high-current FIB of up to 30 nA requires the built-up charge to be dissipated using the methods mentioned above. In the final milling using a low-current FIB of less than 100 pA, on the other hand, little charging up is observed if the milled surface connects to somewhere at ground level. The reason is that the implanted Ga atoms make the surface conductive. Even a high-resistivity surface is generous with the scanning beam of low current.

In the SEM imaging for the insulator samples, on the other hand, the primary electron energy is popularly adjusted to some specified energy in the range of 200 eV to 2 keV (depending on the sample) so as to balance the SE yield. This method is not applied to the SIM imaging using the positive ion probe. The SE emission does not decrease the positive charges building up on the surface, but does increase them.

Deterioration in the SIM image quality due to sputtering

To form a good image, there must be resolution, contrast, and adequate signal-to-noise ratio (S/N). Good images are essentially a system issue, i.e., the instrument plus sample. Unnecessary ion bombardment must be reduced to minimize the unavoidable influence of ion sputtering and redeposition of sputtered atoms.

In order to obtain an image of a given S/N, the beam must dwell at points on the sample for a time t_d to collect enough SE quanta. The number of detected SE quanta per pixel is given by

$$N_e = \gamma I_b t_d \eta / e, \qquad (4.13)$$

where γ is the SE yield, η is the detection efficiency of SEs (taken to be 1 to simplify the model), and e is the electron charge. The average human eye requires an S/N value of at least 5. Considering a two-step process where by primary ions strike a sample and generate SEs under Poisson statistics, the number of incident ions per pixel N_{ion} forms the image at S/N given by [6,10]

$$(S/N) = \{\gamma/(1+\gamma)\}^{1/2} N_{ion}^{1/2}. \qquad (4.14)$$

Assuming $\eta = 1$ and $\gamma = 2$, the image at $S/N = 10$ requires $N_{ion} = 150$, i.e., $t_d = 24 \, \mu s$ at $I_b = 1 \, pA$, or an image acquisition time of 6.3 s for a 512×512 pixel image, and subjects the surface to erosion by $150Y$ atoms. Here, Y is the sputtering yield, i.e., about 1–10 for 30 keV Ga ion bombardment, depending on the target.

Looking closer at the sputtering spot, it can be seen that a scanning FIB forms a local slope on the bombarding spot due to the difference in ion dose between the beam head and tail positions in the scanning direction [69]. The slope angle strongly varies with the scanning speed, the beam density, the beam size, sputtering yield, and redeposition. Since a change in the slope angle corresponds to that in the beam incident angle, the beam scanning conditions also influence the parameters γ and Y. In high-resolution SIM imaging, a single slow scan is usually superior to averaging frames to enhance the *S/N* ratio of the image.

In SEM imaging, we don't usually need to care about the sputtering, but it does meet a contamination problem. Hydrocarbon molecules on the sample surface are cracked and polymerized by the knock-on and/or ionization processes. A carbon-rich contamination film grows over the electron scanning area. At a thickness of the order of 10 nm, the SE emission will be of carbon and the irradiated areas will appear darker in the SEM image.

4.4 Some applications of FIB linked with modern SEM/STEM/TEM

4.4.1 FIB micro-sampling and micro-fabrication

In FIB applications of micro-machining and device transplantation, FIB sputtering plays the role of "cutter" and FIB induced deposition does the role of "attacher" [70]. One of their successful uses is an FIB micro(μ)-sampling for site-specific SEM/STEM/TEM inspection. The micro-samples are dug out by FIB milling without cleaving a wafer or a chip [71,72]. Figure 4.16 shows a schematic flow of the μ-sampling for transmission electron microscope (TEM) / scanning TEM (STEM) inspection. First, the μ-sample (or μ-wedge) of interest (typically $3\,\mu m$ wide $\times 13\,\mu m$ long $\times 7\,\mu m$ deep) is FIB dug-out, leaving a micro-bridge (or a microbeam) to support the μ-sample. Second, a built-in metal needle (or μ-probe) is manipulated into position for lifting the μ-sample. FIB assist-deposition (FIB-AD) is used to bond (or weld) the needle to the μ-sample. Further FIB milling of the micro-bridge separates the μ-sample from the wafer or chip. The separated μ-sample is transferred and then fixed onto the side of a partially cut TEM-grid using FIB-AD. Finally, the μ-sample is FIB thinned to a strip of about $0.1\,\mu m$ thick and $7\,\mu m$ long and is ready for the TEM/STEM inspection. All the steps are accomplished under vacuum in the FIB system. SIM images allow monitoring of each step in live time thereby assisting the operator. Advantages of the FIB μ-sampling for TEM/STEM are as follows: (i) the wafer is left intact, (ii) FIB rethinning is possible after TEM/STEM observation, (iii) it is possible to prepare the

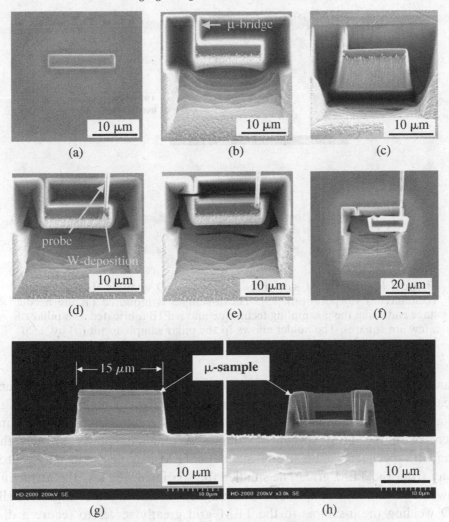

Figure 4.16 Schematic flow of the μ-sampling for TEM/STEM inspection. (a) Tungsten (W) deposition as a protection layer on the site-specific region of interest, (b) FIB trench-milling to dig out the μ-sample, leaving a μ-bridge (or a microbeam) to support the μ-sample, (c) FIB cutting the μ-sample bottom, (d) bonding the μ-sample to the μ-probe, (e) FIB cutting the μ-bridge to release the μ-sample from the wafer or chip, and (f) lifting the μ-sample to transfer onto TEM grid. All pictures in (a)–(f) are SIM images to assist the operator. Pictures of (g) and (h) show 200 kV SEM images of the μ-sample attached on the TEM half-grid edge before and after the FIB rethinning, respectively.

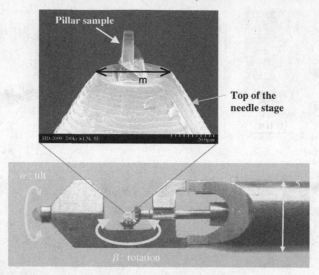

Figure 4.17 FIB/STEM side-entry holder for 3D analysis (Hitachi High-Technologies Corporation) [75]. The μ-sample is mounted on the needle stage end using the μ-sampling technique and is FIB fabricated as a pillar of a few μm squares. The holder allows to the pillar sample to tilt (α) by ±20° and rotate (β) by 360°.

planar/cross-sectional view samples, (iv) multiple sampling from an adjacent area is possible, (v) the method is applicable for magnetic materials, (vi) high reliability of the sampling. The advantage (v) results from a significant decrease in the TEM specimen volume. A magnetic power attracts the magnetic specimen in the strong magnetic TEM-lens field and may pull it away from the TEM grid. The smaller the specimen, the less the attracting power. In addition, disturbance of the magnetic lens field is also smaller. FIB-AD welding the μ-sample to the TEM grid greatly serves to secure a dissipation path for beam heating during the TEM/STEM inspection. The several μm-sizing is also of great use for radioactive samples because their radiation activities are reduced in proportion with the volume to the level of easy handling [75].

Recently, a pillar μ-sampling has been reported for three-dimensional (3D) STEM observation of the same point of interest [74,75]. The FIB/STEM side-entry holder has been specially designed to give the pillar sample a tilt (α) of ±20° and a rotation (β) of 360° (see Figure 4.17) [75]. The conventional-size μ-sample is FIB milled, transferred to the needle stage edge using the manipulated needle, and fixed to it using FIB-AD. The μ-sample is FIB milled to be a 2–5 μm-square pillar that contains the inspection point of interest. Here, the pillar sample is STEM observable at any rotation angle. We can

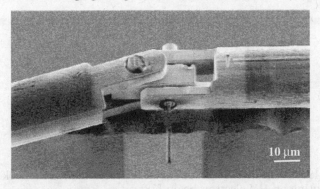

Figure 4.18 Micro universal-joint. Each of the parts is FIB fabricated from a 25 μm diameter aluminum wire and is sequentially assembled with each other using both the manipulated needle and the FIB "cutter" and "attacher" (Hitachi FIB FB-2100 with the micro-sampling system). (Courtesy of T. Tanaka, Osaka Sangyou University.)

reconstruct computerized tomography (CT) images using the STEM images at all rotation angles. Since nothing surrounds the inspection region except along the direction of the pillar base, extreme reduction has been achieved in system peak intensities, i.e., X-rays excited from the surroundings by high-energy stray electron bombardment, in energy dispersive X-ray (EDX) analysis.

Another similar technique to the μ-sampling is an "ex-situ lift-out," where the FIB milled TEM lamella is lift out by the glass needle using an optical microscope under the atmosphere pressure [3,76]. For TEM analysis, the ex-situ lift-out specimen of about 0.1 μm in thickness is usually positioned onto a carbon coated TEM grid. Then, the "ex-situ lift-out" is in general faster than the "μ-sampling," but does not allow to FIB rethin the sample.

The FIB works as micro "cutter" and "attacher" in micro-fabrication. Figure 4.18 shows a typical example, i.e., a micro universal joint. Each of the micro parts has been FIB milled from 25 μm-diameter aluminum wire and assembled with them using the manipulated FIB micro "cutter" and "attacher." This is the very FIB manufacturing system. Both by extracting μ-devices having electrical/mechanical functions and by transplanting them at any desired positions, we will manufacture new hybrid systems, which are difficult to fabricate in the same process. Device transplantation is a task with more challenge in the future line [70].

Another interesting question is "how smooth and flat a sample can FIB fabricate?" TEM 0.3–0.4 μm-thick and very flat specimens of less than several nm in roughness over 10 μm square in size have been successfully FIB milled to observe structures of vortices in high-Tc superconductor material (YBaBiCaCuO) by Lorentz microscopy [77].

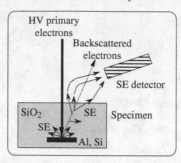

Figure 4.19 Generation of secondary electrons (SE) and backscattered electrons using an HV electron beam [78].

4.4.2 SEM/STEM/TEM microscopy of FIB cross-sectioned specimens

A focused ion beam (FIB) has been actively applied for preparation of about 0.1 μm-thick specimens for transmission electron microscopes (TEMs). For device failure analyses, however, it is difficult to prepare the exact-point TEM specimens. The reason is that the failures are usually beneath the surface and their exact locations are unknown. Then, even step-by-step FIB cross-sectioning may sputter away the failures in the TEM specimen preparation. One of the solutions for exact-point TEM microscopy is high-voltage (HV) SEM imaging in TEM [78].

A TEM has an HV-SEM imaging mode in addition to the conventional transmission imaging mode (e.g., Hitachi 200 kV TEM system, HF-2000). In the HV-SEM imaging mode, SEs emitted from the specimen travel upward through the objective lens bore under the influence of a combination of the forces from the magnetic field of the lens and the electrostatic field of the SE detector (+10 kV) and are collected to form the SEM images. Figure 4.19 shows an electron–sample interaction when a 200 kV electron beam illuminates the sample. Devices such as 4M and 16M dynamic random access memory (DRAM) used in the present study have transistors and capacitors that are composed of Al and poly-Si surrounded by silicon oxide (SiO_2), respectively. When the 200 kV primary electron beam strikes the region of Al or Si, BSEs are more generated in their regions than their surroundings. The reason is that both Al and poly-Si are larger in Z_2 than SiO_2. The SEs are excited not only by the primary electrons, but also by BSEs. The SEs generated deep in the sample, however, are absorbed within the sample because their energies are so small. On the other hand, the BSEs have high energies so that they can travel through 1 μm thick silicon oxide and emerge from the surface. The SEs excited in the surface area of the sample by the BSE and emitted from it reveal the planar position of each BSE's birthplace in the SIM

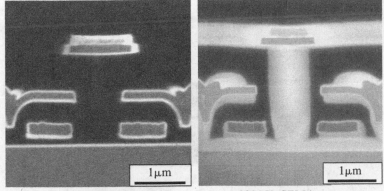

30 kV- SIM image   **200 kV- SEM image**

Figure 4.20 Typical comparison of 30 kV Ga-SIM and 200 kV HV-SEM images of an FIB milled cross section.

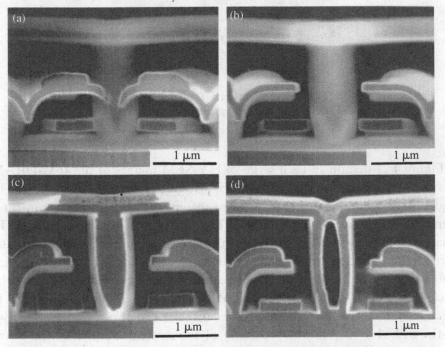

Figure 4.21 Series of HV SEM images taken every 0.1–0.2 μm in FIB milling of a 16M-DRAM specimen (Hitachi 200 kV TEM; HF-2000) [78].

image. Thus, the HV-SEM imaging provides information on sample surfaces as well as inner structures to a depth of about 1 μm. Figure 4.20 shows a typical comparison of 30 kV-Ga SIM and 200 kV HV-SEM images. The information depth for SIM imaging is only a few nm, as mentioned before.

Figure 4.22 (a) STEM view of an FIB milled SRAM and (b) the highly magnified view of metal line showing grain structure. The arrow indicates a barrier metal [79].

Figure 4.21 demonstrates a series of HV-SEM images of the 16M-DRAM device, which have been taken at every 0.1–0.2 μm step in FIB milling. The figure shows that the device structures around the contact change in three dimensions (3D) as milling proceeds. The depth of field is about 1 μm. The image resolution gradually worsens with the depth and reaches about 0.1 μm at a depth of about 1 μm. Fine-step FIB cross-sectioning is repeatedly carried out until the specified region of interest surely appears in both front- and back-side HV-SEM images of the TEM specimen (of about 0.1 μm or less in thickness). Electron energy loss spectroscopy (EELS) of transmitted electrons in the STEM system allows providing of the elemental and chemical information in the site-specific region and has tentatively been applied to the FIB prepared specimen [78].

Recently, the FE-STEM of low voltage (5–30 kV) also has been applied to characterize the device features with ease [79,80]. A standard in-lens FE-SEM (e.g., Hitachi S-5000) was modified by the addition of a STEM detector and a bright field (BF) aperture. Using a "μ-sampling" technique in an FIB system (see Section 4.3.1), a thin specimen including the specific area of interest was dug out, lifted, transferred from the wafer, and fixed onto a TEM grid mounted on a side-entry specimen stage. The μ-sampling technique using the compatible FIB-SEM specimen holder permits SEM/STEM inspection and FIB milling repeatedly, so that the precise area of interest is easily prepared for STEM observation.

The electron straightforward penetration at accelerating voltage $V = 5$–$30\,kV$ is rather short in contrast to that at $V = 200$–$300\,kV$. So, an interesting point is "How fine are the device features that are observable in the low voltage FE-STEM?" Figures 4.22(a) and (b) show the STEM views of the FIB milled static random access memory (SRAM) [79] and its highly magnified view of the metal lines, respectively. It has been found that the low voltage FE-STEM is applicable to obtain rather fine structure STEM images for $0.1\,\mu m$-thick specimens.

References

[1] K. Nikawa. *J. Vac. Sci. Technol. B*, **9** (1991), 2566–77.
[2] T. Ishitani and T. Yaguchi. *Microsc. Res. Tech.*, **35** (1996), 1320–33.
[3] L. A. Giannuzzi and F. A. Stevie. *Micron.*, **30** (1999), 197–204.
[4] T. Ishitani and K. Ohya. *Scanning*, **25** (2003), 201–9.
[5] T. Ishitani, H. Hirose and H. Tsuboi. *J. Electron Microsc.*, **44** (1995), 110–14.
[6] M. Utlaut. *Handbook of Charged Particle Optics*, ed. J. Orloff (New York: CRC Press, 1997), pp. 429–88.
[7] Y. Sakai, T. Yamada, T. Suzuki *et al. Appl. Phys. Lett.*, **73** (1998), 611–13.
[8] Y. Sakai, T. Yamada, T. Suzuki and T. Ichinokawa. *Appl. Surf. Sci.*, **73** (1999), 96–100.
[9] T. Ishitani and H. Tsuboi. *Scanning*, **19** (1997), 489–97.
[10] J. Orloff, M. Utlaut and L. Swanson. *High Resolution Focused Ion beams: FIB and its Applications* (New York: Kluwer Academic/Plenum Publishers, 2003).
[11] J. I. Goldstein, D. E. Newbury, P. Echlin, *et al. Scanning Electron Microscopy and X-ray Microanalysis* (New York and London: Plenum Press, 1992).
[12] L. Reimer. *Scanning Electron Microscopy*, 2nd edn (Berlin: Springer, 1998).
[13] D. C. Joy. *Inst. Phys. Conf. Ser. No. 93*, **1** (1988), 23–32.
[14] T. Ishitani. *Jpn. J. Appl. Phys.*, **34** (1995), 3303–6.
[15] L. Reimer. *Image Formation in Low-Voltage Scanning Electron Microscopy* (Washington, DC: SPIE Optical Engineering Press, 1993).
[16] G. Dearnaley, J. H. Freeman, R. S. Nelson and J. Stephen. *Ion Implantation* (Amsterdam: North-Holland Publishers, 1973).
[17] H. Ryssel and I. Ruge. *Ion Implantation* (New York: John Wiley & Sons, 1986).
[18] Z. J. Ding and R. Shimizu. *Scanning*, **18** (1996), 92–113.
[19] W. O. Hofer. *Scanning Microsc. Suppl.*, **4** (1987), 265–310.
[20] D. Hasselkamp, H. Rothard, K.-O. Groeneveld *et al. Particle Induced Electron Emission II* (Berlin: Springer-Verlag, 1992).
[21] E. S. Parilis, L. M. Kishinevsky, N. Yu. Turaev *et al. Atomic Collisions on Solid Surfaces* (Amsterdam: North-Holland, 1993), pp. 391–473.
[22] R. A. Baragiola, ed. *Ionization of Solids by Heavy Particles* (New York: Plenum Press, 1993).

[23] R. A. Baragiola. *Low Energy Ion-Surface Interactions*, ed. J. W. Rabalais (Chichester: John Wiley and Sons, 1994), pp. 187–262.

[24] J. Schou. *Physical Processes of the Interaction of Fusion Plasmas with Solids*, ed. W. O. Hofer and J. Roth (San Diego: Academic Press, 1996), pp. 177–216.

[25] H. Kudo. *Ion-Induced Electron Emission from Crystalline Solids* (Berlin: Springer, 2002).

[26] H. D. Hagstrum. *Phys. Rev.*, **96** (1954), 325–35.

[27] H. D. Hagstrum. *Phys. Rev.*, **96** (1954), 336–65.

[28] R. A. Baragiola, E. V. Alonso, J. Ferron and A. Oliva-Florio. *Surf. Sci.* **90**, 240–55.

[29] E. J. Sternglass. *Phys. Rev.*, **108** (1957), 1–12.

[30] J. Schou. *Phys. Rev. B*, **22** (1980), 2114–74.

[31] M. Rösler. *Ionization of Solids by Heavy Particles*, ed. R. A. Baragiola (New York: Plenum Press, 1993), pp. 27–58.

[32] P. M. Echenique, F. Flores and R. H. Ritchie. *Solid State Physics: Advances in Research and Applications,* Vol. 43, ed. H. Ehrenrech and D. Turnbull (New York: Academic Press, 1990), pp. 229–308.

[33] I. Nagy, A. Arnau and P. M. Echenique. *Phys. Rev. B*, **38** (1988), 9191–3.

[34] K. Ohya and T. Ishitani. *J. Electron Microsc.*, **52** (2003), 291–8.

[35] D. R. Penn. *Phys. Rev. B*, **35** (1987), 482–6.

[36] J. C. Ashley. *J. Electron Spectrosc. Relat. Phenom.*, **50** (1990), 323–34.

[37] K. Ohya and J. Kawata. *Jpn. J. Appl. Phys.*, **32** (1993), 1244–7.

[38] H.-J. Fitting and J. Reinhardt. *Phys. Status Solidi A*, **88** (1985), 245–59.

[39] R. A. Baragiola. *Interaction of Charged Particles with Solids and Surfaces*, ed. A. Gras-Marti, H. M. Urbassek, N. R. Arista and F. Flores (New York: Plenum Press, 1990), pp. 443–58.

[40] G. Lakits, F. Aumayr, M. Heim and H. Winter. *Phys. Rev. A*, **42** (1990), 5780–3.

[41] J. Lörincik, Z. Sroubek, H. Eder, F. Aumayr and H. Winter. *Phys. Rev. B*, **62** (2000), 16116–25.

[42] K. Ohya and T. Ishitani. *Surf. Coat. Technol.*, **158–159** (2002), 8–13.

[43] W. Eckstein. *Computer Simulation of Ion–Solid Interactions* (Berlin: Springer-Verlag, 1991), pp. 4–32.

[44] J. F. Ziegler, J. P. Biersack and U. Littmark. *The Stopping and Ranges of Ions in Solids* (New York: Pergamon Press, 1985), pp. 109–40.

[45] T. Koshikawa and R. Shimizu. *J. Phys. D: Appl. Phys.*, **7** (1974), 1303–15.

[46] J. P. Ganachaud and M. Cailler. *Surf. Sci.*, **83** (1979), 498–518.

[47] Z. J. Ding, and R. Shimizu. *Surf. Sci.*, **197** (1988), 539–54.

[48] M. Kotera, R. Ijichi, T. Fujiwara, H. Suga and D. B. Wittry. *Jpn. J. Appl. Phys.*, **29** (1990), 2277–82.

[49] A. Dubus, J.-C. Dehaes, J. P. Ganachaud, A. Hafni and M. Cailler. *Phys. Rev. B*, **47** (1993), 11056–73.

[50] J.-Ch. Kuhr and H.-J. Fitting. *Phys. Stat. Sol. A*, **172** (1999), 433–48.

[51] K. Nishimura, J. Kawata and K. Ohya. *Nucl. Instrum. and Meth. Phys. Res. B*, **164–165** (2000), 903–9.

[52] T. K. Olson, R. G. Lee and J. C. Morgan. *Proc. 18th Int. Symp. Testing and Failure Analysis* (1992), 373–380.

[53] M. W. Phaneuf. *Micron*, **30** (1999), 277–88.

[54] T. Ishitani, Y. Madokoro, M. Nakagawa and K. Ohya. *J. Electron Microsc.*, **51** (2002), 207–13.

[55] K. Ohya and T. Ishitani. *Nucl. Instrum. Meth. Phys. Res. B*, **202** (2003), 305–11.

[56] K. Ohya, F. Aumayr and H. Winter. *Phys. Rev. B*, **4** (1992), 3101–4.
[57] K. Ohya and J. Kawata. *Scan. Microsc.*, **9** (1995), 331–53.
[58] M. Kudo, Y. Sakai and T. Ichinokawa. *Appl. Phys. Lett.*, **76** (2000), 3475–7.
[59] K. Ohya. *Proc. 3rd Int. Conf. Atomic Level Characterizations for New Materials and Devices: Nara* (2001), 53–6.
[60] K Ohya and T. Ishitani. *J. Electron Microsc.*, **53** (2004), 229–35.
[61] K. Ohya. *J. Phys. Soc. Jpn.*, **61** (1992), 3013–14.
[62] J. Kawata and K. Ohya. *Radiat. Eff. Def. Sol.*, **130–131** (1994), 131–6.
[63] K. Ohya and J. Kawata. *Nucl. Instrum. Meth. Phys. Res. B*, **90** (1994), 552–5.
[64] P. D. Prewett and G. L. R. Mair. *Focused Ion Beams From Liquid Metal Ion Sources* (Taunton, UK: Research Studies Press Ltd, 1991).
[65] B. A. Brusilovsky. *Vacuum*, **35** (1985), 595–615.
[66] Y. Yahiro, K. Kaneko, T. Fujita, W.-J. Moon and Z. Horita. *J. Electron Microsc.*, **53** (2004), 571–6.
[67] E. S. Mashkova, V. A. Molchanov and D. D. Odintsov. *Doklad. Akad. Nauk SSSR*, **151** (1963), 1074–5; Engl. transl.: *Sov. Phys. Doklad.*, **8** (1964), 806–7.
[68] J. Lindhard. *Mat. Fys. Medd. Dan Vid. Selsk.*, **34** (1965).
[69] T. Ishitani and T. Ohnishi. *J. Vac. Sci. Technol. A*, **9** (1991), 3084–9.
[70] T. Ishitani, T. Ohnishi, Y. Madokoro and Y. Kawanami. *J. Vac. Sci. Technol. B*, **9** (1991), 2633–7.
[71] T. Ohnishi, H. Koike, T. Ishitani *et al. Proc. 25th Int. Symp. Testing and Failure Analysis* (1999), 449–53.
[72] T. Yaguchi, R. Urao, T. Kamino *et al. Mater. Res. Soc. Symp. Proc.*, **636** (2001), D9.35.1–D9.35.6.
[73] Y. Nemoto, Y. Miwa, M. Kikuchi *et al. J. Nucl. Sci. Technol.*, **39** (2002), 996–1001.
[74] T. Yaguchi, M. Konno, T. Kamino *et al. Microsc. Microanal,*, **9** (2003), 118–19.
[75] T. Yaguchi, Y. Kuroda, M. Konno *et al. Hitachi Scientific Instrument News*, **46** (2003), 18–20.
[76] L. A. Giannuzzi, B. W. Kempshall, S. D. Anderseon, B. I. Prenitzer and T. M. Moore. *Microelectronic Failure Analysis Desk Reference Supplement* (2002) 29–35.
[77] A. Tonomura, H. Kasai, O. Kamimura *et al. Phys. Rev. Lett.*, **88** (2002), 237001.
[78] T. Ishitani, Y. Taniguchi, S. Isakizawa *et al. J. Vac. Sci. Technol. B*, **16** (1998), 2532–7.
[79] M. Nakagawa, R. Dunne, K. Koike *et al. J. Electron Microsc.*, **51** (2002), 53–7.
[80] M. V. Moore. *Scanning*, **25** (2003), 159–60.

5

Characterization methods using FIB/SEM DualBeam instrumentation

STEVE REYNTJENS
FEI Company, The Netherlands

LUCILLE A. GIANNUZZI
FEI Company, Hillsboro, OR

5.1 Introduction

In this chapter, we consider some of the characterization possibilities that are typical and in some cases even unique to focused ion beam (FIB) and DualBeam® (FIB/scanning electron microscope (SEM)) instruments. It is well known that FIB based instruments may be used to section and provide analysis of the third dimension for subsequent SEM or FIB imaging. Following the preparation of any cross section, the steps to obtain serial sections are relatively straightforward. Why not, after opening up the inside of the bulk substrate by milling a trench and imaging a polished sidewall of that trench, polish off a thin layer of material, and open up a second surface for imaging? One can continue in the same fashion, and open up a third, a fourth, a fifth, etc., surface, and in so doing create a series of serial sections through a volume. In this manner, one may analyze a selected volume of the sample, instead of "just" a single cross-sectional surface. This type of analysis is "3D characterization," and it constitutes the better part of this section. 3D characterization is more than a series of SEM images; it can equally be applied to FIB images [1], and even complete elemental maps (via energy dispersive spectrometry (EDS)) [2] or 3D crystallographic data (via electron backscatter diffraction (EBSD)) [3,4].

As an option to alternating between polishing and imaging as described above (i.e., automated "slice and view"), collecting images while implementing simultaneous patterning and imaging (SPI mode) can be considered as a

Focused Ion Beam Systems: Basics and Applications, ed. N. Yao.
Published by Cambridge University Press. © Cambridge University Press 2007.

shortcut to 3D SEM imaging. At present, this is limited to SEM imaging while FIB polishing within a DualBeam instrument. As an aside, SPI mode imaging is as much a "process monitoring" or "end point detection" tool (helping the operator decide on the progress of the FIB machining) as a characterization method.

Characterization by scanning transmission electron microscopy (STEM) using the SEM is also possible. In many cases, the preparation method of choice for STEM imaging is FIB or DualBeam. The preparation process and characterization step may be combined very elegantly and conveniently in a DualBeam microscope.

Although less common, FIB based analytical techniques exhibit some very specific possibilities. Near the end of this section, we will briefly elaborate on the merits and difficulties of techniques such as particle induced X-ray emission (PIXE) and secondary ion mass spectroscopy (SIMS).

We conclude the section with some options and extensions to an FIB or DualBeam microscope that can further enhance their characterization capabilities. Electrical probing is possible when using in-chamber probes connected to measurement instruments outside the chamber via electrical feed-throughs. Similarly, micromanipulators in or on the chamber provide means for the investigation of mechanical properties of microstructures.

5.2 3D characterization

With overall increasingly complex man-made or natural devices and objects, the nature of many microscopic investigations is changing from surface information only, to tomographic three-dimensional analysis. In this introduction, 3D characterization using a sequential sectioning technique in the FIB is situated among other existing techniques for 3D imaging. Without attempting a complete overview, we will show that FIB tomography fills an obvious gap in the length scales that are embraced by the existing techniques.

Figure 5.1 shows a graph representing some different techniques for 3D imaging and their associated typical resolution performance and sample size. Three-dimensional imaging techniques have been important in medical technology for quite some time. Magnetic resonance imaging (MRI) is a well-known example used for the 3D imaging of internal organs. The resolution of the MRI technique is of the order of a few hundred micrometers. Confocal microscopy is a technique that uses an optical through-focus series of images of the same sample, and subsequent offline reconstruction into a 3D model.

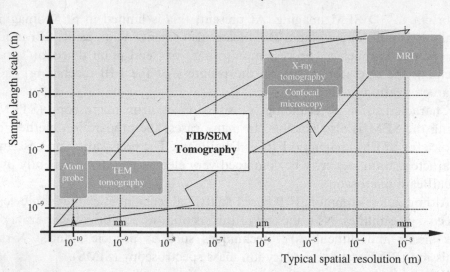

Figure 5.1 Graph showing the typical spatial resolution and sample size of some contemporary techniques for three-dimensional characterization. The obtained resolution scales roughly linearly with the sample size. FIB/SEM tomography clearly fills an obvious gap in the range of available techniques.

Because of its light optical nature, the lateral resolution of a confocal microscope is limited to about 0.2 µm, where its vertical resolution is worse at about 0.5 µm. Sample sizes are on the order of 1 mm. In the same order of magnitude of resolution performance, X-ray tomography is a well-known technique that works with similar-sized samples, and is nondestructive. It is based on recording the attenuation of X-rays as they travel through the sample material, to produce a two-dimensional image. A series of 2D images is recorded along the length of an object, to form a complete 3D representation. In order to obtain the best spatial resolution (1 µm), a synchrotron X-ray source is used. Three to four orders of magnitude lower in resolution as well as sample size, TEM tomography is an extremely powerful technique for analyzing small volumes of material (typically a few µm^2 in surface area and not more than a few hundred nanometers thick). Using dedicated sample holders, a series of images is recorded at various angles of tilt between electron beam and specimen (typically −70 to +70°). The resulting dataset is deconvoluted offline into a 3D model that can be visualized. Finally, atom probe or field-ion microscopy (FIM) techniques enable 3D characterization with true atomic resolution. Using a very thin and sharp needle-like specimen, sample ions are field emitted by a strong electrical field applied to the specimen tip. Three-dimensional elemental information is obtained from combined time-of-flight (TOF) mass spectrometry and a position-sensitive detector. However, the technique is typically limited to conductive samples with small taper angle and a very small tip radius (< 50 nm), which requires extensive preparation. Interestingly, FIB has proven to be an excellent preparation method for this type of sample [5]. However, the analyzed volume is not more than a cubic micrometer.

From Figure 5.1 and the description above, it is clear that these techniques leave an important gap in the accessible dimensional space, corresponding to sample length scales of hundreds of nanometers to hundreds of micrometers. That gap is almost perfectly addressed by FIB/SEM tomography using serial sectioning: SEM imaging resolution ranges down to roughly 1 nm. A practical lower size limit for the FIB "slices" is found to be of the order of magnitude of 10 nm. Analyzed sample sizes range from just sub-micrometer to about 100 μm in length scale. This range of resolution/sample size is found to fit almost perfectly in the existing gap, as can be seen from Figure 5.1.

5.2.1 *From a single cross section to serial sectioning –*
Auto (Slice and View)

It is well known that FIB is an excellent tool for specimen preparation, amongst others, for subsequent SEM or FIB observation. Specifically, the cross-sectioning capabilities of FIB provide a precise, site-specific, and efficient means for looking below the surface, i.e., "into" the bulk of a sample. Even more efficient to this end, is the combination of an FIB and SEM column on one platform (i.e., a DualBeam instrument). This allows the operator to switch from FIB sectioning to SEM observation, and back, in a matter of seconds or, even monitor a machining process directly via simultaneous patterning and imaging (SPI). SPI mode will be described in detail below.

The resulting image of such an FIB prepared cross section yields two-dimensional information about the cross-sectioned plane only. In a Dual-Beam, however, it is possible to extend this technique to multiple serial cross sections in a very straightforward way: why not, after acquiring an image of the original cross section, "peel" off another thin layer of the polished surface by FIB milling, and acquire a second image? The second cross-sectional image yields similar information as the first one, only this time from a slightly shifted position in the volume of the sample. This alternating sequence of cutting and imaging steps can be repeated over and over, until a sufficiently large volume of the sample has been cross-sectioned and imaged. This process is illustrated schematically in Figure 5.2. The result is a series of "images" of a volume, instead of one 2D planar image. The dataset thus inherently contains true three-dimensional information about the sample.

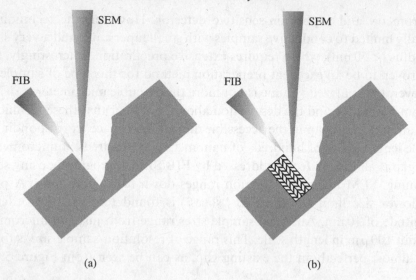

Figure 5.2 A schematic diagram of (a) the concept of a cross section, milled and imaged in a DualBeam and (b) the concept of multiple, serial cross sections where the dashed boxes represent thin layers of material that are polished off sequentially after SEM imaging.

The "images" from the dataset can be (but are not limited to) SEM induced images (secondary electron or backscattered electron images), FIB induced images (secondary electron or secondary ion images), and/or complete EDS and even EBSD maps. Depending on the type of data acquired on a specific class of microscope, manipulation of the sample position may be needed between milling and imaging. The different combinations discussed above are summarized in Table 5.1.

Besides indicating for the need of sample position manipulation, Table 5.1 also gives a typical cycle time. This time encompasses the four steps in a cycle: the actual time spent milling (FIB); the sample repositioning for data acquisition, when applicable; the actual data acquisition time (can vary greatly between just an image, or a complete EBSD map); and the sample repositioning for milling, when applicable. The cycle time given is "typical," i.e., for a cross section of reasonable size (e.g. 10 μm wide and 3 μm deep) and normal milling and imaging parameters. Of course, timing can vary greatly with the cross section size, milling current, imaging speed, and a number of other parameters.

Practical total acquisition times range from a couple of hours to tens of hours, translating into a total number of tens to thousands of slices. Slice thickness and magnification are mainly dictated by the size of the features that are to be analyzed; ideally the slice thickness and lateral image resolution

Table 5.1. *Different types of information can be obtained from serial sectioning. Depending on the instrument class, sample position manipulation will be needed or not. This table summarizes the possibilities and limitations (n/a in the FIB section means this combination is not applicable since it requires a primary electron beam).*

Instrument type	Type of "image" (data) acquired	Sample position manipulation?	Typical cycle time
DualBeam	SEM (SE, BSE)	No	1–5 minutes
	FIB (SE, SI)	Yes	2–10 minutes
	EDS	No	2–15 minutes
	EBSD	Yes	5–30 minutes
FIB only	SEM (SE, BSE)	n/a	n/a
	FIB (SE, SI)	Yes	2–10 minutes
	EDS	Yes	Currently not known
	EBSD	n/a	n/a

are at least ten times smaller than the smallest features. In practice, lateral resolution as well as slice thickness of approximately 10 nm can be obtained in a process that runs for multiple hours. Slice thickness is limited by the precision of ion beam positioning, as well as system drift throughout the entire process. Using thicker slices, a realistic size limit for the analyzed volume is around a length scale of 100 μm (resulting in a total maximum analyzed volume of $1\,000\,000\ \mu m^3$).

Although in principle there is no real limit to the magnification range that can be used, to the size of the volume that can be analyzed, or to the slice density that can be sliced (inversely proportional to the slice thickness), practical limitations are imposed directly or indirectly by time constraints. Assuming a typical and reasonably sized dataset consists of 100–500 slices, the total time that is needed to acquire such a large amount of data is quite long. Therefore, one quickly faces total times on the machine from a couple of hours to tens of hours. This reflection automatically leads to the benefit and even necessity to automate such long tasks. Besides taking away a large burden from the operator, automation will also improve overall speed and reproducibility of the results. In the example below, the utilization of FEI's Auto (Slice and View) automation software to acquire hundreds of serial slices overnight using a Nova NanoLab 600 DualBeam is demonstrated. The automation software allows for compensation of the slightly increasing focal length, as well as for the apparent upward shift of the cross-sectional surface within the field of view, during the slicing process. Both effects are caused by the combination of subsequent material removal and the SEM observation of the cross section under a tilt angle.

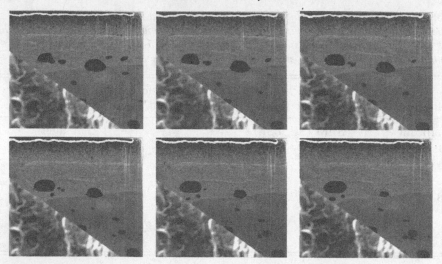

Figure 5.3 A selection of six SEM images after FIB slicing. The complete dataset consisted of 283 images.

Before describing a practical example, it should be mentioned that FIB cross-sectioning is inherently a destructive process, and therefore also serial sectioning is a destructive process. In order for the electron beam (SEM imaging) to access the deeper-lying volume, the material "in between" has to be milled away. Although this material removal is irreversible, the sample modification is only very local, leaving the remainder of the bulk material intact.

An Auto (Slice and View) dataset was acquired overnight on a sample consisting of a copper substrate with a copper coating containing iron oxide inclusions, a nickel coating, and a very thin gold coating. The raw dataset consists of 283 slices with dimensions $16 \times 13.8 \ \mu m^2$. Each slice is recorded as a 1024×884 pixel SEM image. Figure 5.3 shows a selection of six slices from this dataset. The total acquisition time is approximately 14 hours, roughly half of which is spent acquiring images and the other half milling slices.

The size of the complete 3D dataset will have implications when storing it on a hard disk, transporting it over a computer network, and manipulating it offline for further processing and visualization: in the example, the total size of this moderate dataset is about 250 MB.

5.2.2 3D reconstruction

Three-dimensional analysis using FIB tomography is essentially a two-step process. After acquisition of the raw data as described above, this dataset is

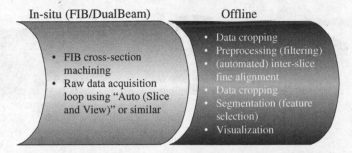

In-situ (FIB/DualBeam)　　　　　　Offline

- FIB cross-section machining
- Raw data acquisition loop using "Auto (Slice and View)" or similar

- Data cropping
- Preprocessing (filtering)
- (automated) inter-slice fine alignment
- Data cropping
- Segmentation (feature selection)
- Visualization

Figure 5.4 A schematic diagram of the FIB tomography workflow.

taken offline for further processing and 3D visualization. This second step is described in this paragraph, by means of the practical example introduced above. We first describe the different steps to be taken, then an example using a specific software package is given. It should be noted that multiple software packages performing similar processing and visualization steps are commercially available.

The entire workflow is illustrated in the schematic diagram of Figure 5.4. After initial cross-section machining and raw data acquisition in the FIB or DualBeam, one ends up with a stack of images. The content of this stack are equidistant cross sections through the analyzed volume.

Offline processing starts with initial data cropping. Though not strictly necessary, this optional step often enables irrelevant data to be discarded (e.g., parts of the image that do not contain cross-sectional information but rather the outer parts of the sample surface), thereby effectively reducing the size of the entire dataset, which in turn results in more efficient manipulation and processing. Then, the image stack may be preprocessed to improve image quality. The preprocessing usually involves application of image filters to the entire data set to remove noise and artefacts, smooth or sharpen the images, or improve contrast and brightness. Inter-slice alignment is usually the next step performed. This can be done by hand (obviously very labour intensive and time consuming on large stacks) or by using an automated algorithm, and is often useful on high-magnification datasets to reduce the effects of drift or small system instabilities. Inter-slice alignment inevitably results in some data points being present on the periphery of the images, where data are not present throughout the entire stack. This is due to shifting of the images relative to one another. Since these data points are no longer useful, they are cropped away in a second cropping step.

In principle, the dataset is now ready for visualization, although depending on the rendering method (see below), it may be needed to highlight certain features in the images, throughout the entire stack, in a segmentation

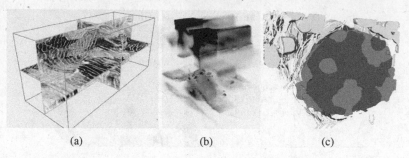

(a)   (b)   (c)

Figure 5.5 Examples of FIB tomography based 3D characterization using different visualization techniques. (a) Orthogonal sections through a steel sample; (b) volume rendering based on voxel transparency of a semiconductor defect; (c) surface rendering after segmentation of a biological cell sample. The length scale of each of these examples is a few micrometers. Amira software was used for the visualization.

operation. This allows, for instance, displaying only some inclusions in a bulk, thereby highlighting only the interesting or relevant parts of the data. Visualization itself can be done in many different ways, depending on the possibilities of the actual software used, the information that one is after, and the quality of the original dataset. The most basic and often very useful way is to show (orthogonal) sections through the volume, thereby enabling virtual cross-sectioning through an arbitrary plane. Alternatively, the dataset can be reconstructed into a 3D volumetric dataset. This is achieved using volume or surface rendering techniques. For volume rendering, the dataset is subdivided into volumetric building blocks called voxels. A voxel is the 3D counterpart of a 2D pixel, and in the case of a stack of SEM images, each voxel carries a gray value based on the original cross-sectional images. For the most powerful and best-balanced visualization possibilities, the voxels should ideally be cubes. In practice, the slice thickness should not be more than ten times larger than the lateral resolution in the individual slice images. The 3D voxel sets are displayed on the computer screen by projecting them onto a planar surface, whereby the gray value can, for example be translated into a transparency value. Volumes rendered in this manner often seem to have the appearance of a translucent suspension of particles in 3D space. In surface rendering, the volumetric data must somehow be translated into a set of surfaces, using segmentation techniques. Again this can be done by hand, or by automated algorithms such as iso-surfacing. The resulting (triangulated) surfaces are then rendered for display using conventional geometric rendering techniques. Figure 5.5 shows examples on different samples of each of the three visualization techniques.

(a)   (b)

Figure 5.6 Examples of SEM/FIB three-dimensional EBSD analysis on a thin-film Ni sample. (a) Main map consisting of 20 nm wide voxels for a total analyzed volume of $1940 \times 1260 \times 500 \, \text{nm}^3$; (b) details of a twinned grain. Grayscale is indicative of crystal orientation.

In practice, we believe the calculation of sections through the volume is the easiest, fastest and simplest way to obtain three-dimensional information. It enables not only true insight into the 3D structure of the sample, but also makes measuring in any direction in space straightforward. Volume rendering requires a lot more user interaction, in order to obtain useful results; however, the final result can be quite rewarding. A disadvantage is the difficulty of working with large datasets, since this technique is computationally very intensive. Surface rendering is probably least used, since it requires a good understanding of the segmentation process, is the most user-intensive, and needs a powerful computer to be efficient.

In the case of EDS or EBSD data collection over multiple slices, not only is the visualization important, but so is the data-mining, i.e., the numerical and statistical analysis of the results. Figure 5.6 shows a visualization of a dataset that was acquired using FEI's "EBS3" package [5,6]. The images are based on a set of EBSD maps from a thin-film Ni sample with a grain size of about 1 μm. For the entire analyzed volume, or more specifically for each of the identified grains, statistical data such as three-dimensional grain size (volume) and orientation, surface area, and the mean grain boundary misorientation, can be automatically extracted.

To conclude this part about 3D characterization, we show some practical results that were obtained on the dataset, originally introduced in Figure 5.3. Figure 5.7 shows orthogonal sections as well as a volume-rendered representation of these data, after alignment and cropping. The visualization software package used for these examples is Amira.

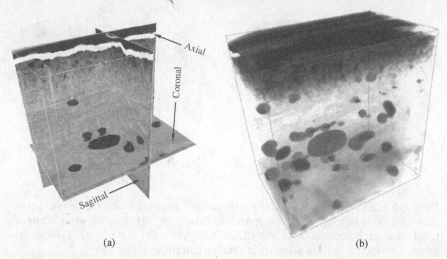

Figure 5.7 3D visualization of the dataset introduced in Figure 5.3. (a) Orthogonal sections clearly reveal the layered structure of the sample; (b) volume-rendering gives an insight into the distribution of the Fe_2O_3 inclusions in the Cu bulk. Visualization realized using Amira software package.

5.3 Simultaneous patterning and imaging (SPI mode)

Serial slicing and viewing, as described in the previous section, alternates between FIB machining and SEM imaging. Essentially, this enables complete optimization of the parameters used for both techniques independently. This is not entirely true for simultaneous patterning (milling or deposition) and imaging. However, simultaneous milling and imaging does have other benefits, not least of all the increased overall operation speed. Therefore, both techniques should be considered complementary techniques, each exploiting their own benefits in specific application areas.

Whereas serial slicing and viewing can be performed on a single-beam FIB instrument, true simultaneous patterning and imaging (SPI) obviously requires two beams to be on at the same time, and is thus strictly limited to, at the present time, DualBeam instruments.

5.3.1 Primary beams and detected signals

Obviously, simultaneous patterning and imaging includes, e.g., both a primary ion beam (patterning) and a primary electron beam (imaging) interacting with the sample material at the same time. A prerequisite for simultaneous patterning and imaging is that the ion and electron beams are coincident on the sample surface, i.e., they should be scanning the identical region of

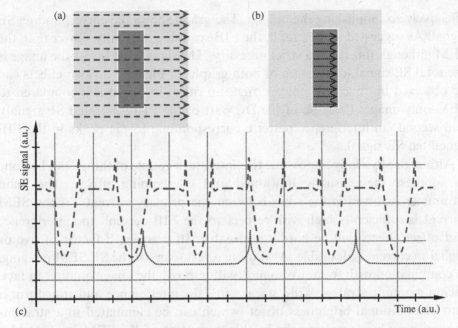

Figure 5.8 Schematic diagram of primary beam scanning and resulting SE signal. (a) SEM scanning across field of view; (b) FIB scanning across milling box; (c) SE signals generated by the SEM alone (gray dashed line) and the FIB alone (black dotted line). The sum of both is the basis for image formation.

interest on the sample. At their intersection point, the primary ion beam interacts with the sample to create secondary electrons, sputtered atoms, secondary ions, and other interaction products. At the same position in the sample, the primary electron beam generates, amongst others, secondary electrons, backscattered electrons, and X-rays. For creating an SEM image, traditionally the secondary electron (SE) or backscattered electron (BSE) signal is correlated with the scanning of the electron beam. This is not different when using SPI mode.

The main difference lies herein, that the FIB also generates an SE signal. As shown in Figure 5.8, the detector cannot or hardly discriminate between secondary electrons generated either by the SEM or by the FIB. The figure shows how the SEM is scanned across a given field of view. In the graph, the generated SE signal is shown as a gray dashed line: as the electron beam scans across the partially milled shape, its edges will generate a peak, while the milled hole itself will appear darker than the original sample surface (topography contrast; we assume a uniform material for simplicity). The figure also shows how the FIB is scanned across the defined pattern, thereby effectively accomplishing the milling. The graph shows the corresponding SE signal. As suggested in the graph, the FIB scanning is usually slower than the SEM although this is not a strict necessity. The signal that forms the image is the total SE signal, or the sum of both graphs in Figure 5.8. Two effects can be observed in "live" SE images: first, an offset in brightness compared to SEM-only images (because of the DC part of the FIB generated SE signal); and second, an interference pattern, corresponding to the peaks in the FIB generated SE signal.

Although physically impossible to eliminate, a couple of things can be done to suppress these usually unwanted and overlapping effects. The main parameter involved is the ratio between the primary currents: if the SEM current is sufficiently high with respect to the FIB current, the interference and offset effects will be hardly noticeable. In practice, a factor of five or higher in current is desirable. Also using a fast scan for the SE SEM imaging, in combination with frame averaging, will improve the image quality. In fact, such a scanning strategy will "smear out" the interference and transform it into an additional brightness offset (which can be eliminated in a straight-forward way by decreasing the brightness setting of the SEM image). Also the scan speed ratio as well as the FIB scan speed itself have an influence; however, it is impossible to go into details of their much more complex behavior within the scope of this text.

A completely different, but very effective approach is to use backscattered electron imaging for SPI mode. Since the FIB obviously does not create backscattered electrons, the offset and interference on the BSE signal will be minimal. A possible disadvantage is that a suitable detector has to be present, and the signal level for imaging is intrinsically lower. Also the need to work at relatively high SEM acceleration voltages may be disadvantageous on some materials, especially considering the better charge neutralization properties of lower-energy electrons.

Finally it should be mentioned that although aesthetically not desirable, the artefacts in SE images can be very informative to the operator, since they offer instant and direct information about the interaction of the ion beam with the sample.

5.3.2 Applications of simultaneous patterning and imaging

As mentioned before, simultaneous patterning and imaging (SPI) is strictly limited to DualBeams. In everyday DualBeam use, SPI mode is used quite often, mainly to monitor the progress of a milling or deposition step. It should be noted that when SPI mode is used to monitor FIB deposition

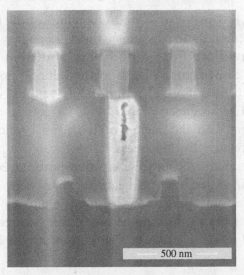

Figure 5.9 Example of an SE SEM image of an FIB milled cross section through a plug in a semiconductor device. SPI mode was used to determine the end point of the polishing process, resulting in a cross section through the center of the plug where the void is located.

processes, smaller amounts of material may be deposited in the raster field as a result of the scanning electron beam. We believe SPI is at its best when used for process monitoring and end-point determination.

When milling a (single) cross section, it is often important to stop the polishing step at an exact position in the sample (e.g. through the center of a unique defect, through the void in a semiconductor plug as shown in Figure 5.9). This end-point determination can be realized very elegantly using SPI, since the operator gets continuous and instant feedback about the polishing process (from an applications point of view, this is the main difference from serial sectioning). As soon as the feature of interest appears in the live SEM image, the FIB polishing process is stopped by the operator.

The last step in FIB TEM preparation is to mill a portion of the specimen thin enough to be electron transparent (e.g., ~ 100 nm or less) [7]. This last step in the process is conveniently monitored by (i) measuring the specimen thickness via top-down FIB imaging in conjunction with (ii) monitoring the remaining thickness of the protecting layer (usually Pt) on the top of the specimen. When this protective layer starts to recede and approach the specimen surface, the FIB thinning process should be stopped. SPI mode proves to be an efficient means to monitor this process (i) without imaging unnecessarily with the FIB and (ii) to precisely stop the FIB milling at the appropriate moment. Furthermore, regions that are FIB milled to electron transparency for TEM will also exhibit changing contrast as observed with the SEM. In particular, as the specimen thins, secondary electrons will not only be emitted from the front surface, but also from the back surface, resulting in higher SEM contrast levels at the thinnest portion of the specimen. SPI mode can be utilized to monitor this contrast change for precise end-pointing of TEM specimen preparation. Of course, this effect is quite complex and depends on the energy of the primary electrons, the target specimen and the relative foil thickness.

The nature of a multi-step process is often advantageous for using SPI mode. In addition, as the process nears its final steps, a more stringent requirement for precision using a lower ion beam current usually goes hand in hand. The former simply justifies the use of SPI, whereas the latter makes it easier to use SPI mode (e.g., lower FIB current means less interference).

5.4 Scanning transmission electron microscopy (STEM) characterization in the DualBeam

Even though, strictly speaking, STEM is realized using the electron beam only (using an SEM in this case), we consider it relevant enough to discuss it here since the preparation method of choice for STEM imaging is often FIB or DualBeam. The preparation process and characterization steps may be combined very elegantly and conveniently in a DualBeam microscope.

We note that for the point of this discussion, STEM refers to SEM-STEM only, i.e., imaging with $\leq$30 keV transmitted or diffracted electrons in an SEM (or DualBeam). This is clearly different to STEM in a TEM, where acceleration voltages of hundreds of keV are typically used, and where higher performance and analytical capabilities are achieved. STEM should not be considered as a replacement for TEM inspection, but rather (i) as a tool to be used prior to TEM, (ii) as an aid in TEM specimen preparation and evaluation, and (iii) an extension of the conventional SEM and DualBeam. In particular, it enables SEM operators to capture images with a spatial resolution of less than 1 nm.

5.4.1 Preparation of specimens for STEM characterization

STEM specimens are prepared from a bulk material using FIB and DualBeam techniques that are identical to TEM specimen preparation [7]. Using DualBeam systems, it is possible to use STEM imaging during the final thinning of the thin foil. This provides a closed-loop feedback system to the FIB/DualBeam operator about the thickness and quality of the foil while it is

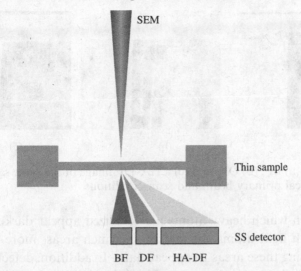

Figure 5.10 A schematic diagram of the concept of STEM imaging. The primary electron beam is scanned through a thin sample. Bright field (BF), annular dark field (DF), and high-angle annular dark field (HA-DF) signals can be collected on the respective segments of a solid-state (SS) detector.

produced. It is possible to use FIB thinning with intermittent STEM inspection, but on some DualBeam systems, these two steps can even be combined in a simultaneous process, providing real-time (live) transmission imaging during the actual thinning using SPI mode.

5.4.2 STEM imaging

STEM imaging is based on the principle of using a focused primary beam of electrons that is scanned across a thin specimen. At each point of the scan, the electron beam interacts with the specimen material, either passing or diffracting through it. The transmitted electrons are collected on a solid-state detector. When energetic electrons strike a solid-state detector (usually made from silicon), electron–hole pairs are formed which produce an electrical current proportional to the energy of the entering electron. This current is then amplified in an external electronic circuit, and translated into gray levels to form an image. The transmission and detection process is illustrated schematically in Figure 5.10.

Collection of the direct beam yields bright field images (labeled "BF" in Figure 5.10). Bright field images yield contrast from mass-thickness effects and diffraction contrast due to crystalline defects in the specimen. The interaction of electrons with heavy atoms is stronger than with light atoms,

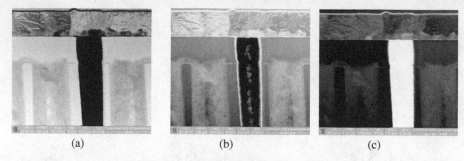

(a)   (b)   (c)

Figure 5.11 (a) BF, (b) DF, and (c) a HA-DF image of the same sample area using identical primary beam and scan conditions.

so that areas in which heavy atoms are localized appear darker in contrast than areas with light atoms. In thicker specimen areas, more electrons are scattered, causing these areas to appear dark. In addition, defects or grains at different orientations can also scatter the primary beam, causing diffraction contrast. In principle, FIB prepared STEM specimens have uniform thickness, so thickness contrast is usually negligible.

For dark field imaging, the opposite result is obtained: heavier and/or thicker areas will yield a brighter contrast, since more scattering of electrons occurs and are detected. The scattered electrons are deflected from their straight paths, and are collected on an off-axis segment, or annular portion of the detector, labeled "DF" in Figure 5.10. Since defects can also scatter the primary beam, dark field imaging can enhance features such as crystal orientation and other crystalline defects.

High-angle dark field detection uses mostly the electrons that are scattered over the largest angles: this detector segment is placed furthest off-axis. Diffracted beams due to crystalline defects are typically scattered at low angles. High-angle scattered electrons depend on mass or Z-contrast. Figure 5.11 shows an example of BF, DF, and HA-DF STEM images of the same sample area. The differences in contrast are evident. In particular, the BF image shows both mass-thickness effects and diffraction effects. The DF image shows some reversal of mass-thickness contrast effects compared to the BF image, but also shows diffraction contrast effects. However, the HA-DF image shows almost entirely Z-contrast effects; i.e., the heavy W plug shows the brightest contrast in the image, but some diffraction contrast is also noted.

5.4.3 STEM elemental analysis (EDS)

One of the most appealing features of the SEM(DB)-STEM technique is the improved spatial resolution that it creates for X-ray energy dispersive

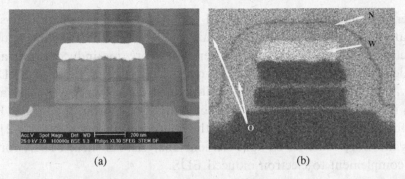

(a)   (b)

Figure 5.12 (a) DF STEM image and (b) XEDS map of a semiconductor device showing high-resolution elemental analysis. The arrows indicate the presence of the respective N, O, and W.

spectrometry (XEDS). Since the specimens are electron transparent, the beam spread and interaction volume normally associated with bulk SEM samples will be greatly reduced. Therefore, significantly higher spatial resolution can be obtained using the STEM detector while performing X-ray analysis. Typically, a resolution improvement from hundreds of nanometers or even some micrometers, to about 30 nm can be achieved. It also provides less background in the spectrum and allows for better separation of peaks as well as lower count rate mapping conditions. An example of a high-resolution X-ray map on a semiconductor device is shown in Figure 5.12. A nitride line with thickness well below 50 nm is clearly resolved in the XEDS map.

5.5 FIB analytical techniques: particle induced X-rays and secondary ion mass spectroscopy

5.5.1 Focused ion beam induced X-rays (FIBIX)

As previously mentioned, ion–solid interactions yield a complex combination of particle sputtering and signal generation, among which is the ejection of characteristic X-rays. This physical phenomenon is the basis for particle induced X-ray emission (PIXE). PIXE has generally been performed using light ions or protons accelerated at energies in the MeV range. To distinguish that FIB induced X-ray analysis is performed with heavy ions of moderate energy (i.e., 30 keV), we shall use the term FIBIX. FIBIX can be used in a DualBeam as a complement to SEM based XEDS analysis [8].

The advantages of ion beam induced X-ray emission are twofold. First, the FIB induced X-rays originate from a region close to the surface of the sample material – typically not more than a few tens of nanometers deep (e.g., a smaller interaction volume compared to SEM). Thus, surface analysis or accurate depth profiling could be achieved. Second, the Bremmstralung is theoretically orders of magnitude lower. Therefore, ion induced spectra may provide superior signal to noise ratios over electron beam induced XEDS. A disadvantage to FIBIX is that a well-defined cut-off energy exists, above which it is impossible to obtain characteristic X-rays from the target. However, the technique is most suited for the collection of soft X-rays ($< 2\,keV$), and is therefore sensitive to light element analysis. As such, FIBIX forms a good complement to electron induced EDS.

5.5.2 Secondary ion mass spectroscopy (SIMS)

The fundamental process of FIB sputtering is the basis of SIMS. While most of the sputtered species are neutrals, secondary ions are also emitted. These secondary ions are detected in an ion mass analyzer such as a quadrupole, magnetic sector, or time-of-flight (TOF). Correlating the analyzer signal with the scanning of the ion beam results in ion images with lateral element distribution. Taking the sputtering yield into account, depth profiles can be obtained as well. Since FIB process mix a small cascade interaction with a small probe size, lateral resolution of elemental specific ion images as low as 20 nm may be obtained, while local concentrations of the order of a few percent can be detected. In comparison, lateral resolution for non-FIB SIMS ranges from 50 to 200 nm, depending on the primary beam element used.

Ga^+ ion beams offer a very small probe size and are most widely used in commercial FIB systems. However, they offer a relatively low yield of secondary ion production. The secondary ion yield for Ga^+ is typically about a factor of 40 less than O_2^+, which is used in some non-FIB SIMS equipment [7]. Although superior lateral resolution can be obtained with commercial FIB instruments, the relatively low secondary ion yield combined with low collection efficiency in quadrupole mass spectrometers limits the utility of elemental mapping with FIB-SIMS. Enhancement of the secondary ion yield or detection efficiency is needed to increase the sensitivity of FIB-SIMS.

5.6 Other opportunities for in-situ characterization

Focused ion beams, and DualBeams in particular, enable further opportunities for in-situ characterization. Like SEM, FIB enables the use of mechanical micromanipulators and electrical probes, for characterization of mechanical and electrical properties, inside the microscope. Furthermore, it is also possible to combine in-situ fabrication or modification of mechanical or electrical structures with live monitoring of the relevant parameters.

Acknowledgements

Thanks to Hans Mulders and Mark Wall, FEI, for added contributions.

References

[1] D. N. Dunn and R. Hull. *Appl. Phys. Lett.*, **75** (1999), 3414–16.
[2] P. G. Kotula, M. R. Keenan and J. R. Michael. *Microsc. Microanal.*, **9**, Suppl. 2, (2003), 1004–15.
[3] S. Reyntjens. Dualbeam (FIB/SEM) sample preparation for sub-micrometer EBSD analysis. Presented at *Advances in Focused Ion Beam Microscopy: FIB 2003*, Cambridge, UK (2003).
[4] M. K. Miller, K. F. Russell and G. B. Thompson. *Ultramicroscopy*, **102** (2005), 287–98.
[5] J. J. L. Mulders and A. P. Day. Three-dimensional texture analysis. *Proc. ICOTOM 14*, Leuven, Belgium (2005), 237–42.
[6] J. J. L. Mulders and H. L. Frazer. *Microsc. Microanal.*, **11**, Suppl. 2 (2005), 506–7.
[7] L. A. Giannuzzi, F. A. Stevie, eds., *Introduction to Focused Ion Beams: Instrumentation, Theory, Techniques and Practice* (Berlin: Springer, 2005), pp. 201–28.
[8] L. A. Giannuzzi. *Scanning*, **27** (2006), 165–9.

6

High-density FIB-SEM 3D nanotomography: with applications of real-time imaging during FIB milling

E. L. PRINCIPE

Carl Zeiss SMT Inc., Redwood City, CA

6.1 Introduction

The ability to acquire, display, and interrogate multi-dimensional volumetric data sets has been well established through various scientific disciplines. The medical field in particular has exposed the public to tomographic methods through now common medical procedures such as computed axial tomography (CAT), magnetic resonance imaging (MRI), and positron emission tomography (PET). In an analogous fashion the focused ion beam (FIB) and scanning electron microscope (SEM) can combine to generate tomographic data using the FIB-SEM platform.

Early work completed by a small number of researchers gives an indication of the variety of information that can be obtained by FIB based tomography: Uchick *et al.* [1] and Kotula *et al.* [2] have produced secondary electron (SE) and X-ray images (XEDS) of 3D structures; Dunn *et al.* [3] have created 3D mass spectral images using planar FIB etching (FIB-SIMS); Inkson *et al.* [4] and Sakamoto *et al.* [5] have published 3D reconstructions based on secondary electron images, and Principe [6] has demonstrated the application of FIB-Auger spectroscopic 3D reconstruction at 10 nm resolution. Konrad *et al.* [7] have demonstrated sequential automated acquisition of 3D electron backscatter pattern (EBSP) tomographic data. This prior work has shown that FIB based tomographic methods have volumetric data resolution down to 20 nm or less (i.e., FIB-SE, FIB-Auger) and have tremendous potential for a variety of investigations in both material science and biology.

Hardware has existed for over a decade to allow collection of volumetric data from a set of sequential FIB serial sections yet FIB based nanotomography

Focused Ion Beam Systems: Basics and Applications, ed. N. Yao.
Published by Cambridge University Press. © Cambridge University Press 2007.

has until now remained impractical for a routine user. Wider utilization of FIB based tomographic methods required improvements in the ease of acquisition, data collection speed, and density of raw data collection. Advanced FIB-SEM systems such as the CrossBeam® family now permit unattended automated acquisition of volumetric data sets combining XEDS, EBSP, SE, and backscatter electron (BSE) data. Another enabling factor is integration of software tools that allow reconstruction, visualization, and analysis of the data sets including transparency mapping, surface meshing, exploration of sub-volumes, and animation. This recent evolution in both hardware and software on FIB-SEM platforms coupled to advanced data analysis is driving the rapid growth in the field of FIB based tomography by providing accessibility to the routine user.

With the advent of high-resolution simultaneous secondary electron (SE) and backscatter electron (BSE) imaging during the FIB sectioning process on the CrossBeam platform it has become practical to acquire several hundred data frames in the span of less than one hour in a continuous and automated fashion for volumes of $500\,\mu m^3$ or greater. The approach of using simultaneous and continuous SEM imaging during the FIB milling process to create a 3D reconstruction and the method of 3D quantification described in this chapter was developed and first demonstrated by the author. Real-time imaging coupled with automated image recording facilitates the data acquisition process while providing volumetric resolution at the nanoscale. High throughput is also a direct result of being able to capture high quality image data at high density in real-time during the milling process.

As will be described through the examples presented, the information content is not limited to powerful visualization and failure analysis; it is a research tool that can be applied to quantitatively correlate microstructure, material properties, and system performance. Aspects of data acquisition, data processing, and quantification will be described.

6.2 Overview of 3D tomography

The word tomography comes from the Greek words *tomos* ("to cut or section") and *graphein* ("to write"). FIB-SEM nanotomography fits this literal definition quite well. In this process the sample is sectioned with the FIB and the result is "written" in the form of digital data. A three-dimensional reconstruction results when the individual tomograms are arranged in proper order to reconstruct representation of the original object. The terminology for a three dimensional digital image element is known as a voxel, which is

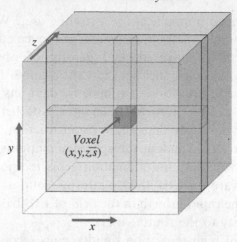

Figure 6.1 The voxel is the term applied to the volume element in a 3D reconstruction, analogous to a pixel in a 2D image. The voxel defines the location in three-dimensional space of the signal s. Note the orientation of the axes. The x–y plane is the same reference frame seen in the SEM. (Adapted graphic courtesy of Eric Lifshin.)

described in reference to Figure 6.1. Each voxel is defined in terms of its x, y, and z position in the volume as well as by an information vector **s** representing the signal. The information contained in **s** can represent any number of appropriate signals including a secondary electron (SE) signal, an X-ray signal (XEDS), an electron backscatter diffraction signal (EBSD), a backscatter electron (BSE) signal, or an Auger electron signal to name a few. In the context of this chapter FIB-SEM 3D reconstruction is an inclusive term that refers to the platform enabling all the various forms of tomographic data sets described above. Depending upon the information vector and operating conditions the method classifies as **FIB-SEM** nanotomography when the minimum analytical volume is sufficiently small to be considered nanoscale.

Spatial resolution of CAT, MRI, and PET data can range from 1 μm (for specialized equipment) to several millimeters (most commercial applications). Spatial resolution of the FIB-SEM nanotomography methods described in this context is 5–20 nm, but it is also a destructive technique. X-ray tomography, MRI tomography, and PET scan methods can also be applied to much larger volumes than is practical with the FIB-SEM 3D reconstruction method. The niche for FIB-SEM nanotomography using a liquid-metal ion gun (LMIG) lies in obtaining 3D detail on the nanometer scale over tens of micrometers or less.

6.2.1 Ultra-high-resolution tomographic techniques

Other tomographic techniques familiar to electron microscopists and material scientists include transmission electron microscopy (TEM) tomography and atom probe spectroscopy. TEM tomography is an established technique involving acquisition of a tilt-series of TEM images or acquisition of hundreds of images of identical structures, in the case of a single-particle reconstruction. The single-particle reconstruction technique is usually applied to biological structures. TEM tomography requires preparation of an electron transparent lamella or preparation and imaging of a large number of identical structures. Therefore the sample preparation requirements are generally more restricted and the volumes are typically much smaller in TEM tomography than the FIB-SEM tomographic method. However, TEM tomography permits deep sub-nanometer or even angstrom level resolution.

Atom probe tomography is an emerging commercial analytical technique that yields quantitative 3D compositional data for all elements and isotopes with 0.2 nm resolution. The instrumentation is a combination of a field ion microscope (FIM) and a time-of-flight (TOF) spectrometer [8]. The position and identification of each extracted atom is mapped to the original volume of the sample, which consists of a sharp needle less than 10 nm in diameter. The atom probe method offers very high spatial resolution but is limited to the pre-defined volume of the tip. The most versatile and dominant method to prepare atom probe tips is via FIB. In this sense, atom probe tomography can be also considered an FIB dependent tomographic technique. Thus while TEM tomography provides higher spatial resolution in terms of image data and atom probe yields a nearly ultimate compositional spatial resolution; FIB-SEM 3D reconstruction methods are in a unique range of application, versatility, and resolution.

6.3 Electron beam imaging and signal detection for FIB-SEM 3D reconstruction on a CrossBeam

Parameters that define 3D reconstruction data will be discussed in terms of the signal origin and information content. Paramount is optimization of the electron microscope conditions and understanding the operation theory of various detectors. This foundation of knowledge begins with an understanding of the origin of the various signals. Due to the importance of the subject a section will be dedicated to outline the basic aspects of electron interactions with solids. A following section will describe the requisite hardware with particular emphasis on detection principles and unique elements of the electron column design.

6.3.1 Electron–solid interactions and resolution in
FIB-SEM tomography

The elastic and inelastic interactions of the electron beam with the solid determine the lateral spatial resolution, depth resolution, and the information content conveyed by signals used in the FIB-SEM 3D reconstruction method. A description of the origin of the various signals employed will provide the background to interpret the FIB-SEM nanotomography results presented in this chapter and allow the reader to appreciate the detection principles employed. This background will also help the analyst select data acquisition parameters such as electron probe voltage, probe size, probe current, the type of signal collection, and the frequency of image data collection. As the following is an abbreviated discussion to define our strategy for FIB-SEM 3D reconstruction, ·the interested reader is directed to reference resources for further study [9,10].

Electron beam energy, interaction volume and information depth

As a first-order estimate of the resolution of the FIB-SEM nanotomography we will examine the concept of electron interaction volume. An electron interaction model is not a strictly rigorous estimate of the tomographic resolution, but it serves as a useful visual aid to open discussion. In Figure 6.2 are shown two images of the same cluster of catalyst nanoparticles which individually are on the order of 15–20 nm in characteristic dimension. The image on the left was recorded at 0.8 kV while the one on the right was imaged at 0.37 kV. Both images were acquired at an original screen magnification of 500 k× and a field width of 750 nm. The 0.37 kV image was acquired following the 0.8 kV image. The brightness setting is the same in both images while the contrast setting is 2.8% higher in the lower voltage image. Image data are paired with a Monte Carlo simulation of the electron interaction volume at each respective voltage. The simulation was completed using a freeware package, WinCasino [11], and employs the modified Bethe expression of Joy and Luo [12]. The electron interaction volume data modeling yielded an interaction volume for 0.8 kV electrons of approximately 21 nm. At 0.8 kV the interaction volume is similar to the characteristic dimension of the average nanoparticle. At 0.37 kV the interaction volume is on the order of 7 nm, which is less than the characteristic dimension of the average nanostructure. Comparing the image data, more surface detail is evident in the individual nanostructures in the image recorded at 0.37 kV. The reduced interaction volume and change in secondary yield leads to improved surface imaging sensitivity on the nanoparticle and demonstrates

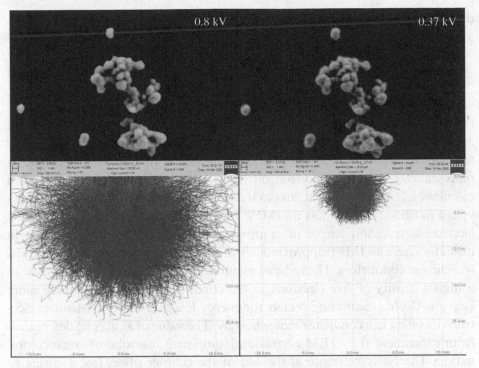

Figure 6.2 The influence of probe voltage on the interaction volume. The interaction volume provides a first-order estimate of the equilateral resolution limit in an FIB-SEM 3D nanotomographic reconstruction. The model data suggest a resolution limit of ∼5–20 nm for a 0.8 kV beam. The Monte Carlo simulation employed a package called WinCasino using the modified Bethe expression and assumes a silicon material. See text for details on the image data.

the benefit of extreme low voltage. The interaction volume simulation used silicon as the model material but the interaction volume will vary significantly depending upon probe voltage, electron beam geometry, and target material. A high atomic mass material such as tungsten yields an interaction volume of approximately 7 nm at 1 kV and less than 3 nm at 0.4 kV.

A more detailed examination of interplay between electron energy, interaction volume, and resolution requires introduction of a term called the inelastic mean free path, or IMFP. This fundamental material parameter carries units of length and is given the designation λ^0. The IMFP is formally defined as the average distance that an electron of a given energy may travel between successive inelastic collisions in a given material. The IMFP thus determines the maximum depth below the surface from which secondary electrons can originate and still be omitted from the surface without energy loss. The intensity distribution of emitted electrons at a given energy as a function of depth follows an exponential decay according to the Beer–Lambert law:

$$I(E, z) \cong I^0 \exp\left(\frac{-z}{\lambda^0 \cos\theta}\right), \tag{6.1}$$

where I^0 is the initial arbitrary intensity, z is the depth variable and θ represents the emission angle of the electron with respect to the sample normal. A value of 3λ accounts for 95% of the total electron distribution and is a common value for the electron attenuation length to estimate practical analysis volumes while 5λ represents 99.3% of the electron distribution and is preferred for numerical calculations. As an example, an electron of approximately 1.4 kV energy traveling in silicon dioxide has an IMFP of approximately 3.3 nm to yield a 3λ electron attenuation length of approximately 10 nm. Electrons traveling in insulators have an IMFP approximately three to five times larger than electrons traveling in conductors. The reduced attenuation length in conductors is due to a higher density of free electrons in the conduction band that supply more electron–electron scattering centers for energy loss and thereby attenuate electrons traveling in a conductor more strongly. The value of λ^0 may be determined from experiment (i.e., TEM correlation), empirical formula, or some combination. The NIST reference at the end of the chapter offers free software to generate IMFP data as well as effective attenuation length data from theoretical models or from built-in database compilations (NIST database 82) [13].

Secondary and backscatter electrons

While the primary beam energy may range from 100 eV up to 30 kV the secondary electrons always comprise a low energy continuum which by definition extends from 0 to 50 eV. But 90% of the secondary electron distribution is below 10 eV. This would suggest that FIB-SEM nanotomography employing secondary electrons may have spatial and depth resolution down to a few nanometers or even less, regardless of the primary beam energy. But the representation thus far is an incomplete scenario.

A typical secondary electron is the product or by-product of multiple random scattering events with each event involving an energy loss interaction. Further, the IMFP of an electron is a function of energy and the IMFP so-called "universal curve" reaches a minimum at approximately 50 eV for metals and then steadily increases again with decreasing electron energy. This means a 10 eV electron may travel further between scattering events than a 50 eV electron. The IMFP increases below approximately 50 eV because lower energy electrons lack the energy for a significant interaction probability (a plasmon interaction for metals). Thus, random scattering processes and the low energy dependence of the IMFP collectively smear the spatial distribution of secondary electrons.

Resolution of the low energy secondary electrons is further reduced from the ideal because the total distribution of low energy secondary electrons consists of two types of secondary electrons, given the designations SE_1 and SE_2 electrons. While SE_1 and SE_2 electrons have similar energies they are derived from two distinct processes. The former (SE_1) represent secondary electrons directly excited by the incident electron beam and carry high spatial resolution information. The latter (SE_2) are secondary electrons indirectly excited by backscatter electrons (BSEs).

A BSE is defined as the primary electron that scatters through an angle $>90°$ and escapes the surface with an energy greater than 50 eV and up to the primary electron beam energy. The BSE family also consists of two broad categories of electrons that collectively have experienced a mixture of elastic and inelastic scattering events. Within the first category are those BSEs that have undergone one or more large angle elastic scattering events through interaction with atomic nuclei. The primary electron essentially recoils directly out of the material. The category I BSEs (BSE^I) will have energy close to the primary beam energy and include a distribution that comprises the so-called elastic peak. The BSE^I display a small angular spread with respect to the primary incident beam geometry and thereby retain high *lateral* spatial resolution. The *depth* resolution of the BSE^I will depend upon the primary beam energy and the escape depth in the material. Category II BSEs span a broader continuum of energy and lose energy through multiple inelastic scattering events transferring energy to produce SE_2 electrons in the process prior to escaping from the surface. The BSE^{II} electron may emerge relatively far from the incident primary beam after multiple scattering events and therefore this detected signal also carries a lower lateral spatial resolution than either SE_1 or BSE^I signals.

Therefore the total SE signal is a high resolution SE_1 signal convolved with a delocalized contribution from SE_2 electrons generated by BSEs. At a voltage greater than approximately 5 kV the contribution of the SE_2 signal to the total detected SE signal may be 2–3 times greater than the intensity contribution of the SE_1. The SE_2 also follow the character of the BSE^{II} signal of deeper origin. Thus the SE_2 signal carries information influenced by the distribution of subsurface features and has inherently lower resolution. The influence of BSE^{II} on the SE image data is particularly evident where the composition is varying with depth (i.e., tungsten embedded in silicon) or in the case of buried edges which display enhanced secondary electron production. It would seem that the physics of electron scattering has conspired to smear out the achievable resolution. Fortunately there are strategies that can be introduced to gain back some ground that will be employed in the FIB-SEM 3D reconstruction. This strategy is related to the low voltage scattering regime.

Low voltage electron scattering regime

As the energy of the primary beam is reduced toward 1 kV the volume from which the SE signal is generated and the volume from which the BSE signal is generated become very similar. It is experimentally problematic to distinguish SE_2 generated by BSEs and SE_1 excited by the primary beam below 1 kV. In many respects this is a welcome outcome because the limitation imposed by the BSE production of SE_2 evaporates. At the same time the total BSE signal depth resolution improves at low voltage as the penetration decreases and the BSE^I signal in particular remains a viable high spatial resolution signal, approaching the resolution of the SE_1 signal. The most important attribute of the BSE signal has not yet been emphasized. The BSE signal intensity increases in a nearly monotonic manner with the atomic number of the nuclei from which it scattered [14]. Hence, the local intensity in a BSE image is proportional to the mass of the target atom and can be employed to form a Z-contrast image. This is a very useful contrast discrimination scheme for 3D reconstruction, as will be demonstrated.

We've now established that one possible strategy to optimize 3D resolution is to acquire SE_1 and BSE^I signals in a low voltage scattering regime. This phase space must be balanced with the achievable performance of the electron column in this regime to maintain a sufficiently bright focused primary electron beam at low energy coupled with effective electron detection schemes. These topics will be covered in a description to follow on hardware and detection principles.

Electron beam – sample geometry on a CrossBeam

Prior discussion has assumed an electron beam with normal incidence to the sample surface. On an FIB-SEM platform there is a fixed angle between the ion and electron column which is close to 35°. If an FIB cut is made with the FIB axis normal to the surface then the incident electron beam will form an angle with the cut face that is close to this 35° fixed angle (the precise value varies a few degrees depending upon the instrument manufacturer). The incident beam geometry means that the electron range will be reduced roughly in proportion to the cosine of the incident angle. Penetration of the electron beam normal to the cut face will then be approximately 20% less than the normal incidence condition but the beam is also elongated in the sample plane a complementary amount. Ultimately the beam geometry does not dramatically change the depth resolution estimate at low voltage or the information content. However, electronic tilt compensation must be employed to ensure that the scan area yields a correct dimension in both the x and y planes of the cut face to allow accurate quantification. This compensation is easily accommodated in all modern instruments.

6.3.2 Electron probe formation and electron detection principles in the Gemini column and CrossBeam platform

Ideally we would like to enlist the SE_1 electron exclusively as this is an inherently high-resolution signal. While it is not possible to purely separate SE_1 and SE_2 electrons because they span the same energy range, it is possible to collect a heavily SE_1 weighted signal using a proper in-lens detection system. Likewise, in the strategy identified above to optimize 3D depth resolution it is desired to form a high-brightness low voltage finely focused electron probe to reduce interaction volume and minimize the delocalization of SE and BSE signals. It is also desirable to extract a BSE^I signal under these same conditions. Finally, it would be very beneficial if we could collect these signals simultaneously in real-time during the milling process in an automated fashion to maximize data density and gain high throughput. With reference to the schematic cross section of the Gemini® column in Figure 6.3 the principles of electron probe formation and electron detection will be explained in the following paragraphs that allow those goals to be realized.

Figure 6.3(a) shows the schematic cross section overview with the optical elements of the Gemini column labeled. The source is a Schottky field emitter that combines high brightness, long lifetime, and excellent stability. The aperture changing system is electromagnetic and requires no mechanical adjustments. Note the position of the in-lens BSE detector which resides above the in-lens secondary detector. Along the length of the column is a liner tube element labeled as the beam booster. Beneath the in-lens secondary is a combination electrostatic-electromagnetic objective lens, detailed in Figure 6.3(b). Figure 6.3(c) depicts the crossover free beam path of the electron column. The following paragraphs relate how the crossover free beam path, the beam booster, the in-lens detection system and the electrostatic-electromagnetic objective lens combine to extract the data signals for optimized 3D reconstruction at low voltage.

Crossover free beam path

Figure 6.3(c) shows a representation of the condenser lens, but the strength of this lens is never operated in a crossover condition (i.e., brought to a focal point). Traditional electron columns contain one or more crossovers in the electron beam path. The crossover free beam path column minimizes the

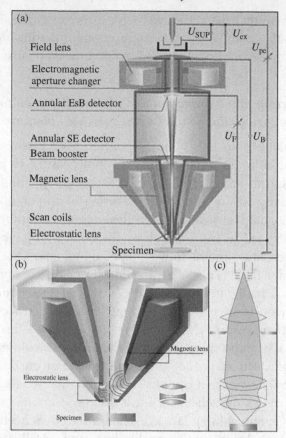

Figure 6.3 Schematic representation of key elements of the Gemini column. (a) Overview of the optical elements. (b) Detail of the combination electrostatic–electromagnetic Gemini objective lens. (c) The crossover free beam path with optical representations for the condenser lens and the effective triplet formed by the objective lens.

Boersch effect, which is a term describing energy broadening of the electron beam created by stochastic coulombic interactions between electrons. The Boersch effect leads to chromatic aberration in the electron column and is significant in areas of high electron density such as at a crossover condition and at low voltage. By minimizing the Boersch effect with a crossover free beam path the column reduces chromatic aberration and enhances low voltage performance.

Beam boosted column The beam booster has two functions: (i) it is part of the primary electron beam optics and (ii) it is part of the optical detection system. The beam booster is an 8 kV voltage that is applied on top of the selected primary beam landing energy. For instance, if the selected landing energy is 200 volts the actual primary beam energy coming down the column is 8.2 kV. By maintaining a high potential at the source regardless of the selected landing energy the tip brightness is kept high under all conditions. One of the pitfalls of low voltage operation on traditional electron columns is the severe reduction in brightness that follows when an emitter is operated at low voltage. The beam booster obviates this problem and contributes to excellent low voltage performance. The 8 kV potential is reduced to ground potential at the final electrostatic lens element thereby creating an electrostatic *deceleration* field gradient as seen by the primary beam that reduces the electron beam to the selected landing energy as it reaches the sample surface. This deceleration field also has a negative aberration lensing effect on the probe. However, the SE electrons produced on the sample surface experience an electrostatic *acceleration* field gradient of equal but opposite sense as that experienced by the primary electron beam. Due to the electrostatic acceleration field gradient the secondary electrons are pulled up the column across the 8 kV potential. Therefore the beam booster is acting as part of the optical collection and detection system. The beam booster potential also "stretches" the differences in small surface potentials, analogous to a contrast stretching operation. Even for a low voltage primary beam the effect of the beam booster also allows direct detection of both SEs and BSEs, as opposed to conversion detection.

On-axis in-lens versus Everhart–Thornley secondary electron detection

As the secondary electrons are accelerated up the column by the beam booster they pass through an electrostatic focus then follow a toroidal path around the primary beam and impact an *on-axis* in-lens secondary electron detector. On-axis in-lens detection maintains true optical projection geometry to maximize contrast and detection efficiency. Alternative designs employ variations of a "snorkel" lens where the detector is off-axis on one side of the column and an electrostatic grid bends secondary electrons off the optic axis and into the detector. The nonaxial collection field of a snorkel lens design can distort the primary beam, particularly at low voltage. A second SE detector, an Everhart–Thornley (ET) style scintillator is mounted on the side of the main chamber. This detector generates topographic contrast due to its asymmetric position on one side of the chamber but it also collects a larger proportion of SE_2 and SE_3 secondary electrons (SE_3 are electrons produced by interaction with the pole piece and surrounding chamber components). Figure 6.4 shows a pair of images taken of a cross section from an uncoated semiconductor device. The images were recorded at 890 eV and a 2 mm

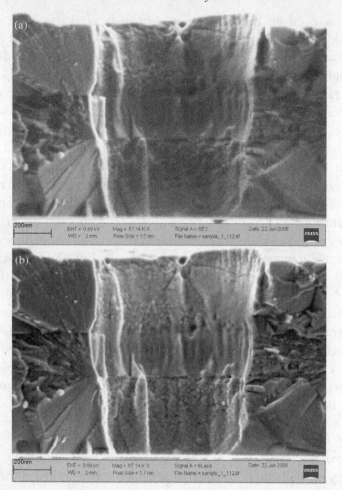

Figure 6.4 Semiconductor cross section image acquired at 890 eV at an original screen magnification of 215 000× and a horizontal field width of 1.74 μm. Both images were acquired simultaneously under identical beam conditions. Panel (a) was recorded with the Everhart–Thornley detector and panel (b) was recorded with the in-lens SE detector. The improved resolution in the lower in-lens image is due to emphasis on SE_1 detection and the reduced contribution of SE_2 and SE_3 image signals.

working distance with an original screen magnification of 215 000× (67.14k× in 4×5 format) and a field width of 1.7 μm. Using dual channel mode both images were acquired simultaneously during a single scan where the image in Figure 6.4(a) was captured on the Everhart–Thornley SE detector and the image in Figure 6.4(b) was recorded using the in-lens SE detector. As both images were acquired simultaneously under identical primary beam

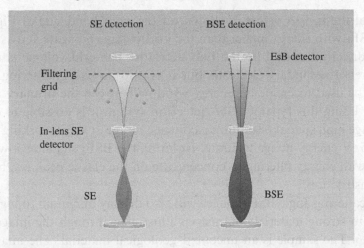

Figure 6.5 Schematic representation of the electron projection paths defining the in-lens SE detection and in-lens EsB detection. Secondary electrons are projected onto the in-lens SE detector while the BSE^I follow an axial projection onto the upper in-lens energy selective backscatter detector.

conditions the images compare the character of the different signals captured by the two detectors. The signal-to-noise ratio is somewhat lower in the ET image relative to the in-lens image due to the short working distance. It is clear that the in-lens detector reveals greater detail and improved resolution with respect to the ET detector. This example visually illustrates the effect of electron–solid interactions discussed above and the difference between a high spatial resolution SE_1 in-lens detection signal and an ET detector convolved with a contribution of SE_2 and SE_3 signals in the image. In the 3D reconstruction application we will primarily employ the in-lens detectors to take advantage of the superior spatial resolution.

Low voltage on-axis in-lens energy selective backscatter (EsB) detection

The principle of on-axis in-lens electron projection detection is extended to partition and extract low energy BSE^I from SEs. The following discussion is with reference to Figure 6.5. It is possible to separate the BSE^I signal from SEs and BSE^{II} electrons because the BSE^I are distinct both in terms of energy distribution and trajectory distribution. Secondary electrons pass through an electrostatic focus and diverge en route to the on-axis in-lens SE detector, as shown on the left-hand side of Figure 6.5. A small number of secondary electrons pass through the annular opening in the SE detector and are repelled by an adjustable field generated by an electrostatic grid. Low loss BSE^I electrons which follow a nearly *axial trajectory* pass through the annular portion of the in-lens detector, shown on the right-hand side of Figure 6.5. Those BSEs with energy higher than the grid voltage navigate across the field and are directly detected by the EsB detector. The grid voltage is not only used to reject secondary electrons but it is also an energy filtering element that can be used to create an energy window between the primary and grid voltages. Using this type of BSE^I detection system it is possible to collect a low voltage high spatial resolution z-contrast image at short working distance with a 25 eV energy image window. Isolating the BSE^I signal at low voltage coupled with energy filtering to concentrate on the elastic peak is a benefit in 3D reconstruction.

High scattering angle BSE^I images will also display minimum topography in addition to strong material contrast, as illustrated through the image pair in Figure 6.6. The sample is an uncoated geological material. The images were acquired at 750 V with a horizontal width of 7.7 μm. Both images in Figure 6.6 were acquired simultaneously during a single scan and captured using two detectors. The side chamber mounted SE_2 detector was used to collect the image shown in (a) and the energy selective backscatter (EsB) detector was used to collect the image in (b). The SE_2 detector image shows strong topographic contrast while the EsB image displays minimum topography and strong z-contrast, in a dramatic example of the different information content extracted by the two detectors. Other contrast mechanisms such as Schottky barrier contrast, voltage contrast and charging contrast are also suppressed from an EsB image. Enhanced z-contrast with minimum topography is a benefit in FIB-SEM 3D reconstruction to separate different materials and phases throughout the 3D volume during data reduction. Similarly, the elimination of edge contrast, charging contrast, and other contrast mechanisms deleterious to FIB-SEM nanotomography is also a significant benefit. The ability to extract this information at low voltages of 1 kV and less means that depth resolution is also optimized. Moreover, this type of in-lens BSE detector requires no insertion, alignment, or adjustment. Thus, in many respects the in-lens EsB is an ideal detector for the purposes of FIB-SEM 3D nanotomography. It is expected that future work will continue to highlight the benefits of FIB-EsB nanotomography in both materials science and emerging biological applications.

Electrostatic–electromagnetic objective lens

The Gemini column derives it name from the "twin" element configuration of the objective lens. With attention to Figure 6.3(b), the upper portion of the objective lens is an electromagnetic element while the lower portion of the objective lens is an electrostatic element. Collectively, these elements combine

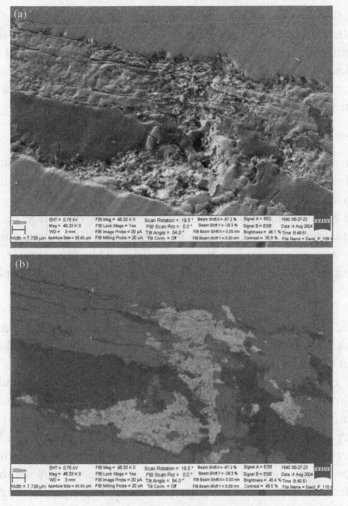

Figure 6.6 Comparison of images acquired simultaneously on an EsB CrossBeam highlighting the difference in information content from an SE$_2$ detector and an in-lens BSE detector. The images were acquired at 750 V and collected simultaneously using two different detectors. The image depicted in (a) was formed by the side chamber mounted SE$_2$ detector and shows strong topographic contrast. The lower image (b) produced with the in-lens EsB detector displays minimum topographic contrast and strong z-contrast.

to produce the electron optical analogy of an achromatic lens triplet, represented schematically in the lower right portion of Figure 6.3(b). The first lens of the triplet is an electromagnet element and the final two lenses of the triplet are electrostatic. Traditional electron optics systems consisting only of focusing elements lead to a lens system with positive aberration coefficients.

However, the action of the decelerating electrostatic field functions as a negative aberration lens, represented by the central element of the triplet. The final electrostatic lens element is the pole piece focusing element. This unique objective lens design leads to excellent low energy performance because the chromatic and spherical aberration coefficients actually decrease with primary electron voltage.

Real-time SEM imaging during FIB milling

The electrostatic component of the Gemini objective lens results in the absence of any appreciable magnetic field at the sample surface and this leads directly to a unique ability to image in real-time at ultra-high-resolution during the FIB milling process. Blurring and beam distortion result during any attempt to simultaneously operate a gallium FIB column and an SEM column in the presence of a magnetic field. This interference is due to interaction between the magnetic field with the isotopes and multiply charged ions within the gallium ion beam. The electrostatic elements on the lower portion of the Gemini objective lens contain the vast majority of the electromagnetic field generated by the upper portion of the objective lens, creating an opportunity for operation of simultaneous focused ion beam milling and scanning electron beam imaging.

Another notable consequence derived from the ability to simultaneously image with the SEM during the FIB milling process is the formation of a hybrid contrast mechanism that combines the SE image contrast produced by the primary electron beam and the channeling contrast produced by the ion beam on crystallographic samples. This operation mode is often desirable due to strongly enhanced orientation contrast that complements the electron beam induced SE contrast in the collected image. This hybrid contrast is only present while both beams are operating simultaneously.

It should be also noted that the real-time imaging during the ion milling process will not influence the BSE image contrast generated by the primary electron beam because ion bombardment does not produce backscatter electrons. Therefore collecting the in-lens EsB signal can also be advantageous when it is desired to suppress orientation contrast. Dual channel operation allows the operator access to display both these signals simultaneously on two separate computer screens during the milling operation.

6.4 Data acquisition parameters for FIB-SEM 3D reconstruction

The preceding established a rationale for using low voltage operation in order to optimize depth resolution and a strategy to extract SE_1 and BSE^I

information content to achieve high-resolution FIB-SEM nanotomography. Electron detection principles that allow ultra-high-resolution imaging at low voltage and *simultaneous* collection of both in-lens SE and in-lens BSE[I] signals have been described. Finally, high density data collection of these signals is facilitated by the ability to image in real-time during the FIB milling process by taking advantage of the electrostatic objective lens design.

The following is a general guideline for data specific acquisition parameters described for various types of FIB-SEM 3D reconstruction including FIB-SE, FIB-BSE, and FIB-EBSD. Obviously there are nuances to the techniques and methods described and the analyst is encouraged to adapt the methods described here according to their own circumstances. In practice, once the operation parameters are considered acquisition is executed with relative ease by an operator oriented with both SEM and FIB instrumentation. The setup process can also be assisted through a computer-aided expert system to facilitate the automated procedure.

6.4.1 Electron beam imaging current

An important parameter under the operator's control to affect the signal-to-noise ratio in the recorded image data is the electron probe current. The Gemini column does not suffer the same tradeoff between probe current and probe size as encountered with a traditional electron column. The details are related to the lack of crossover in the beam path and the fact that "spot size" is not adjusted by changing the condenser lens setting as on a system with crossover electron optics. So it is possible to maintain high resolution and high current while enhancing depth of focus. The high current/high depth of field mode on the Gemini column approximately doubles the current while also increasing the depth of field due to a commensurate reduction in the convergence angle. Therefore, the high current mode is generally a good choice for a given aperture as an increased depth of field is a benefit that is in addition to improved signal intensity. Probe currents can be in the range of less than 10 pA to greater than 20 nA, as desired. Software also allows an option to automatically track working distance during the milling process to maintain focus over large cut depths.

The real-time SEM imaging conditions during ion milling used to produce the 3D reconstruction data in most of this work yielded an incoming count of approximately 20 000 electrons averaged per pixel per frame and represents a moderate current. The calculations are derived from the dwell time per pixel, the probe current, and the frame averaging parameters.

6.4.2 FIB milling current

The FIB milling current will dictate the quality of the cut face and the amount of time required to mill the selected volume. The current and dwell time of the ion beam should be designed to ensure that each milling element cuts completely through all materials in the cross section to the desired depth. A "mill to depth" option allows a cut depth to be numerically input. The milling calculation is based upon a materials file that calibrates milling rates derived from measured or theoretical values input by the user.

The amorphous damage depth due to the milling process is usually not an issue for SEM imaging. However, be aware of the amorphous damage layer associated with the FIB milling process and ion beam sensitivity of the material. A lower FIB voltage will reduce the amorphous depth while a reduced milling current may reduce the damage rate, particularly on organic or biological specimens. Amorphous damage is a potential issue when collecting FIB-EBSD tomographic data. The analytical volume of EBSD is estimated to be in the range of 30 nm and good EBSD have been generated from 30 kV FIB using a 2 nA beam. However, the analytical volume of the EBSD data and the tolerable range of amorphous damage will be highly material dependent and even grain orientation dependent. In the case of FIB-EBSD it is most practical to run tests to confirm good quality patterns for a given FIB milling current.

An acceptable current range for FIB-SEM data collection can be anywhere from 10 pA–20 nA. The specific ion beam current will depend upon the milling properties of the material, the physical volume of the data cube and the time allotted for the acquisition. Assuming a nominal data volume consisting of 10 μm in width (x-axis), a z-depth of 5 μm and a y-depth of less than 10 μm, the acquisition can be completed in approximately one hour or less on a silicon based material with a current of 20–50 pA (recall with reference to Figure 6.1, that the z-axis represents the axis along which the cutting plane (x–y plane) will advance).

6.4.3 Acquisition time

One hour is quite reasonable for the typical sample volumes of approximately 500 μm^3, but there is no serious compromise associated with longer acquisition time. Some FIB-EBSD experiments have operated continuously for up to four days. Sample drift also is not an issue since the data set requires image alignment even with zero image drift, as described in the data reduction section. In sequential acquisitions, as in FIB-EBSD and FIB-XEDS, automatic image registration is employed between data acquisition cycles for both FIB and SEM alignments.

6.4.4 Image collection frequency

The depth resolution along the cutting plane has been estimated as 5–20 nm while the data acquisition time has been defined as 3600 seconds. Together with knowledge of the linear dimension along the cutting plane (x–y plane and z-axis, with reference to Figure 6.1) it is possible to determine the appropriate image acquisition rate. It is a good practice to acquire the data at a slice thickness that corresponds to half the estimated resolution limit. Assuming a 20 nm depth resolution, an image will be collected every 10 nm. Assuming this 10 nm value and a linear z-axis dimension of 1.5 μm, the number of required frames is 150. The total number of required frames to achieve a 10 nm resolution in this example corresponds to an image recorded every 24 seconds. Using the frame averaging noise reduction mode the system can automatically average approximately 200 frames every 20 seconds. It is a sound practice to collect an image at twice the frame averaging cycle rate. Continuing with this example, collecting an image every 10 seconds yields a ∼4 nm slice thickness and a total of approximately 350 frames. In this instance the data acquisition parameters are such that we are over sampling by nearly a factor of three to five relative to the estimated resolution limit. In the data reduction process one may choose to work with all the data frames or cull the data using any index parameter, i.e., every other frame. Thus, in general, an image recorded every 10–20 seconds will produce a high quality FIB-SEM nanotomographic reconstruction at a 10–20 nm resolution over volumes of 250–500 μm³.

If very high milling rates are desired for milling larger volumes or smaller volumes in a shorter time period then faster AVI capture rates may be used. The system software on the CrossBeam has a built-in AVI capture capability that allows the images to be recorded as frequently as every 100 ms.

6.4.5 Pixel size

The image pixel size is a function of both the SEM magnification and the image store resolution. At a screen magnification of ∼10 000× it is reasonable to employ a 1 K × 1 K store resolution but it is possible to save the images at a 3 K × 2 K store resolution. It is interesting to note that the maximum achievable resolution can be recorded without any a priori knowledge of the structures contained in the data volume. Following data acquisition it is of course possible to post-process any sub-volume of interest and digitally zoom in on any feature in the volume, but that operation won't ultimately increase the information content contained in the data.

6.4.6 Storing images

The storage volume required will depend upon the number of AVI movies recorded over time, the store resolution, and the length of those movies. A typical movie will require 200–300 Mb of disk storage and a comparable amount disk space for post-data reduction files. As a practical matter it is a good idea to have a separate storage volume to accommodate the movie data if one intends to complete a large number of reconstructions, process the data, and create inspiring animations. Fortunately, data storage is cheap.

6.5 Quantification of FIB based 3D tomographic data

The continuous nature of the milling process, in combination with a method devised to determine the slice thickness between image frames, facilitates direct quantification [15]. The continuous nature of the process also means that the data slices are more uniform in nature and less subject to thickness variation, charge fluctuation, and re-alignment errors associated with stopping and restarting the tomographic slice process.

The procedure involves patterning one or more features into or on the surface of the region of interest that can also be viewed in cross section during the FIB tomographic process. This feature may be etched into the surface by either the FIB itself or by another means. The feature could also be deposited onto the surface through the FIB or other method. In other words, the feature (s) can be either recessed (i.e., etched into) or raised (i.e., deposited) with respect to the sample plane. If deposited, the pattern can consist of any suitable material such as an oxide, metal, organic, inorganic, or any combination thereof. The feature must simply have a known geometric relationship between the top view and the cross section. The most direct example is a pattern consisting of at least two straight lines forming a known angle etched into the surface perpendicular to the cross section, as depicted in Figure 6.7.

Since the angle defined by the lines is known, a mathematical relationship can be established that is generally described by:

$$\triangle t_n = L_n \left(\frac{A_n - A_{n+1}}{A_n} \right) = \left(\frac{A_n - A_{n+1}}{\tan \theta} \right) \tag{6.2}$$

On the right-hand side of Figure 6.7 is an image showing actual fiducial lines etched into a platinum protective layer on top of a multilayer semiconductor

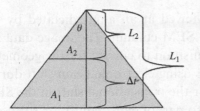 

Figure 6.7 The geometrical relationship above is applied to quantify the average slice thickness in each frame of the FIB-SEM reconstruction volume. On the left is a schematic related to Eqn. (6.2). On the right is an example of the fiducial lines etched using the FIB into a platinum layer created using the deposition capabilities of the FIB. The dashed box around the central fiducial line is used to perform automated image stack alignment. As the cut face progresses the angled lines move toward the central line allowing quantification of the slice thickness based upon the known geometrical relationship.

device. As the cut face progresses the angled line(s) move toward the center line. If the angle is 45° then the change in distance between the center line and the angle line is equal to the slice thickness. This fiducial system allows quantification through the mathematical relation above based upon either the average image slice thickness over the entire acquisition or the thickness of individual image slices. The ability to image the quantitative mark on the cross section in a single view improves speed and accuracy of the data. The data can be processed after the image data acquisition and do not require any prior knowledge of sample sputter rates (or other properties of the samples) nor a calibration or measurement during the data acquisition process. The fiducial marks are not seen in the processed final data set as they are usually placed to one side of the image just outside the region.

6.6 Data reduction

The typical steps involved in data reduction are summarized as follows: (i) image alignment; (ii) selection of the sub-volume representing the region of interest from the original data volume; (iii) image processing (optional); (iv) reconstruction of the volume; (v) any additional volumetric analyses (i.e., porosity, phase density, etc.); (vi) volume visualization, which could include application of color and transparency mapping; and (vii) animation of the results as desired.

As mentioned above, image alignment is required even if there is no sample drift because the image plane is translating in the $+y$-direction in the SEM view as the cutting plane progresses. This geometric effect occurs because the stage is tilted and the cutting plane is viewed at an angle dictated by the angular separation between the FIB and SEM column. The image data are properly tilt compensated during the acquisition to correct for the geometric tilt. The image data may also be adjusted using electronic beam shift during the acquisition in order to compensate for the translational shift in the SEM image. Automatic image registration is also part of data acquisition software.

Image processing may involve contrast enhancement, segmentation, threshold operations, or any other combination of standard image processing steps. For the results presented in this chapter image processing was not extensive and consisted of contrast adjustment or pass filters to facilitate transparency mapping. However, in future more sophisticated image processing will likely be employed to develop quantitative metrology applications as the field of 3D FIB-SEM nanotomography continues to emerge.

A variety of commercial software and freeware is available to complete 3D reconstructions that vary in sophistication and capability, including the integrated Smart3D software. Reconstructions of the type depicted on the following pages can be processed within one hour. The quality and ease of data processing is strongly dependent upon the quality of the original image data volume. Strong contrast and good signal intensity translate into an easily processed and high quality data set.

6.7 High density quantitative FIB-SEM nanotomography results

A series of examples follow that demonstrate the application of FIB-SEM 3D nanotomography including real-time data acquisition during the milling process. There is an inherent challenge capturing the experience of a dynamic process in the static format of a text. The reader should bear in mind that one can interrogate the data interactively including the ability to rotate the object freely, zoom in or out and even generate a "fly-through" path on the region of interest. It is common to animate the final results for presentation purposes.

6.7.1 FIB-SEM 3D reconstruction of a superconducting thin film

Figure 6.8 depicts a High-Tc superconductor thin film with second phases (dark areas in cross section). The top panel is a frame from the original AVI capture and the second panel is the volumetric reconstruction with selective transparency applied to display only the second phase material. This example demonstrates the ability to investigate a very complex and irregular structure. Interpolative schemes are generally not necessary due to the high data density

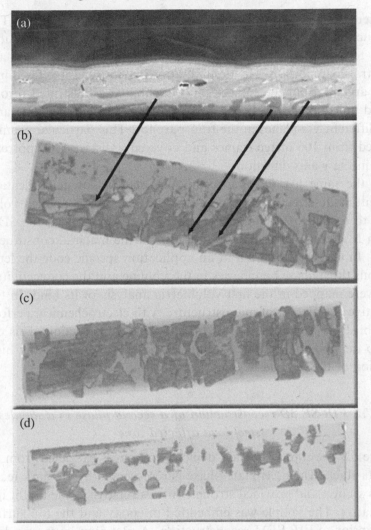

Figure 6.8 3D FIB-SEM nanotomographic reconstruction of a High-Tc superconductor thin film with second phases (dark areas in cross section). Panel (a) is a frame from the original AVI capture and the panel (b) depicts the volumetric reconstruction with selective transparency applied to highlight the second phase material. Common features in each image are marked by the arrows. Panel (c) is a different view but also shows a different section plane from the full transparency. The section plane is indicated by the dashed marker. Panel (d) is a transparency section plane close to the bottom of the reconstruction volume and highlights the nucleation density of second phases along the interface. (Sample courtesy of Dean Miller and Jon Hiller, Argonne National Laboratory.)

and a user specified slice thickness in the typical range of 5–20 nm. The solid reconstruction (no transparency applied) is not shown. In addition, it is possible to "slice and dice" either the solid reconstruction or data with applied transparency any way desired, allowing considerable freedom in the analysis and study of the subject. This capability is illustrated through the third and fourth panels of Figure 6.8 where the transparency data are taken at two different x–z planes in the transparency. This particular example was completed from 100 image frames and covered a distance of approximately 0.75 μm in the y-axis direction.

Using the same live imaging techniques, method of quantification, and equipment as described here a team of researchers have applied volumetric analysis to a solid oxide fuel cell sample [16]. They applied the 3D reconstruction to determine volumetric fractions of the material constituents and porosity. In addition, with aid of an application specific code the lengths of contiguous three phase boundaries in the volume and the connectivity of sub phases were mapped in the first volumetric analysis of its kind. In this way quantitative correlation of microstructure with electrochemical performance was enabled via a quantitative FIB-SEM 3D reconstruction. This example points to the potential of the technique and future directions and highlights application of the method beyond visualization.

6.7.2 *FIB-SE 3D reconstruction of a second phase in a stainless steel heat affected zone*

In Figure 6.9 the top two panels depict two images extracted from the original data stack. The precipitate phase is represented by the dark features in the cross section and provided strong contrast relative to the bright intensity of the matrix. The sample was embedded in epoxy and the top surface was polished prior to FIB-SEM reconstruction. A platinum layer was deposited on the surface to protect the portion of the structure at the surface. The fiducial marks for quantification (not shown) were also etched into the platinum layer outside the field of view. Again, the FIB-SEM nanotomographic reconstruction faithfully reproduces details in the original structure. The reconstruction is 563×135 pixels and consists of 200 frames over a y-axis linear dimension of approximately 1.4 μm to yield a 4 nm slice thickness encompassing a volume of approximately 25 μm³. Growth, shape, and nucleation parameters can thus be studied in three dimensions. Several data sets were obtained from this material at different sections of the weld. The different shapes and structures of the phases frozen into the material reflected

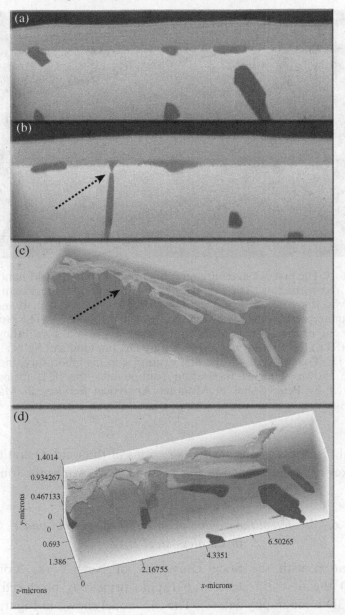

Figure 6.9 3D FIB-SEM reconstruction with transparency applied to highlight phases formed in a heat affected zone of a stainless steel weld. The top two images (a, b) are frames extracted from the original movie. Reconstruction views are shown in (c) and (d). A common feature detail is identified by the arrows. Axes units are micrometers in (d). (Sample courtesy of Mahesh Chaturvedi, University of Manitoba.)

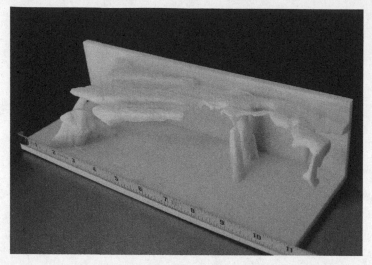

Figure 6.10 The first 3D stereolithographic print of an FIB-SEM nanotomographic reconstruction. The original object dimension was less than 10 μm across while the macroscale model of that original object is over 10 inches in length and represents the same structure shown in Figure 6.9. The project to convert one of the 3D nanotomography data sets into a stereolithography format (STL) for printing was motivated by Dr. Jim Quinn at SUNY, Stony Brook. Dr. Quinn produced the first examples in his laboratory, including the one shown. The file translation routine into the STL format was completed by Peter Sobol of Monona Analytical Services in Madison, Wisconsin.

the thermal histories from the various regions of the weld. Quantification of all three axes has been applied to this volume, as shown in the fourth panel.

6.7.3. *Stereolithographic printing of the second phase in stainless steel HAZ*

In cooperation with the State University of New York, Stony Brook (SUNYSB) the first 3D stereolithographic print was produced from an FIB-SEM 3D tomography dataset. The 3D model shown in Figure 6.10 was produced from the data set shown in Figure 6.9. Originally approximately 10 μm in length the physical model measures over 25 cm to allow a virtual nanoscale to macroscale conversion. Using a routine to convert the original intensity matrix into a format compatible with the 3D printer (STL format) it was a fairly straightforward process to produce a quantitative three-dimensional solid model. Following a similar approach it is also possible to scale replicate an object using the deposition capabilities within the CrossBeam.

Integral software on the CrossBeam platform allows the user to reslice the 3D reconstruction following data reduction and then rebuild the object at a chosen scale layer by layer using any of the selected deposition materials, including metals, oxides or carbon precursors. A potentially interesting application of this technology is to scale replicate pattern templates using FIB-SEM reconstructions produced from natural structures. Objects could range from diatoms to the interlaced ladder-like structure of the cross section from a butterfly's wing. The stereolithographic 3D models·also make lovely bookends.

6.7.4 3D reconstruction of a multi-layer semiconductor device

Figure 6.11 depicts a multi-layer semiconductor device with copper inter-' connects. The top panel is a cropped portion of a frame from the original movie. Voltage contrast is apparent in the SEM image data where some interconnects are bright while others are dark. Since these reconstructions are intensity-based, the dark interconnects do not display clearly throughout the selective transparency volume. Collection of a BSE image is one approach to eliminate the voltage contrast when desired. The second and third panels represent solid reconstructions depicting different section planes. Recall that the top surfaces are planes *perpendicular* to the cutting plane. Note the detail of the copper grain structure in the third panel highlighted by the arrow. It is possible to trace the copper grain structure throughout the volume.

FIB-EDS 3D reconstruction of a multi-layer semiconductor device

Figure 6.12 depicts an FIB-EDS reconstruction of the same device shown in Figure 6.11. The information vector in this reconstruction is associated with X-ray intensity data for the copper L line. These data were acquired at 5 kV in order to excite the Cu_L X-ray. The reconstruction is based upon 18 data slices and has an inherently lower resolution than the FIB-SE data due to the larger analytical volume associated with the X-ray signal at 5 kV. The key benefit of FIB-EDS is the opportunity to combine high-resolution image data with compositional information.

6.7.5 FIB-BSE 3D reconstruction using the in-lens EsB backscatter detector

Figure 6.13 is an FIB-BSE nanotomographic reconstruction of a multi-layer semiconductor device. Figure 6.13(a) shows a single frame from the image

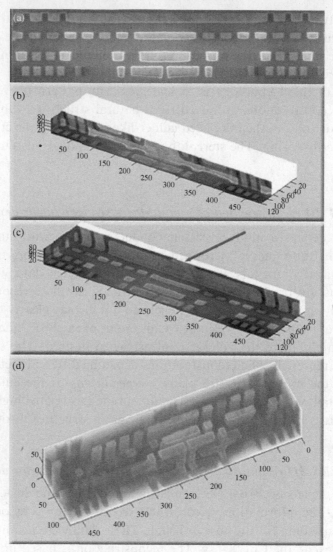

Figure 6.11 3D FIB-SEM reconstruction of a multi-layer semiconductor device. Panel (a) is a cropped frame extracted from the original AVI movie. Panels (b) and (c) are solid 3D reconstructions sectioned along different planes and *perpendicular* to the FIB cutting plane. Note the detail in the cross section, particularly in the panel (c), where the copper grains are clearly visible (marked by the arrow). Note also the voltage contrast seen in the image. Panel (d) presents the volume reconstruction with transparency applied to highlight the copper interconnects. (Sample courtesy of Bernie Levine, SUNY at Albany Nanotechnology Center, College of Nanoscale Science & Engineering.)

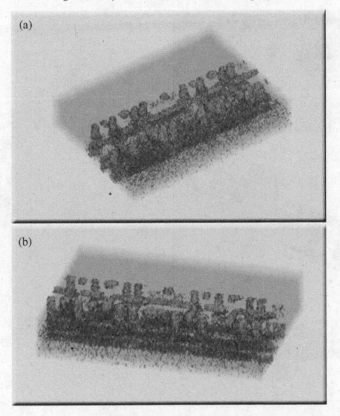

Figure 6.12 FIB-EDS reconstruction of the structure shown in Figure 6.11. Original EDS data maps were acquired by James Evertsen, SUNY at Albany Nanotechnology Center using a PGT EDS spectrometer.

data stack showing the total cross section, which extends approximately 40 μm in width. The data set was collected using a 20 pA milling current and acquired in less than 2 hours. The highlight area in Figure 6.13(a) is the sub-volume from which the 3D reconstruction shown in Figure 6.13(b) and (c) was generated. The 3D reconstruction data set consists of approximately 350 frames with a 15 nm slice thickness encompassing a volume of 22 μm × 3.5 μm × 4 μm in the *x*, *y*, and *z*-axes respectively. The data were acquired using the in-lens backscatter (EsB) at a primary voltage of 0.8 kV. The brightest contrast regions in the EsB image stack, such as the eight areas seen in Figure 6.13(a) correspond to the via material. The next brightest contrast is from the metal interconnects and the lowest contrast level is from the interlayer dielectric. In the reconstructions in Figure 6.13(b) and upper image in Figure 6.13(c) the via material is mapped to blue while the interconnects are mapped to red. In Figure 6.13(b) transparency mapping is applied to remove the

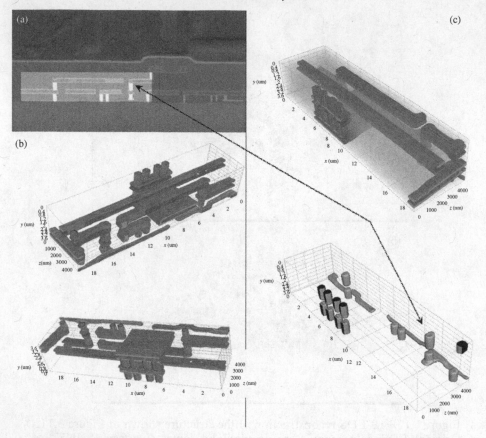

Figure 6.13 FIB-BSE 3D nanotomography of a multi-layer semiconductor device using in-lens EsB detection. The backscatter image data stack was captured using a primary voltage of 0.8 kV. The EsB detector produces high-resolution z-contrast, which is favorable for separating different materials in the 3D reconstruction while suppressing voltage contrast and edge enhanced contrast effects.

interlayer dielectric completely. Upper panel in Figure 6.13(c) is a three color composite with a lower degree of transparency applied to the dielectric (green) to provide additional depth contrast. The view in the upper image of Figure 6.13(b) is at positive elevation viewed from the rear of the volume with respect to the orientation in Figure 6.13(a) while the lower image is a negative elevation view (bottom view) as seen from the front. The lower image in Figure 6.13(c) is the same view as the upper image in Figure 6.13(c) represented as an isoclinic surface that maps the 204 intensity level from the 0–255 range in the data and thereby shows exclusively the via material throughout the volume. The image in Figure 6.13(a) represents a single plane of the data stack that

Figure 6.14 FIB-BSE 3D nanotomography example from a 65 nm node dual
damascene structure. Note that a volume reconstruction can be completed
by a surface mesh over the same region of interest. Several reconstructions
also can be combined from different sub-volumes within the same dataset.

bisects the approximate center of the eight via structures seen along the back
plane region of the isoclinic view in Figure 6.13(c). The arrows identify a
corresponding structure in each image. Figure 6.14 provides another example
of FIB-BSE reconstruction representing a BW of 65 nm dual damascene
structures. Various views in the figure illustrate that volume reconstructions
can be combined with a surface mesh mapped to specific intensity values.

6.7.6 FIB-EBSD

The EBSD technique is too rich in information and complexity to give a full
accounting in the scope of this text. A general overview is given to facilitate
presentation of the FIB-EBSD methodology and results. The interested reader
should consult references for further study [17]. In the EBSD technique an
incident electron beam forms an angle of approximately 20° with the sample
plane, as shown in Figure 6.15. The beam voltage is generally 15–20 kV to
obtain a properly scaled pattern. The glancing angle geometry of the electron
beam facilitates production of strong intensity electron channeling patterns
(commonly called electron backscatter diffraction patterns) which escape from
the near surface region of the sample. The projected patterns are captured onto
a phosphor screen. The EBSD pattern consists of multiple pairs of Kikuchi
lines. Each pair forms a Kikuchi band that corresponds to a specific plane

within the crystallographic lattice. The band width is proportional to the lattice spacing. The intersection of the several bands represents a zone axis. Collectively, the Kikuchi pattern contains information on the interplanar angles in the crystal and its orientation. The spatial resolution of the technique is on the order of 10–20 nm under ideal conditions with a field emission microscope while the information depth is material dependent and of the order of 30 nm. In addition to grain orientation, crystal phase, and crystallographic texture the EBSD patterns are sensitive to strain and lattice deformation.

In order to apply the EBSD technique in 3D it is necessary to address issues of sample preparation and geometry. Preparation of the surface is typically a stringent requirement in EBSD because the quality of the patterns produced is highly sensitive to the morphology and amorphous damage depth. If the residual damage from mechanical polishing or other preparation is too severe no patterns will result. Electropolishing is often employed as a final preparation step to yield a smooth surface with minimum damage. Using the FIB it is possible to generate a suitably smooth surface for high quality EBSD patterns. However, it is still important to properly prepare the sample prior to FIB and that will often require mechanical polishing, but for reasons primarily associated with geometry rather than surface finish. With reference to Figure 6.15 the EBSD pattern can only be produced if there is a line of sight from the sample surface to the camera. One method to meet this requirement is to create the FIB cut on the edge of a sample. If an FIB cut is made too far into the interior of the specimen then most of the cut face will be blocked from the EBSD camera and, unless very large cuts are made, only a very small area from the top of the cut face will generate an EBSD pattern. Aside from locating the FIB cut at the edge it is also very helpful to work on a polished edge. A smooth and polished edge minimizes the amount of FIB milling time required to mill away rough edges prior to beginning the actual 3D FIB-EBSD. Once a suitable sample is produced it can be mounted and set up for the automated FIB-EBSD acquisition.

The schematic in Figure 6.16 shows the electron beam, FIB, and sample geometry for generating 3D FIB-EBSD data on the standard CrossBeam hardware configuration. For the FIB-EBSD technique the sample will oscillate between two positions, as shown in Figure 6.16. One position is the FIB milling position, designated position A and the second position is the EBSD data collection position B. However, instead of producing an FIB cut normal to the sample surface at 54° tilt, a cut face is generated with the sample stage tilted at 17°. Alternatively, a 17° pre-tilt holder may be used. After the sample is mounted with a smooth edge perpendicular to the stage and the stage is tilted at 17° the user then selects the area of interest and the volume to be analyzed

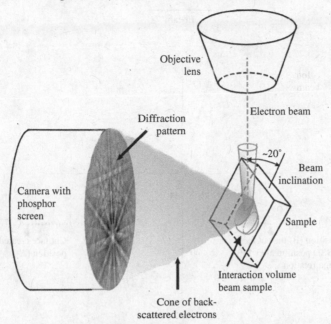

Figure 6.15 Schematic showing the electron beam and sample geometry for EBSD. The electron beam forms an angle of approximately 20° with the sample plane. The glancing angle geometry of the electron beam facilitates production of strong intensity electron channeling patterns (commonly called electron backscatter diffraction patterns) which escape from the near surface region of the sample. The projected patterns that are captured on a phosphor screen form a family of Kikuchi bands that contain crystallographic information from the sample.

with the aid of a setup wizard. A fiducial mark (i.e., an "X") is milled into an appropriate location on the sample surface next to the analysis region and the location is recorded by the FIB image registration routine. This is the FIB milling position (position A). The stage is then rotated 180° about the same 17° tilt axis into position B and that location is recorded by the SEM image registration. As illustrated in Figure 6.16, the geometry dictates that the cut face is now oriented properly for acquisition of the EBSD data.

The user is guided through the remainder of the setup process including the EBSD parameters and then the automated routine is initiated. The system will mill the selected area then rotate into position B and collect the EBSD data and then return to position A and begin milling the next data slice. This process repeats until the volume is completed. The time overhead transitioning between the two stage positions is less than 20 seconds, allocating most of the time to the EBSD acquisition. The milling time will depend upon the volume to be analyzed. Very large acquisitions have run continuously for up to four days.

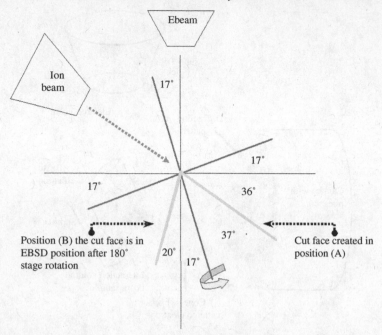

Figure 6.16 Schematic showing the electron beam, FIB and sample geometry for generating 3D FIB-EBSD data on the standard CrossBeam hardware configuration. Instead of producing an FIB cut normal to the sample surface at 54° tilt, a cut face is generated with the sample stage tilted at 17°. Alternatively, a 17° pre-tilt holder may be used. This is the FIB milling position designated as position A. After the cut face is created the stage is rotated 180° about the tilt axis into position B. The geometry dictates that the cut face is now oriented properly for acquisition of the EBSD data.

The example shown in Figure 6.17 is taken from the work of Konrad *et al.* described in [7]. The analysis was conducted on the area surrounding a hard particle (laves) in an Fe$_3$-Al alloy. The purpose of this study was to characterize the strong orientation gradients surrounding the hard particle produced during the warm rolling process. Figure 6.17(a) shows the cut face where the highlight area is the approximate region from which the 3D reconstruction was completed. The hard particle is the bright area in the central portion of the region of interest. Figure 6.17(b) shows an inverse pole figure generated from one slice of the EBSD data stack showing the hexagonal particle in the center surrounded by four orientation gradient zones developed during rolling. Note the EBSD data is tilt corrected. Figures 6.17(c) and (d) depict the 3D rendering of the EBSD data showing large angle grain boundaries.

There is an exception to the general rule stated earlier that a low voltage is required to achieve good depth resolution. The explanation to this exception is

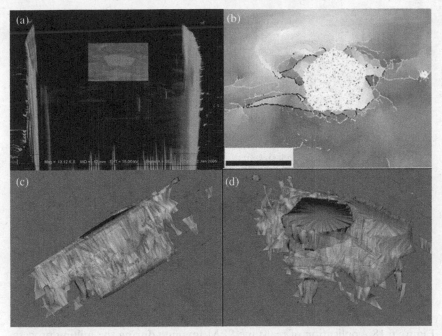

Figure 6.17 3D FIB-EBSD data investigation orientation gradients surrounding a hard laves particle in an Fe_3-Al alloy. (Data courtesy of S. Zaefferer [7].)

related to both the electron beam–sample geometry and electron detection schemes. When forward scattered low loss electrons are collected using an appropriate beam–sample geometry and detector it is possible to maintain a high spatial resolution signal with a relatively small interaction volume even at high voltage. The effects of collector take-off angle on low loss BSE imaging is reviewed nicely by Wells [18]. This situation is quite relevant in FIB-EBSD tomography because the electron beam typically is incident at a 20° angle with respect to the sample, which is also geometry suited for low loss forward scattered electrons. When forward scattered electrons are collected in this beam–sample geometry it is possible to collect a relatively high depth resolution signal even at high voltage. Commercial EBSD systems typically have an option to attach a forward scattered detector to the EBSD camera and this provides a means to collect the high depth resolution signal at high voltage. It is also an excellent accessory to provide orientation contrast imaging.

6.8 Applications in end-point detection

Certain benefits of simultaneous imaging during the milling process are intuitive. It is easy to understand the advantage of being able to observe the

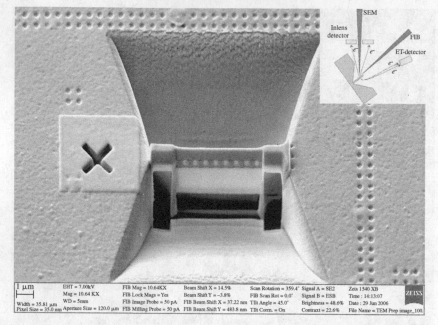

1 μm	EHT = 7.00kV	FIB Mag = 10.64KX	Beam Shift X = 14.5%	Scan Rotation = 359.4°	Signal A = SE2	Zeis 1540 XB	ZEISS
	Mag = 10.64 KX	FIB Lock Mags = Yes	Beam Shift Y = −3.8%	FIB Scan Rot = 0.0°	Signal B = ESB	Time : 14:13:07	
Width = 35.81 μm	WD = 5mm	FIB Image Probe = 50 pA	FIB Beam Shift X = 37.22 nm	Tilt Angle = 45.0°	Brightness = 48.6%	Date : 29 Jun 2006	
Pixel Size = 35.0 nm	Aperture Size = 120.0 μm	FIB Milling Probe = 50 pA	FIB Beam Shift Y = 483.8 nm	Tilt Corrn. = On	Contraxt = 22.6%	File Name = TEM Prep image_100	

Figure 6.18 Application of real-time SEM imaging during the milling process in TEM sample preparation. The image formed by the Everhart–Thornley (ET) detector creates a bright area in the thin portions of the lamella. The ET detector geometry collects the secondary electrons emitted from the back face as the lamella becomes thin. The ET detector intensity in the thin region of the lamella is a function of the material, primary electron voltage and the primary beam current, establishing a premise for automated EPD.

milling process via simultaneous SEM imaging at any magnification and at ultra-high resolution. This advantage clearly applies to end-point detection (EPD). Beyond manual EPD based upon the operator's vision and judgment new possibilities exist to automate EPD through the SEM imaging conditions in conjunction with various detectors and software control. Nearly any application in failure analysis will also gain advantage from the enhanced accuracy offered through real-time SEM observation during the milling process. It is hoped the reader will see other possibilities related to their own interests.

6.8.1 TEM sample preparation end-point detection

Consider the TEM sample shown in Figure 6.18. With regard to TEM preparation, EPD based upon real-time FIB-SEM imaging relies upon the different geometry and interaction volume between the in-lens secondary and the ET secondary electron detectors. The image in Figure 6.18 was acquired with the ET detector. Note that the thin portion of the lamella is brighter than the surrounding thicker portions of the lamella. The ET detector forms this bright image because as the lamella becomes thinner a point is reached where secondary electrons begin to emit from the back face of the lamella. The lamella thickness corresponding to emission of secondary electrons from the back face of the lamella is a function of the material and the primary beam voltage. This effect is due in part to the geometry of the lamella and the orientation of the primary electron with respect to the ET detector, as illustrated by the inset in the upper right corner of Figure 6.18. The in-lens SE detector will always display an "opaque" image while the ET SE detection will generate an image with a "transparent" appearance as the lamella becomes thin. This difference in the character of images generated by the in-lens and SE detectors establishes a method for automated EPD. Thus, it is possible to monitor the lamella thickness in real-time and establish an end-point detection based upon the secondary electron signal when both signals are displayed and monitored independently.

6.8.2 Patterned device prototyping

The final example illustrates the advantage of real-time imaging during the FIB milling process to produce patterned devices in a study of micrometer-scale ferroelectric capacitors. Conventional micro-fabrication techniques are not amenable to complex oxide devices because standard chemical or reactive ion etching processes are ineffective. However, it is possible to etch well-defined regions with an FIB to produce capacitors with dimensions far below 1 μm. In Figure 6.19 a ferroelectric capacitor 15 μm in diameter is connected to two contact pads, using leads formed from the top electrode material. The size (25 μm × 35 μm) and pitch (100 μm) of the pads was chosen to match the specifications of a commercial high-frequency electrical probe. The entire structure including the pads and the capacitor are electrically isolated from the remainder of the sample by 700 nm-deep trenches. The left contact pad is milled down to the bottom electrode, which is 500 nm from the surface and only 100 nm thick. The top capacitor electrode is removed in a region close to the capacitor in order to create a well-defined capacitor area. The top electrode is connected to the right pad by an undercut bridge structure (5). Fabricating the bridge required milling deep holes from both sides with an angle between the ion beam and the sample surface of approximately 26° instead of the typical 90°. Real-time FIB-SEM imaging was vital as EPD in the fabrication of this structure, as ion beam based EPD showed little or no variation through the various layers.

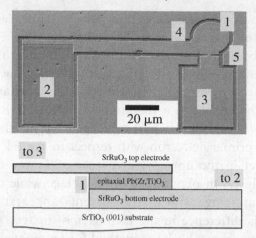

Figure 6.19 A ferroelectric capacitor structure designed for fast polarization switching experiments. The device size is about 15 micrometers (1); its bottom electrode and top electrode are connected to the left (2) and right (3) contact pad, respectively. The top electrode is cut (4) between the device and the bottom electrode contact pad. The top electrode is connected to the top electrode contact pad by a bridge structure (5). (Sample courtesy of Paul Evans, Alexei Grigoriev, Dal-Hyun Do, and Chang-Beom Eom of University of Wisconsin, Madison. FIB work completed by Jon Hiller, ANL.)

6.9 Summary

High density FIB-SEM 3D tomography examples have been produced through the benefit of simultaneous SEM imaging during the milling process and automated image capture. A methodology was established based upon low voltage imaging and in-lens SE_1 and in-lens BSE^I detection principles to optimize resolution and information content. The estimate of the achievable resolution of the technique is in the range of 5–20 nm depending upon the material. A method of quantification of 3D reconstruction data has also been devised and presented. Data acquisition parameters for high quality 3D reconstructions have also been outlined. Typical samples require approximately one hour of milling time and comprise data sets consisting of 100–400 image slices captured every 10–20 seconds over a volume of 250–500 μm^3.

Examples have also been shown that extend FIB based tomography to FIB-EDS and FIB-EBSD. The ability to automate the acquisition process and run unattended for extended periods of time makes this approach practical. In future it is likely that data formats will allow multiple data sets to be incorporated into a single file and could include FIB-SE, FIB-BSE, FIB-EDS, and FIB-EBSD data stacks. Future work will also likely extend the application to biological and life science. Within all fields, that potential to fuse data from various data sets will drive progress in FIB-SEM 3D reconstruction data analysis and allow powerful correlation between morphology, microstructure, and material properties in three dimensions at the nanoscale.

End-point detection schemes that also take advantage of simultaneous SEM imaging during FIB milling have been described. Applications of EPD based upon SE imaging include TEM sample preparation, failure analysis, and prototyping of complex patterned devices.

Acknowledgements

The author would like to thank Dr. Dean Miller, Jon Hiller, Dr. Eric Lifshin, and James Evertsen for their partnership in developing advanced applications including FIB based 3D volume reconstruction on the Carl Zeiss CrossBeam platform. Acknowledgements to Stefan Zaefferer and his colleagues at the Max Planck Institute for Ion Research for their contribution in the early development of the automated FIB-EBSD technique on the Carl Zeiss CrossBeam platforms. Acknowledgements are also owed to the team from EDAX/TSL who contributed to the FIB-EBSD development project including Damian Dingley, Paul Scutts, and Andy Fisher. Contributors from Zeiss on the FIB-EBSD project include Peter Gnauck, Hidde Wallart, and David Hubbard. Other Zeiss contributors to the overall 3D reconstruction hardware automation and software capabilities on the CrossBeam platforms include Richard Moralee and Patrick Cooper. A special thanks to Paul Evans and his research group at University of Wisconsin, Madison for permission to present their results on ferroelectric devices. Jim Quinn at the State University of New York, Stony Brook holds the honor of printing the first stereolithographic model of an FIB-SEM 3D reconstruction. Peter Sobol of Monona Analytical in Madison, Wisconsin created a routine to convert the 3D intensity matrix data from the FIB-SEM reconstruction into the STL format required for stereolithographic printing and also aided in development of an early version of custom 3D reconstruction visualization software.

References

[1] M. Uchick *et al. Proc. Microscopy and Microanalysis*, Savannah, GA (2004), 330–3.
[2] P. G. Kotula, M. R. Keenan and J. Michael. *Micros. Microanal.*, **12** 1 (2006), 36–48.
[3] D. N. Dunn and R. Hull. *Appl. Phys. Lett.*, **75** (1999), 3414–6.

[4] B. J. Inkson. *Scripta Mater.* **45** (2001), 753–8.

[5] T. Sakamato. *et al. Jpn. J. Appl. Phys.* **37** (1998), 2045–51.

[6] E. L. Principe. *Introduction to Focused Ion Beams: Instrumentation, Theory, Techniques and Practice*, ed. L. Giannuzzi and F. A. Stevie, (Berlin: Springer-Verlag, 2004), pp. 301–27.

[7] J. Konrad, S. Zaefferer, D. Raabe *et al. Acta Materialia*, **54** (2006), 1369–80.

[8] M. K. Miller. *Mater. Charact.* **44** (2000), 11–27.

[9] L. Reimer. *Scanning Electron Microscopy: Physics of Image Formation and Microanalysis*, 2nd edn. (Berlin: Springer, 1998).

[10] J. Goldstein, D. Newbury, D. Joy *et al. Scanning Electron Microscopy and X–Ray Microanalysis*, 3rd edn. (Berlin: Springer, 2003).

[11] WinCasino version 2.42. Universite de Sherbrooke, Sherbrooke, Canada. *www.gel.usherbrooke.ca/casino/What.html*

[12] D. C. Joy and S. Luo. *Scanning*, **11** (1989), 176–80.

[13] NIST Standard Reference Database 82. NIST Electron Effective-Attenuation-Length Database, *www.nist.gov/srd/nist82.htm*.

[14] The monotonic increase in BSE intensity with increasing atomic mass is valid at moderate and high voltage. At very low voltage nonlinear behavior may occur as the BSE coefficient of light elements can increase and that of heavy elements decrease. Low voltage BSE can even result in contrast reversal of light and heavy elements. This interesting behavior is not fully understood and is beyond the scope of this text but the analyst should be aware this can occur, as it will certainly influence image interpretation.

[15] Patent pending.

[16] J. R. Wilson, W. Kobsiriphat, R. Mendoza *et al. Nature Materials*, 11 June (2006), 1–4.

[17] A. J. Schwartz, M. Kumar and B. L. Adams, eds. *Electron Backscatter Diffraction in Material Science*, (New York: Kluwer Academic/Plenum Press 2000).

[18] O. C. Wells. *Scanning*, **2** (1979), 199–216.

7

Fabrication of nanoscale structures using ion beams

AMPERE A. TSENG

Arizona State University

7.1 Introduction

Nano-fabrication aims at building nanoscale structures, which can be used as components, devices, or systems, in large quantities with potentially low costs. Here, a nanoscale structure is characterized by a feature size in the range of 0.1 to 100 nm. Recently, ion beams have become increasingly popular tools for the fabrication of various types of nanoscale structures for different applications. In this chapter, the capabilities of the ion beam (IB) technology for nano-fabrication using the projection printing and direct writing approaches are discussed and examined.

The IB technology has many advantages over other energetic particle beams in nano-fabrication. For example, when compared to electrons, ions are much heavier and can strike with much greater energy density on the target at relatively short wavelengths to directly transfer patterns on hard materials (such as semiconductors, metals, or ceramics) without producing forward- and backscattering. Thus, the feature size of the patterns is directly dictated by the beam size and the interaction of the beam with the material considered. On the other hand, the electron beam or photon beam can only effectively write on or expose soft materials (such as photo or e-beam resists), and the corresponding feature sizes are determined by the proximity of the backscattered electrons or wave diffraction limits. Furthermore, the lateral exposure in IB is very low; thus, just exposing the right areas. As a result, a fine beam of heavy ions can produce very narrow line widths in the substrate and has a better capability to directly fabricate nanoscale structures. The IB technology has become not only a powerful fabrication tool adopted by the

Focused Ion Beam Systems: Basics and Applications, ed. N. Yao.
Published by Cambridge University Press. © Cambridge University Press 2007.

semiconductor industry for mask repairing, device modification, failure analysis, and integrated circuit (IC) debugging, but also a popular tool in making high-quality and high-precision nanostructures [1–3].

Two basic schemes, projection printing and direct writing, are used in IB nano-fabrication. In projection printing, a collimated beam of ions passes through a stencil mask and the image of the mask is projected onto the substrate using electrostatic lens systems, while in direct writing, a small spot of the focused ion beam (FIB) is struck directly onto the substrate without a mask. The technique of ion projection printing, also known as ion projection lithography (IPL), which originates from the semiconductor industry, enables parallel production of a large number of devices and has been one of the major candidates for next-generation lithography (NGL) to replace the currently used optical lithographic technique. Other types of printing schemes, including proximity printing and contact printing, are not included in this chapter because they offer limited flexibility and no rigorous development efforts have been reported for future industrial applications. The IB direct writing is a collection of several techniques, including milling, implantation, ion induced deposition, and ion assisted etching, which remove materials from, change properties in, or deposit materials on a substrate by direct impingement of the ion beam on the target with or without chemical reactions.

In this chapter, both projection printing and direct writing techniques are explored for their abilities in fabricating nanostructures. In projection printing, the current developments in its equipment, ion source, masks, and resists with focus on their applications for NGL are assessed. The nanostructures resulted from different exposures and trial tests are presented to illustrate their feasibility and resolution. In direct writing, the ion source and equipment for FIB are first assessed; the principles and techniques of four FIB-related processes are then introduced and evaluated separately. The associated nanostructures made by different procedures are also presented in order to illustrate the versatility and advancement of the FIB-related techniques. Finally, concluding remarks with recommendations are provided.

7.2 Ion projection lithography (IPL)

Lithography is the process of transferring patterns from one medium to another. In this section, recent development related to the technology of IPL is presented.

7.2.1 Ion source

Two types of ion sources, either point or volume-plasma sources, have been developed to produce nanometer resolution patterns. The volume-plasma sources are normally used for IPL, where a collimated ion beam is formed and projected on a substrate or resist through a mask with or without demagnification. On the other hand, the point source is used to shape an IB for direct writing.

In general, the axial energy spread of the ion beam, when coupled with the chromatic aberration in the ion optical column, can lead to blurring in the printed pattern on the target. The multicusp volume-plasma source has replaced the duoplasmatron volume source for ion projection printing because its ability to provide a lower axial-energy spread of ions minimizes the chromatic aberration of the projected image. Also, the multicusp source can be used to produce large volumes of uniform, quiescent, and high-density plasmas with high gas and electrical efficiencies. The multicusp source is based on electron-impact ionization, in which the energy transferred to a gas molecule from an energetic electron exceeds the ionization energy by means of ionizing collisions. Both the electrons and excited ions are accelerated by a dc field or rf power at a frequency of a few MHz (up to 13.56 MHz) and confined by an imposed magnetic field. The multicusp source is generally used to produce hydrogen and helium ion beams. Other ion beams of Ne, Ar, and Xe can also be generated.

An international IPL development program, called MEDEA, led by Infineon and Sematech, has used a coaxial multicusp ion source for its newly built IPL system, also known as the process development tool (PDT) for testing the virtual source size, energy spread, homogeneity, total extract current, and life time of the source [4]. With a specially designed extraction system, the PDT can extract a 5 keV He^+ beam to yield an axial energy spread as low as 0.6 eV. Consequently, this ion source can be expected to produce sharp features and to achieve the 50 nm resolution target. The light ions (H^+, H_2^+, H_3^+, He^+) are particularly suitable for projection printing because they have very little forward scattering and give off very little energy to the secondary electrons in the polymeric resist.

7.2.2 IPL system

In an IPL system, a collimated beam of ions passes through a mask and is accelerated by an electrostatic lens to create a demagnified image of the mask on the wafer. The wafer is stepped chip by chip. Many efforts have been

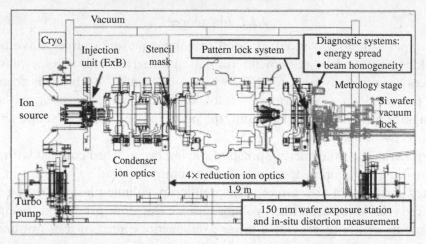

Figure 7.1 Schematic of PDT system. (Reprinted from [13] with permission from American Institute of Physics.)

devoted to the advancement of IPL technologies for the applications below 100 nm. In the MEDEA project, a 4×reduction ion optics has been developed for its prototype process development tool (PDT), in which a printing area of 12.5 mm × 12.5 mm is designed to be printed on the wafer. To obtain a full image of the wafer, each printing area is stitched one-by-one via synchronization of the beam and wafer moment [5]. The PDT has been used to investigate the industrial suitability for mass production of next-generation chips and is one of the major candidates for NGL to replace the conventional photolithography.

The co-axial multicusp ion source used in PDT has a very low energy spread at the 1 eV FWHM (full-width-half-maximum) level. Multi-electrode electrostatic ion optics has been used as the diverging electrostatic lens while an online diagnostic system and field composable lens are used to compensate for mechanical manufacturing inaccuracies. Figure 7.1 shows a schematic of the PDT ion-optical system. The multicusp ion source is equipped with a co-axial Wien (EXB) mass filter. A multi-electrode electrostatic condenser lens illuminates the 150 mm stencil mask with a nearly telecentric ion beam of 115 mm in diameter. After exciting the stencil mask, the ion beamlets are further accelerated to energies in the 70–150 keV range and then demagnified into a parallel beam whose image is focused at the wafer. A nine-element multi-electrode reduction optics is used to accelerate and demagnify the beamlets, while a pattern lock system monitors the position of 12 reference beamlets as they travel through the ion-optical system. Diagnostic elements are provided to measure the energy spread, beam uniformity, and distortion.

The uniformity of the current density can be controlled within 3%. By system optimization, it is expected that the uniformity can be achieved within 1%.

The Lawrence Berkeley National Laboratory has developed a maskless ion projection printing system called maskless microion-beam reduction lithography (MMRL). As reported by Ngo *et al.* [6], MMRL consists of a co-axial multicusp ion source, a multibeamlet pattern generator, and an all-electrostatic reduction ion-optical column. The pattern generator is used to create lithographic patterns to eliminate the need for masks. During processing, each individual ion beamlet can be switched on or off to form the lithographic pattern by biasing the extraction electrode with respect to the plasma electrode. Removing the use of stencil masks and the first stage, normally required by the other IPL systems (such as PDT), can eliminate the cost and efforts for mask development and fabrication as well as having a great potential to reduce the equipment and operation cost. Jiang *et al.* [7] and Ngo *et al.* [6] have used nine 50 μm switchable apertures to generate the beamlets with 10× reduction ion optics to demonstrate the proof-of-concept of MMRL and concluded that the maskless system is a vital candidate for NGL.

7.2.3 *Mask and resist*

Most of the IPL systems use two types of masks for pattern transformation as indicated by many investigators [5,8,9]. The mask needs to be made from an ion-transparent substrate with a pattern made up of an absorber surface. The first one can be made of a single layer of silicon that uses a varying thickness of the silicon acting as the absorber and the transparent membrane. Using the same single silicon crystal minimizes the problem created due to the thermal distortion and proton scattering in conventional masks. This is relatively easy to manufacture using the conventional IC processes. The second type, known as the stencil or open mask technology, is made up of open masks, in which the patterns are etched on the metal foils. This type of mask has excellent contrast, as the ions are not affected by passing through the open hole before striking on the resist. However, since there is no sub-layer to hold the etched metal pattern, every section in the etched pattern has to be connected to each other and no island or detached sections can be included in the pattern. As a result, the second type can only be used for very limited geometries.

Deep-ultraviolet (DUV) chemical-amplified resists and polymethyl methacrylate (PMMA) are the two most popular types of IPL resists [5,6]. They have the best combination of resolution, minor sensitivity, and etch resistivity. Other resists used for IPL include Hoechst's AZ series, Olin's HPR 506, OCG's HPR and ARCH [9,10]. By applying the Top Surface Imaging (TSI)

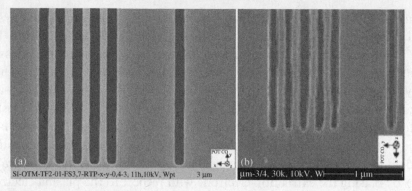

Figure 7.2 Mask-to-wafer transfer by PDT. (a) SEM image of 400 nm resolution patterns of overlay test mask (modified from [5]), (b) 100 nm line-space pattern printed in 240 nm thick Shipley XP9946-D resist. (Courtesy of Hans Löeschner of IMS Nanofabrication in Vienna.)

principle and using Ga ion beam exposure associated with silylation and oxygen dry etching, a DNQ (diazonaphthoquinone)/novolak-based resist pattern can be obtained to be as small as 30 nm, yet maintaining a high aspect ratio of up to 15 as indicated by Arshak *et al.* [11].

For a PDT demonstration, 150 mm stencil masks have been developed and fabricated on SOI (silicon-on-insulator) mask-wafer blanks using established semiconductor fabrication processes. The silicon membrane is 3 μm thick and coated with a 500 nm thick carbon protection layer to ensure adequate mask life. The design of these masks contains an array of overlay measurement structures. Through PDT's 4×reduction ion-optics, one of the overlay test masks having an array of I-marks with a linewidth of 400 nm (Figure 7.2(a)) has been successfully used to print a 100 nm wide I-marks pattern in a 240 nm thick Shipley XP9946-D resist as shown in Figure 7.2(b) [5]. A 37.5 keV He$^+$ ion beam at an exposure dose of $1.35\,\mu C/cm^2$ is used to project the I-marks pattern on this Shipley DUV chemical amplified resist, which exhibits a sensitivity of $0.8\,\mu C/cm^2$ for 37.5 keV He$^+$ ions [12]. This and other tests have demonstrated that mask-to-wafer transfer for isolated and arrayed lines/space can be performed within the required accuracy. Also, as reported by Löeschner *et al.* [13], 75 nm feature patterns can be achieved and pattern collapses are the limiting factor for determining the ultimate resolution of the system.

Moreover, exposures have been performed using the ion projector IPLM-02 manufactured by Ion Microfabrication Systems (IMS) on the Shipley UV II HS resist [14]. A 3.5 keV He$^+$ ion beam is used and accelerated behind the mask to 75 keV. The resist used is 180 nm thick and has been diluted so that a smaller resist thickness of 180 nm could be obtained. For a 75 keV H$^+$ ion exposure, the

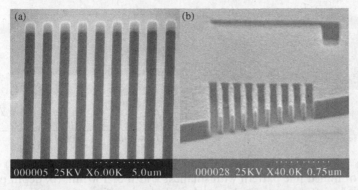

Figure 7.3 Ion projection resolution evaluation. (a) SEM image of stencil mask with 650 nm wide lines and spaces, (b) SEM image of 75 nm line pattern printed by 8.7×reduction in 180 nm thick Shipley DUV resist. (Courtesy of W. H. Bruenger of Fraunhofer Institute in Berlin.)

resist sensitivity is 10^{12} ions/cm^2, which corresponds to 0.15 μC/cm^2. An open stencil mask with arrayed patterns fabricated on a SOI wafer is used for a resolution study. Figure 7.3(a) shows the stencil mask with a linewidth of 650 nm, while, through an 8.7 demagnification by ion optics, a pattern of 75 nm wide lines and spaces is printed on the DUV resist without pattern collapse as shown in Figure 7.3(b). As indicated in Figure 7.3, the mask shows good edge quality, while no proximity effect is visible at the line ends in the exposed resist pattern, even though the resist has been removed by a second exposure of equal dose in the front part of the picture to allow SEM side view. It is noteworthy that the higher dose used in the test could effectively reduce the edge roughness of the resist.

7.2.4 Resistless IPL

Ions have a unique feature of directly modifying a wide range of materials without the need of a resist. In IPL, a whole surface area can be treated in parallel and can be patterned into a substrate in a single-step process that should be much faster than the series process or direct writing using FIB. IPL has been used for the resistless patterning of semiconductors (Si, GaAs, poly-Si), insulating layers (SiO$_2$, Si$_3$N$_4$), and metals (Al, Ni, Mo, Au) in addition to patterning organic thin films and resists. In many conditions, it is desirable to avoid the use of a resist. For instance, in making high-temperature superconductor nanostructures, undesirable chemical reactions could occur with the resist.

Bruenger *et al.* [15] have used the IPLM-02 ion projector to perform direct milling on a 35 nm thick Au film using an open stencil mask with 8.7

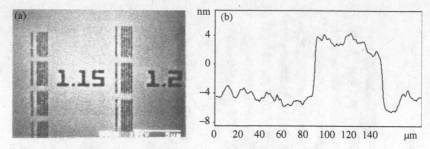

Figure 7.4 IPL milling with Xe^+ ion in polycrystalline Au film. (a) SEM image of milled line pattern with smallest width of 130 nm, (b) depth profile of milled pattern measured by white light interferometry. (Courtesy of W. H. Bruenger of Fraunhofer Institute in Berlin.)

demagnification ion optics. To increase the milling rate, the light ions normally provided by the multicusp ion source used in IPLM-02 are replaced by the heavier Xe^+ species without any change in emission stability or uniformity. Figure 7.4(a) shows an array of line patterns milled with a dose of 2×10^{15} Xe^+ ion/cm^2 at 75 keV having the smallest linewidth of 130 nm, while, based on the profile measurement by white light interferometry shown in Figure 7.4(b), the milling depth is 8 nm. The apparent roughness of the gold film is believed to be due to its grain structure.

IPL milling should be especially attractive for fabricating magnetic nano-dots, since other techniques, including photolithography, e-beam lithography, and nanoimprint lithography have to use resist-based processes that are not preferable to keep the topography of a surface unchanged. This is extremely important for magnetic disks with a surface roughness of a few nanometers. Through a stencil mask, Dietzel *et al.* [16,17] used the IPLM-02 projector with a 3×10^{14} Xe^+ ion/cm^2 dose at 73 keV to mill magnetic Co-Pt multilayers. Based on the magnetically altered areas measured by magnetic force microscopy (MFM), Dietzel *et al.* [16,17] reported that magnetic islands with an averaged diameter of less than 100 nm are formed. Also, the atomic force microscopy (AFM) measurement indicates that the surface roughness of the topography after the IPL process is 1.1 nm in rms, which confirms that the topography change by IPL is negligible. Thus, IPL is acceptable for magnetic media applications. Earlier, Bruenger *et al.* [15] had also used the IPLM-02 projector to mill a magnetic Fe-Pt film with a 10^{16} He^+ ion/cm^2 dose at 75 keV and reported that the averaged magnetic island size was 340 nm. By experimental evidence and numerical simulation, Dietzel *et al.* [16] and Bruenger *et al.* [15,18] noticed that Xe^+ and Ar^+ are more effective by two orders of magnitude compared to He^+ and believe that with an appropriate mask design, IPLM-02 can make magnetic dots as small as 50 nm in diameter, which would result in a storage density greater than $10\,Gbit/cm^2$.

7.3 Direct writing

FIB is normally used for the direct writing of nanostructures. IB direct writing is a process of transferring patterns by using an energetic FIB to directly hit the target to cause physical and/or chemical changes in the target materials. Four major direct writing processes will be considered in this section milling, implantation, ion assisted etching, and ion induced deposition. While the first two are governed by physical alterations, the latter two are dominated by chemical transformation.

7.3.1 Ion source

FIBs use liquid-metal ion sources (LMIS), which are high brightness ion sources because they can produce a beam of heaver ions that can be focused into fine spots of the order of 10 nm with adequate current densities for direct writing. Almost all metals that have relatively low melting temperatures and low reactivity can be used as the sources. The range of materials being used in FIB systems is also expanding to further increase the extent of their applications. The ion sources that are currently available include Al, As, Au, Be, Bi, Cs, Cu, Ga, Ge, Er, Fe, In, Li, Ni, Pb, Pd, Pr, Pt, U, and Zn [1]. Among these, Ga is the most popular ion species used in FIB for direct writing. In order to lower the melting point and to control the reactivity, alloy sources, such as PdAs, PdAsB, AuSi, and AuSiBe, are frequently used to deliver the dopants for semiconductors [19].

A typical LMIS consists of a capillary tube with a needle through it, an extraction electrode, and a shielding. The capillary acts as a reservoir that feeds liquid-metal to the tip. The interaction of the strong electrostatic force generated by the extraction electrode and the surface tension causes the liquid-metal meniscus to form a sharply peaked cone, also known as the Taylor cone. The application of the critical Taylor voltage on the liquid-metal cone extracts positively charged ions. The ions are condensed into focused parallel beams by the lens and ground electrodes, also called the upper or condenser lens [1].

7.3.2 FIB systems

The basic components of an FIB system consist of an ion source, ion optics, a substrate stage, and a vacuum chamber with auxiliary equipment. Figure 7.5

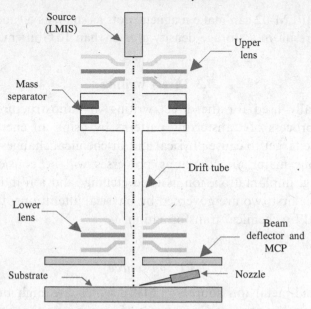

Figure 7.5 Schematic of FIB system.

schematically shows an FIB system with an LMIS. The ions are focused and collimated into parallel beams by the upper (condenser) lens. Then, the ion beam is passed through a mass separator and a drift tube. A mass separator is only necessary for high-energy systems equipped with liquid alloy ion sources. It is set up to filter out unwanted ion species emitted from the alloy ion sources by allowing only the required ions with a fixed mass–charge ratio to pass through. Below the mass separator, there is a long collinear drift tube equipped with a stigmatic focusing lens for stigmatism and collimation, which eliminates the ions that are not directed vertically. The lower (objective) lens is located below the drift tube and helps in reducing the spot size and in improving the focus. Following the objective lens is an electrostatic beam deflector that controls the final trajectory or landing location of the ions on the substrate.

Many accessories, including the nozzle shown in Figure 7.5, are equipped for FIB assisted etching and induced deposition. Also, in processing insulated substrates, a nozzle type of equipment is needed to provide low energy nonfocused electrons to compensate the charging on the substrate surface and to neutralize the substrate. Often, a multi-channel plate (MCP) is located above the target. The MCP helps in recording the secondary electron emission and thereby, helps in viewing the substrate. All the components are usually placed in a low-pressure chamber evacuated to the 10^{-7} Torr regime.

This is done so that the mean free paths of the ions are increased and the strength of the beam is not reduced due to the interference of the particles in the chamber [1,20].

7.3.3 FIB milling

Essentially, milling is a process combining physical sputtering, material redeposition, and amorphization. In this section, following the evaluation of sputtering, the effects of redeposition and amorphization are discussed. The combined effects in milling are then illustrated by evaluating the milling characteristics of three basic nanostructures: holes, trenches, and 3D stacked structures. An understanding of these milled patterns can lead to enhancing the material removal rate in milling and further improving the dimensional controllability for milling high-precision, more complex nanostructures.

Sputtering and sputter yield

When an energetic ion bombards the surface of the target solid, a variety of ion–target interactions can occur. The most important interaction for milling is sputtering, in which the energy transferred from the ions to the targeted substrate is high enough leading to a collision cascade involving substrate atoms on or near the surface to cause material removal. For the majority of the ion sources used in FIB, the optimized ion energy leading to a surface collision cascade of most engineering target materials is in the range from 10 to 100 keV. For energy higher than the order of 100 keV, the ions can easily penetrate into and be trapped inside the target substrate where the implantation occurs. For ion energy higher than the order of 1 MeV, inelastic interactions can become dominant and many high-energy particles can be generated. In milling, only the sputtering interaction is desirable.

Normally, the sputter yield is used to measure the efficiency of the sputtering process and is defined as the number of atoms ejected from the target surface per incident ion. A software package, called SRIM (Stopping and Range of Ion in Matter), has been widely used for predicting the sputter yield for many different ions over a wide range of energies. SRIM is a comprehensive program that uses a Monte Carlo treatment of ion–atom collisions to calculate the stopping range of ions (10 eV to 2 GeV/atomic mass unit) into matter [21,22]. It provides the distribution of the ions and the kinetic phenomena associated with the ion's energy loss, including target damage, phonon production, ionization, ion reflection, implantation, and sputtering.

The dependence of the sputter yield on ion energy, incident angle, and the target material can be predicted by SRIM. Figure 7.6 shows the sputter yield

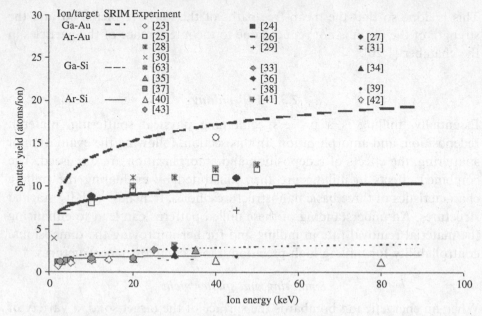

Figure 7.6 Energy dependence of sputter yield of Au and Si substrates by Ga and Ar ions at normal incidence.

predicted by SRIM for two types of target substrates (Si and Au) using two types of ion sources (Ga and Ar) at normal incidence as a function of the ion energy. Ga and Ar are the two most popular ion species in FIB milling, while Si and Au are the two most useful substrates used in nano-fabrication. They will be used as the example ions and target materials for later discussions. As shown, the sputter yield grows as the ion energy increases, but its rate of increase reduces. Although the data are not shown, in most of the cases, the sputter yield either levels off or decreases for ion energies higher than 100 keV, where the implantation becomes dominant, as ions penetrate into the substrate and are trapped in the lattices. Since SRIM is based on the Monte Carlo method, the number of ions used in the simulation can have an effect on the results. Earlier studies indicate that the simulations become converged when the ion number surpasses 500. To be conservative, one thousand ions are used for all of the SRIM simulations presented here.

Figure 7.6 also shows that the sputter yield of the Au substrate is much higher than that of the Si substrate, and that Ga can produce higher sputter yields than Ar. In general, heavier ion sources or lower surface binding energies (or sublimation energy) of target materials can produce higher sputter yields [1]. Also, a decrease in the target sublimation energy causes higher sputter yields due to the fact that sputtering is a phase change process: an atom

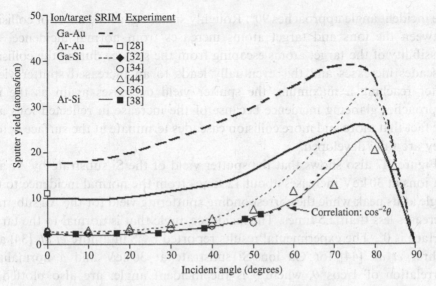

Figure 7.7 Angular dependence of sputter yield of Ar and Ga ions on Au and Si substrates.

in a solid phase being ejected into a gas phase. The atomic (or ion) weights of Ar and Ga are 39.95 and 69.72, respectively, while the binding energies of Au and Si are 3.80 and 4.69 eV/atom, respectively.

To gauge their reliability, the SRIM predictions are further compared to a large group of experimental data. As shown in Figure 7.6, the data for Ga ions on Au substrates are reported by Blauner *et al.* [23] and Xu *et al.* [24], while the data from Almen and Burce [25], Robinson and Southern [26], Carter and Colligon [27], Nenadovic *et al.* [28], EerNisse [29], Müller *et al.* [30], and Poate [31] are for the case of Ar ions on Au substrates. For Si substrates, the results of Yamaguchi [32,63], Pellerin *et al.* [33], Santamore *et al.* [34], Bischoff and Teichert [35], and Frey *et al.* [36] are shown for the case of using a Ga FIB, while the data using Ar FIB are obtained from Southern *et al.* [37], Sommerfeldt *et al.* [38], Andersen and Bay [39], Blank and Wittmaack [40], Kang *et al.* [41], Morgan *et al.* [42], and Zalm [43]. As shown in Figure 7.6, the SRIM predictions agree very well with the experimental data for all the cases considered. Since the data presented by Lehrer *et al.* [44] are almost identical to those of Frey *et al.* [36], only the latter is cited.

Figure 7.7 shows the SRIM simulation of the dependence of the sputter yield on the incident angle at two different ion energies. Again, Ar and Ga, in the sputtering of Au and Si substrates, are considered. As shown, for all the cases considered, increasing the incidence angle increases the sputter yield until it reaches its maximum near 80° then, it decreases very rapidly to zero as the incident angle approaches 90°. Roughly speaking, as the angle of collision between the ions and target atoms increases from normal incidence, the possibility of the target atoms escaping from the surface during the collision cascades increases and this eventually leads to an increased sputter yield. After reaching a maximum, the sputter yield decreases again as the ion approaches glancing incidence because of the increase in reflected ions and the fact that more and more collision cascades terminate at the surface before they are fully developed.

Figure 7.7 also shows that the sputter yield of the Si substrate by Ar and Ga ions at 30 keV increases about 12 times from the normal incidence to the angle at its peak while the corresponding sputtering yield for the Au substrate increases less than 2.5 times. The incidence angle that is normal to the target surface is 0°. The experimental results reported by Santamore *et al.* [34] and Lehrer *et al.* [44] for Ga ions/Si substrate at 30 keV and a normalized correlation of $1/\cos^2\theta$, where θ is the incident angle, are also plotted in Figure 7.7. As shown, a good agreement has been found between the SRIM predictions and the experimental results, especially for incident angles less than 60°. In fact, the experimental results fall in between the SRIM predictions and the correlation. For larger incident angles, the SRIM prediction is much higher than the experimental observation. At glancing angles, surface channeling plays an important role and causes the sputtering yield to decrease. The behavior of the angular dependence of sputter yields has been widely observed by many other researchers, including Sommerfeldt *et al.* [38], Yamamura *et al.* [45], and Vasile *et al.* [46].

Based on the above discussion, the material removal rate by sputtering, or the sputter yield, is dependent not only on the substrate material, but also on many processing parameters, including the ion energy, angle of incidence, and scanning procedures.

Redeposition and amorphization

It has been found that the majority of the sputtered atoms (neutrals and ions) can be ejected from the solid surface into a gas phase. Since the ejected atoms are not in their thermodynamic equilibrium state, they tend to condense back into the solid phase upon collision with any solid surface nearby. Consequently, a portion of the ejected atoms can bump into the already sputtered surface and redeposit onto it. For example, in trench milling, the atoms ejected from the bottom of the milled channel have a certain possibility to collide with the milled sidewalls and become redeposited.

The milling of different micro- and nanostructures from simplified 2D (such as nanoholes and nanochannels) to complicated 3D geometries has been recently reviewed to characterize the basic sputtering, amorphization, and redeposition phenomena [20]. For instance, Tseng [1] indicated that redeposition could be greatly reduced if multiple passes instead of a single pass are used in milling under the same amount of total ions. In multiple passes, each successive pass removes redeposited material from the previous pass. In general, redeposition is affected not only by the scanning pattern of milling, but also by the milled geometry, the dynamics of the ejected atoms, and the sticking coefficient of the target material. Normally, redeposition reduces the milling rate and the aspect ratio of the milled structures. It also makes the amount of the material milled difficult to be controlled and its final geometry hard to predict because a certain portion of the ejected atoms can be removed by the vacuum pumping system or can be redeposited outside the milled area.

If the energy or dose level of the incident ions is not high enough for sputtering, amorphization may occur in the bombarded area of a crystalline substrate and induce the substrate to swell. For example, in the case of a crystallized Si substrate bombarded by Ga ions, the dose level to cause amorphization is in the order of 10^{15} ions/cm^2, while the effective sputtering dose should be at least two orders of magnitude higher than the amorphization dose [36,47]. In amorphization, the incident ions in most cases are buried in the target material and may also displace the target atoms from their lattice sites so that the displaced atoms are relocated on the nearby region. In the case of silicon, its density at the amorphous state is much lower than that of crystalline silicon as reported by Custer *et al.* [48]. Also, since the volume of swelling is much larger than the volume of the buried or implanted ions, swelling should be mainly caused by the density changes or the amorphization rather than the volume increase caused only by the buried or implanted ions. The magnitude of the swelling due to amorphization can be as high as tens of nanometers. Thus, amorphization can diminish the dimensional accuracy of nanostructures and should be important in nanofabrication. An understanding of swelling or amorphization should be the first step to minimizing or controlling it.

Nanostructure milling

Li *et al.* [49] have used a 30 keV Ga$^+$ FIB to mill an array of nanoholes (dots) with a single pass on highly n-doped single-crystal Si(100) substrates. One of the dot-array profiles milled at 1 pA with a dwell time of 2 s for each dot using a 10 nm (FWHM) diameter beam is shown in Figure 7.8. As shown, the hole (dot) has a V-shaped cross section and is approximately 20 nm in depth and 50 nm in diameter with a pronounced ring-shaped structure

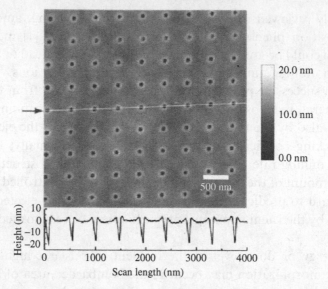

Figure 7.8 AFM image of hole (dot) array milled at 1 pA on Si substrate. (Reprinted from [49] with permission from Institute of Physics and IOP Publishing.)

surrounding the crater. In fact, the protruded ring structure is the swelling of the substrate due to amorphization and is the inherent shape obtained by single-pass FIB milling. Since most of the FIB roughly resembles a Gaussian ion distribution, the intensity at the fringe (tail) of the beam is much smaller than that at the core (center region). Thus it is not strong enough to sputter materials but is sufficient enough to cause amorphization that induces substrate swelling [1]. The ridge or protrusion height shown in Figure 7.8 should mainly be a result of the swelling of the substrate and can be as high as on the order of 10 nm. Redeposition can add materials in the ridge, but the amount of material added should be much less than that of swelling.

Tseng *et al.* [50,52] studied the single-pass milling of several nano- and microchannel patterns on a gold-coated substrate using a 90 keV As^{2+} FIB with a beam FWHM diameter of 50 nm at a current of 5 pA with a dwell time varying from 5 to 50 ms. As shown in Figure 7.9(a), the AFM image depicts the pattern milled with 5 ms dwell time and indicates ridges being formed along the channel banks (shoulders). Figure 7.9(b) delineates the corresponding cross-sectional profile, displaying a V-shaped contour similar to that of the milled hole-array shown in Figure 7.8. Figure 7.9(c) shows the variations of four profile features with respect to the dwell time indicating that although all of the features increase with the dwell time, the increase rates are gradually reduced as the dwell time increases. This may indicate that

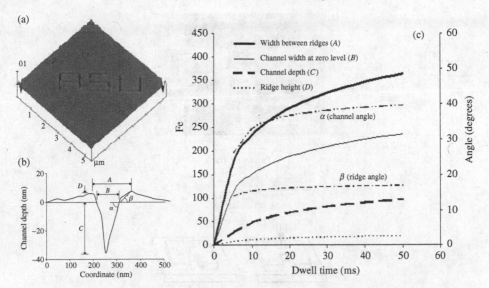

Figure 7.9 "ASU" nanochannel pattern on gold layer milled by 90 keV As^{2+} FIB: (a) channel cross section profile and feature definition, (b) AFM image at 5 ms dwell time, (c) profile measurements for various dwell times.

a large amount of sputtered materials are redeposited into the channel since the mouth width and the milling depth should increase proportionally with the dwell time without redeposition. The four features are defined in Figure 7.9(b): *A* is the ridge width, which is the distance between the ridge peaks, *B* is the mouth width, which is the channel width with respect to the original surface, *C* is the depth from the original surface, and *D* is the ridge height.

Also, the angles to characterize the steepness of the channel walls and the ridge, which are defined in Figure 7.9(b), are calculated and reported in Figure 7.9(c). The wall slope angle (α) approaches 90° if the wall is vertical or parallel to the incident beam direction, while the ridge angle (β) is zero when no swelling occurs. As shown, the effects of the dwell time or the amount of doses on the angles are insignificant except in the initial transient stage. The ridge or β angle varies from 15° to 17° with the dwell time changing from 10 to 50 ms, while the corresponding wall slope or α angle increases from 33° to 39°, which is about twice the ridge angle. It is to be noted that the swelling angle is highly dependent on the amount of the density change by amorphization, while the wall angle can be dictated by the number of the repetitive passes following the pattern that the larger the number, the higher the wall angle.

Frey *et al.* [36] and Li *et al.* [49] have used a 30 keV Ga^+ FIB to mill nanoscale channels with a single pass on Si substrates. They used different

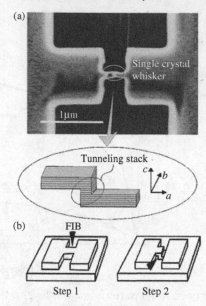

Figure 7.10 Stacked nanostructure. (a) SIM image of intrinsic Josephson junction with $0.01\,\mu m^2$ ($\sim 100\,nm \times 100\,nm$) tunneling stack, (b) FIB fabrication steps. (Reprinted from [51] with permission from American Institute of Physics.)

beam diameters at different beam currents and all learned that the channel profiles milled with a single pass have a "V" shape. Their results further confirm that the V-shaped channel profile is the inherent shape obtained by single-pass FIB milling. Also, the measurements shown in Figure 7.9 and in [49] reveal that the mouth width of the V-shaped channel can be much larger than the beam diameter by almost one order of magnitude. This may suggest that at a higher dwell time, the ion intensity outside the core region of the FIB is sufficiently high enough to produce a sizable amount of sputtering.

Kim *et al.* [51] used a 30 keV Ga^+ FIB to mill an intrinsic Josephson junction (IJJ) structure used as the basic element for quantum computation. The IJJ is a Z-shaped stacked structure made by $Bi_2Sr_2Cu_1O_{6+\delta}$ (Bi-2201) single crystal whiskers. As shown in the Figure 7.10(a), the intrinsic Josephson junction has a $0.01\,\mu m^2$ ($\sim 100\,nm \times 100\,nm$) tunneling stack area. The 3D tunneling stacks were patterned on a special rotation stage. The steps of the FIB milling process are shown in Figure 7.10(b). A width, depending on the required junction size, was milled in the whisker thickness from the perpendicular direction. By tilting the sample stage up to 90°, two grooves of the bridge were then milled completely from the lateral sides in order to create the required junction size. This technique, FIB in concert with a

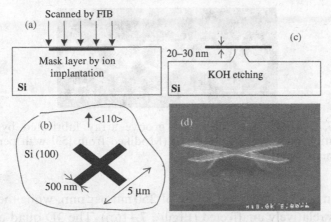

Figure 7.11 Nanoquad-cantilever. (a) Implantation by FIB scanning (cross section), (b) top view of FIB scanned surface with quad-cantilever layout, (c) KOH etching for nonimplanted substrate (cross section), (d) SEM image of fabricated quad-cantilevers, 30 nm thick, 500 nm wide, and 5 μm long. (Courtesy of J. Brugger of Swiss Federal Institute of Technology.)

rotation stage, can fabricate nanoscale stacks from thin films and single crystals.

7.3.4 FIB implantation (FIBI)

The FIB system not only has capabilities of performing subtractive (milling and etching) processes, but it can also perform additive (implantation and deposition) processes. In this subsection, the implantation ability of FIB will be explored by showing the specific procedures in creating different nano-structures.

A combination of FIB implantation (FIBI) and wet etching has been used to fabricate a nanoscale quad-cantilever on a Si (100) substrate having a 10 Ω n-type background dope. As shown in Figure 7.11, the quad-cantilever is first patterned by ion implantation using a 30 keV Ga$^+$ FIB at a current of 100 pA in the Si substrate at a sufficiently high dose, i.e., higher than 10^{15} ions/cm^2. At this level of dose concentration, i.e., higher than 10^{15} Ga$^+$ ions/cm^2 in silicon, the etch rate of certain etchants, including KOH (potassium hydroxide), can dramatically be reduced and the doped silicon has been used as an etch stop or a mask in IC fabrication. The thickness of the cantilever structure is dictated by the penetration depth (or the range) of the implanted ion, which can be controlled by the incident energy of the ions. Therefore, a selective etching is accomplished as the substrate is dipped into the 40% KOH solution at 60°C, in which the silicon layer with no

Figure 7.12 SEM image of round groove array fabricated by Si-FIB implantation on AlGaAs substrates. (Modified from [53] with permission from American Institute of Physics.)

implantation is etched away at a rate of 250 nm per min, while the implanted region stays relatively unaffected (Figure 7.11(c)). The 3D quad-cantilever is 20–30 nm thick, 500 nm wide, and 5 mm long, as shown in the SEM image (Figures 7.11(b) and (d)). Using this implantation/wet etching combined method, other types of nanoscale freestanding structures, including a nano-cup and a U-shaped nanocantilever, have also been reported [20].

Utilizing a similar procedure, Hiramoto *et al.* [53] fabricated an array of grooves on epitaxial AlGaAs layers by implantation combined with etching. The implantation is performed by line scanning of an Si^+ FIB with a 100 nm FWHM diameter at 200 keV. They found that the cross sections of the doped area could be varied from U to round shapes by reducing scanning speeds or increasing the doses. Figure 7.12 shows the round-shaped grooves that are wider in the interior and narrower near the surface. The grooves are formed by selectively etching the highly doped region that is implanted at a scanning speed of 0.1 mm/s with a line dose of 1.0×10^9 ions/mm using a hot HCl etchant at 70°C. The round grooves are about 350 nm in diameter and the distances between the etched grooves are about 30 nm. Since the groove diameter is much larger than the beam diameter (100 nm), the implanted ions are spread into the substrate and the grooves can be implanted very close at a distance much less than the beam diameter. Hiramoto *et al.* [53] also applied a similar technique to make very small GaAs conducting wires ranging from 20 to 100 nm in diameter.

FIB implantation has been used to irradiate polymeric resist materials to increase their resistance to subsequent physical etching. Arshak *et al.* [11] used a 30 keV Ga^+ FIB with 1.17×10^{-2} C/cm^2 doses to make implanted patterns in Shipley SPR660 positive photoresist layers, which can form a Ga_2O_3 compound against O_2 reactive ion etching (RIE). This process, also known as graphitization, can harden the photoresist to be conveniently used as an etching mask for O_2 dry etching. As shown in the FIB image in Figure 7.13, two 100 nm negative resist lines with a 100 nm space (left) and two 30 nm

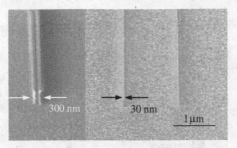

Figure 7.13 Cross-sectional FIB image of double 100 nm (left) and 30 nm (right) line patterns in SPR660 photoresist after O_2 RIE. (Courtesy of Khalil Arshak of University of Limerick of Ireland.)

resist lines (right) were patterned after RIE in the SPR660 resist layer. This demonstrates that anisotropic patterns can be achieved by graphitization and can be directly incorporated into the standard photolithographic processes.

Ion implantation has also been used to change the electrical, optical, and magnetic properties of different materials for making nanodevices. In the case of changing electrical properties, Sumita *et al.* [54] fabricated a CDW (charge-density-wave) device with a thickness of 40 nm and a width of 100 nm, and found that the device shows collective sliding and negative resistance above some threshold electric field in the low temperature region. This suggests that the quantum tunneling phenomena occur in the fabricated CDW, which can only result from a condition that the structure is a true one-dimensional conductor or is narrow enough to cause the collective sliding of electrons without interferences. Shinada *et al.* [55] have improved FIB optics for single-ion implantation and this high precision implantation instrument can perform one-by-one doping of impurity ions into nanoscale semi-conductor devices for various applications.

7.3.5 FIB induced deposition (FIBID)

The FIB can be used as a deposition tool for fabricating nanostructures. FIB induced deposition (FIBID) uses ion energy to initiate and localize chemical vapor deposition (CVD) in a specific location by a direct writing technique, also known as FIBCVD. The FIB system should be equipped with a gas nozzle to inject a variety of organo-metallic precursor gases on the target surface in the vacuum chamber (Figure 7.14(a)). The high-energy ion beam can decompose the organo-metallic molecules adsorbed onto the surface of the substrate. This leads to the release of metal atoms, which are incorporated onto the surface, while the remainder resulting from the decomposition

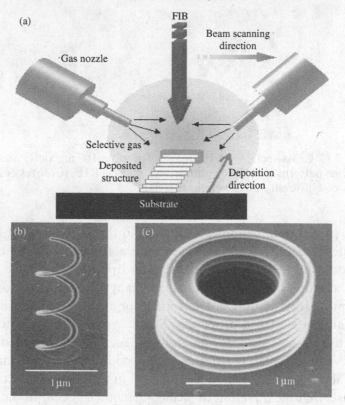

Figure 7.14 Nanostructures by FIBID. (a) Schematic of FIB induced deposition, (b) SEM image of microcoil with a coil diameter of 600 nm, coil pitch of 700 nm, and wire diameter of 80 nm, (c) microbellow with a thickness of 100 nm, and a pitch of 800 nm. (Courtesy of Shinji Matsui of Himeji Institute of Technology of Japan.)

reaction is normally volatile and can be eventually removed from the vacuum chamber by the pumping apparatus. Many types of precursor gases have been used for FIBID of a variety of metal and ceramic structures, such as precursor gases of WF_6, $C_7H_7F_6O_2Au$, $(CH_3)_3NAlH_3$, $C_9H_{17}Pt$, and TMOS+O_2 being used to produce W, Au, Al, Pt, and SiO_2, respectively [3]. Normally, the gas nozzle is controlled at a height of 0.1 to 1 mm above the target surface with an angle of 30° to 60° having the precursor gas evaporated from a heated container.

Matsui *et al.* [56] fabricated several nanostructures using a 30 keV Ga^+ FIB with an aromatic hydrocarbon precursor gas for creating diamondlike amorphous-carbon nanostructures. Figure 7.14(b) shows an SEM image of a microcoil structure with a 600 nm coil diameter, a 700 nm coil pitch, and a wire diameter of 80 nm. The microcoil was made with an FIB diameter of

7 nm and a dwell time of 0.11 s with an exposure time of 40 s and a beam current of 0.4 pA. The basic vacuum chamber pressure is 2×10^{-5} Pa before the gas is introduced, and 5×10^{-5} Pa after. By changing the FIB scanning speed or the deposition rate, the desired pattern with different coil pitches can be fabricated. Figure 7.14(c) shows another nanostructure, a microbellow having a thickness of 100 nm, a pitch of 800 nm, an external diameter of 2.75 μm, and a height of 6100 nm. The total exposure time was 300 s at a beam current of 16 pA. The realization of these structures implies that FIBID is a promising technique for the fabrication of 3D parts at nanoscales, although its mechanical performance has yet to be investigated. Using a similar technique, a number of other nanostructures have also been fabricated by Watanabe *et al.* [57].

7.3.6 FIB assisted etching (FIBAE)

Focused ion beam assisted etching (FIBAE), also known as chemical assisted or gas assisted ion etching, uses a chemical reaction between the substrate surface and gas molecules adsorbed on the substrate. By injecting a precursor reactive gas into the milling process, the material removing mechanism can be changed and the associated milled shape can be altered. The major advantages of FIBAE over FIB milling are higher material removal rates, lower redeposition on sidewalls, and reduced implantation damages. However, while ion milling can be used to remove almost all kinds of solid materials, FIBAE can only be applied to certain substrates since a precursor gas must exist that forms volatile products with the substrate. Furthermore, to have the desired dimensional control, a spontaneous reaction between the precursor gas and the substrate should be avoided or minimized.

The machine setup for FIBAE is similar to that for deposition shown in Figure 7.14(a), except that the precursor gas is different. Frequently used precursor gases in FIBAE include Br_2, Cl_2, I_2, and XeF_2 for etching various metals and insulators relevant to semiconductor processing, and water vapor for the etching of carbon based materials. Also, mixtures have been developed for FIBAE such as Cl_2/NH_3 for copper [58]. However, no effective etch chemicals have yet been developed for etching certain metals, including Pt, Au, and most metal-oxide compounds used in superconductors or ferroelectrics [3].

Normally, to fabricate nanoscale structures, the corresponding material removal rate should have subnanoscale resolutions, and the best way to achieve this is to "minimize" the material removal rate in a controllable manner. Since FIBAE has relatively high material removal rates which may be too high to be used for nanoscale material removing, FIBAE has seldom been used in nano-fabrication. In fact, the smallest feature sizes for the structures made by FIBAE have been found in the range between 200 and 300 nm, which include an egg-crate pattern etched on a Cu layer using a Ga FIB with $W(CO)_6$ gas [59] and parallel trenches on diamond substrate by Ga ions using XeF_2 gas [60]. For a structure smaller than this range, the irregularities induced from etching, including pits, ripples, ridges, and roughness, which can be much larger than 10 nm even in the best-controlled conditions, can overwhelm the intended shape or result with unacceptable precision [60–62]. To claim the legitimacy or ability of nano-fabrication, the feature sizes of the irregularities have to be controlled at a level much smaller than 10 nm. Consequently, no further evaluation or discussion will be provided on FIBAE.

7.4 Concluding remarks

The recent developments in the ion beam (IB) technology for nano-fabrication have been studied including the ion projection lithography (IPL) and the focused ion beam (FIB) approaches. In general, the IB technology has been rapidly developing recently to produce more complicated nanoscale structures necessary in various fields. The key to IB nano-fabrication is the ability to operate an IB with a proper beam size, current, and energy to remove or add a required amount of material at a subnanometer or near-atomic scale from a pre-defined location in a controllable manner. Consequently, the ion beam has to be stable and of high quality, and be controlled within a few nanometers in diameter for an FIB or within subnanometer resolutions for an IPL system. Also, the system equipment has to be improved with ultra-precision stage and feeding operations under a controlled environment, including the direction and position of the ion beam on the target synchronizing with the substrate movement as well as with the chemical feeding instruments for deposition and etching. The accuracy of the substrate stage to perform translation and rotation should be within subnanometer ranges with respect to the target position of ion beams.

To achieve near-atomic scale accuracy, it is inevitable to have a feedback system to control the material removal rate or deposition rate. A feedback signal that detects or monitors the fabricated dimensions with subnanoscale accuracy during operation becomes a necessity. A reliable IB system with the capability for end-point detection or in-situ monitoring should be developed further, especially to improve the quality of these online inspection or monitoring techniques, mainly the scanning electron microscopy, scanning ion microscopy, and secondary ion mass microscopy. It has been found that the feedback system in the current commercially available IB system is still in its infant stage. Other product qualities, such as flatness, texture, and surface finish, can also be important in many applications and the in-situ detection systems should also be worthwhile for investigation development.

In IPL, the prototype systems surveyed have indicated that further fine-tuning and improvement in their resolution and precision are needed so that the critical feature sizes of the printed patterns can reach to the level of 50 nm to make the system a vital candidate for next-generation lithography (NGL) for the semiconductor industry. Also, the accuracy of the stitching scheme requires further refinement to demonstrate the system level integrity, since a large field image has to be projected with minimum distortion and high reliability. Furthermore, it could be critical that ion-beam resists with sharper resolution and higher sensitivity as well as with better processing character-istics need to be developed. Since the IPL tool is much more complicated than the FIB system, the trade-off between pattern generation flexibility and resolution with pattern writing speed will be a major concern for future development of the IPL technology.

It is believed that FIB direct writing will still play a major role in nano-fabrication because of its combined superiority in flexibility, resolution, precision, and cost and of its combined ability for material addition (deposition and implantation) and material removing (milling and etching). Among the four techniques used in FIB direct writing, milling, implantation, induced deposition, and assisted etching, FIBAE is the most efficient and its processing rate is about one-order of magnitude faster than the other three processes. Currently, however, the material removal rate of FIBAE may be too large or unstable to be controlled within the required subnanoscale level. As a result, a better precision in control of the material removal is needed for FIBAE to be suitable for nano-fabrication. Furthermore, although the four FIB related processes are all used in the semiconductor industry for different applications, these FIB techniques, especially FIBID and FIBAE, have been developed for a limited number of materials directly related to IC fabrication. For example, no etch or deposition chemicals have been reported for effec-tively etching or deposition for most metal-oxide compounds used in superconductors or ferroelectrics. More effective chemicals suitable for FIBID and FIBAE should be developed to extend their applications.

In milling, the effects of sputtering, redeposition, and amorphization are all important and have to be understood for enabling better control over the process. For example, the material swelling due to amorphization can be as high as 10 nm, which may not be significant in micro-fabrication, but it can certainly diminish the dimensional accuracy of nanoscale fabrication. Cur-rently, if the throughput is not the issue, an ultra small amount of material can be milled from the substrate using extremely small beam diameters (of the order of 10 nm) and currents (of the order of 1 pA). Furthermore, if no information is available on the milling of special materials and structures, it can be roughly assumed that the milling yield is close to the value of the normal-incident sputtering-yield because many studies have shown that the effect of redeposition and amorphization often counterbalances the increase in milling yield due to the changing of the incident angle.

Future research emphasis will focus on making heterogeneous nanostructures and multiscale systems. Both approaches need to develop the ability to hierarchically integrate nanosystems with larger scale systems. Again, this requires a high degree of process control in the sensing and actuation of matter at the nano- and subnanoscales, as well as capabilities for scaling-up. Eventually, in the post-nano-fabrication or post-miniaturization era, the nano, micro and macro approaches will have to cross boundaries, i.e., if the nano-fabrication needs to be successful, it has to be integrable with its counterpart in the micro and macro sizes, and vice-versa.

Acknowledgements

The author gratefully acknowledges the support for this study by the US National Science Foundation under Grant No. DMI-0002466, CMS-0115828 and DMI-0423457 and by Walsin Lihwa Corp. The support and encouragement in preparing this manuscript from Dr. Wilhelm H. Bruenger of Fraunhofer Institute in Berlin (Germany), Professor Jürgen Brugger of Swiss Federal Institute of Technology (Switzerland), Dr. Ka-Ngo Leung of Lawrence Berkeley National Laboratory (USA), Professor Shinji Matsui of Himeji Institute of Technology (Japan), and Professors Zhao-Ying Zhou and Xiong-Ying Ye of Tsinghua University (China) should be specifically acknowledged.

References

[1] A. A. Tseng. *J. Micromech. Microeng.*, **14** (2004), R15–34.
[2] A. A. Tseng. *Small*, **1** (2005), 594–608.
[3] A. A. Tseng. *Small*, **1** (2005), 924–39.
[4] H. Löeschner, E. J. Fantner, R. Korntner *et al. Mat. Res. Soc. Symp. Proc.*, **739** (2002), 3–12.
[5] R. Kaesmaier, A. Ehrmann and H. Löeschner. *Microelectron. Eng.*, **57–58** (2001), 145–53.
[6] V. V. Ngo, B. Akker, K. N. Leung *et al. J. Vac. Sci. Technol. B*, **21** (2003), 2297–303.
[7] X. Jiang, Q. Ji, L. Ji, A. Chang and K. -N. Leung. *J. Vac. Sci. Technol. B*, **21** (2003), 2724–7.

[8] U. S. Tandon. *Vacuum*, **43** (1992), 241–51.
[9] J. Melngailis, A. A. Mondelli, I. L. Berry and R. Mohondro. *J. Vac. Sci. Technol. B*, **16** (1998), 927–57.
[10] W. H. Bruenger, M. Torkler, M. Weiss *et al. J. Vac. Sci. Technol. B*, **17** (1999), 3119–21.
[11] K. Arshak, M. Mihov, A. Arshak, D. McDonagh and D. Sutton. *Microelectron. Eng.*, **73–74** (2004), 144–51.
[12] S. Hirscher, R. Kaesmaier, W. -D. Domke *et al. Microelectron. Eng.*, **57–58** (2001), 517–24.
[13] H. Löeschner, G. Stengl, R. Kaesmaier and A. Wolter. *J. Vac. Sci. Technol. B*, **19** (2001), 2520–4.
[14] W. H. Bruenger, M. Torkler and L. -M. Buchmann. *J. Vac. Sci. Technol. B*, **15** (1997), 2355–7.
[15] W. H. Bruenger, M. Torkler, C. Dzionk *et al. Microelectron. Eng.*, **53** (2000), 605–8.
[16] A. Dietzel, R. Berger, H. Grimm *et al. IEEE Trans Magnetics*, **38** (2002), 1952–4.
[17] A. Dietzel, R. Berger, H. Loeschner *et al. Adv. Mater.*, **15** (2003), 1152–5.
[18] W. H. Bruenger, C. Dzionk, R. Berger *et al. Microelectron. Eng.*, **61–62** (2002), 295–300.
[19] K. Edinger. *Direct-Write Technologies for Rapid Prototyping Applications: Sensors, Electronics, and Integrated Power Sources*, ed. A. Pique and D. B. Chrisey (San Diego, CA: Academic Press, 2002), p. 347–83.
[20] S. Reyntjens and R. Puers. *J. Micromech. Microeng*, **11** (2001), 287–300.
[21] J. P. Biersack and L. G. Haggmark. *Nucl. Inst. Meth. Phys. Res. B*, **174** (1980), 257–69.
[22] J. F. Zeigler, J. P. Biersack and U. Littmark. *The Stopping Range of Ions in Solids* (New York: Pergamon, 1985).
[23] P. G. Blauner, Y. Butt, J. S. Ro and J. Melngailis. *J. Vac. Sci. Technol. B*, **7** (1989), 609–17.
[24] X. Xu, A. D. D. Ratta, J. Sosonkina and J. Melngailis. *J. Vac. Sci. Technol. B*, **10** (1992), 2675–80.
[25] O. Almen and G. Burce. *Nucl. Inst. Meth.*, **11** (1961), 257–78.
[26] M. T. Robinson and A. L. Southern. *J. Appl. Phys.*, **38** (1967), 2969–73.
[27] G. Carter and J. S. Colligon. *Ion Bombardment of Solids* (New York: Elsevier, 1968), Chapter 7.
[28] T. M. Nenadovic, Z. B. Fotiric and T. S. Dimitrijevic. *Surf. Sci.*, **33** (1972), 607–16.
[29] E. P. EerNisse. *Appl. Phys. Lett.*, **29** (1976), 14–17.
[30] K. P. Müller, U. Weigmann and H. Burghause. *Microelectron. Eng.*, **5** (1986), 481–9.
[31] J. M. Poate, W. L. Brown, R. Homer and W. M. Augustyniak. *Nucl. Inst. Meth.*, **132** (1976), 345–9.
[32] A. Yamaguchi and T. Nishikawa. *J. Vac. Sci. Technol. B*, **13** (1995), 962–6.
[33] J. G. Pellerin, G. M. Shedd, D. P. Griffs and P. E. Russell. *J. Vac. Sci. Technol. B*, **7** (1989), 1810–12.
[34] D. Santamore, K. Edinger, J. Orloff and J. Melngailis. *J. Vac. Sci. Technol. B*, **15** (1997), 2346–9.
[35] L. Bischoff and J. Teichert. *Forschungszentrum, Rossendorf*, Germany, FZR-217 (1998); 36 pp.
[36] L. Frey, C. Lehrer and H. Ryssel. *Appl. Phys. A*, **76** (2003), 1017–23.

[37] A. L. Southern, W. R. Willis and M. T. Robinson. *J. Appl. Phys.*, **34** (1963), 153–63.

[38] H. Sommerfeldt, E. S. Mashkova and V. A. Molchanov. *Phys. Lett.*, **38A** (1972), 237–8.

[39] H. H. Andersen and H. L. Bay. *J. Appl. Phys.*, **46** (1975), 1919–21.

[40] P. Blank and K. Wittmaack. *J. Appl. Phys.*, **50** (1979), 1519–28.

[41] S. T. Kang, R. Shimizu and T. Okutani. *Jpn. J. Appl. Phys.*, **18** (1979), 1717–25.

[42] A. E. Morgan, H. A. M. de Grefte, N. Warmoltz and H. W. Werner. *Appl. Surf. Sci.*, **7** (1981), 372–92.

[43] P. C. Zalm. *J. Appl. Phys.*, **54** (1983), 2660–6.

[44] C. Lehrer, L. Frey, S. Petersen and H. Ryssel. *J. Vac. Sci. Technol. B*, **19** (2001), 2533–8.

[45] Y. Yamamura, Y. Itakawa and N. Itoh. *Angular Dependence of Sputtering Yields of Monatomic Solids*, IIPJ-AM-26, Nagoya University, Nagoya, Japan (1983).

[46] M. Vasile, J. Xie and R. Nassar. *J. Vac. Sci. Technol. B*, **17** (1999), 3085–90.

[47] A. Lugstein, B. Basnar, J. Smoliner and E. Bertagnolli. *Appl. Phys. A*, **76** (2003), 545–8.

[48] J. S. Custer, M. O. Thompson, D. C. Jacobson *et al. Appl. Phys. Lett.*, **64** (1994), 437–9.

[49] H. W. Li, D. J. Kang, M. G. Blamire and W. T. S. Huck. *Nanotechnology*, **14** (2003), 220–3.

[50] A. A. Tseng, I. A. Insua, J. S. Park, B. Li and G. P. Vakanas. *J. Vac. Sci. Technol. B*, **22** (2004), 82–9.

[51] S. -J. Kim, Y. I. Latyshev and T. Yamashta. *Appl. Phys. Lett.*, **74** (1999), 1156–8.

[52] A. A. Tseng, B. Leeladharan, B. Li, I. A. Insua and C. D. Chen. *Int. J. Nanoscience*, **2** (2003).

[53] T. Hiramoto, K. Hirakawa and T. Ikoma. *J. Vac. Sci. Technol. B*, **6** (1988), 1014–17.

[54] T. Sumita, T. Nagai, H. Kubota, T. Matsukawa and I. Ohdomari. *Synthetic Metals*, **103** (1999), 2234–7.

[55] T. Shinada, H. Koyama, C. Hinoshita, K. Imamura and I. Ohdomari. *Jpn J. Appl. Phys.*, Part 2, **41** (2002), L287–90.

[56] S. Matsui, T. Kaito, J. Fujita *et al. J. Vac. Sci. Technol. B*, **18** (2000), 3181–4.

[57] K. Watanabe, T. Morita, R. Kometani *et al. J. Vac. Sci. Technol. B*, **22** (2004), 22–6.

[58] K. Edinger. *J. Vac. Sci. Technol. B*, **17** (1999), 3058–62.

[59] J. D. Casey, M. Phaneuf, C. Chandler *et al. J. Vac. Sci. Technol. B*, **20** (2002), 2682–5.

[60] J. Taniguchi, N. Ohno, S. Takeda, I. Miyamoto and M. Komuro. *J. Vac. Sci. Technol. B*, **16** (1998), 2506–10.

[61] A. Stanishevsky. *Thin Solid Films*, **398–399** (2001), 560–5.

[62] D. P. Adams, M. J. Vasile, T. M. Mayer and V. C. Hodges. *J. Vac. Sci. Technol. B*, **21** (2003), 2334–43.

[63] H. Yamaguchi. *J. de Physique, Colloque C6*, **48** suppl. 11 (1987), C6.165–70.

8

Preparation for physico-chemical analysis

RICHARD LANGFORD

University of Manchester

8.1 Introduction

Focused ion beam (FIB) and focused ion beam/scanning electron micro-scope (FIB/SEM) systems are invaluable tools for the preparation of specimens for physical and chemical analysis. Both systems are routinely used to prepare site-specific specimens for transmission electron microscopy (TEM), SEM and FIB microscopy of a wide range of materials including polymers [1], steels [2,3], surface coatings [4,5], catalysts [6], and semiconductors [7–11]. These systems are also used for the preparation of specimens for other analytical techniques such as Auger [12] and atom probe (AP) field ion microscopy [13].

The increase in use of FIB and FIB/SEM systems during the 1990s for the preparation of specimens for material analysis is due to the significant advantages they offer over other methods such as broad ion beam (BIB) milling, electro-polishing, or mechanical polishing. The main advantage is that the cross sections and TEM lamellae can be prepared to within 50 nm of a feature or region of interest (ROI), making it possible to analyze specific defects, phases, or interfaces. Such positional accuracy is very difficult to achieve using other sample preparation techniques. Although polishing can be used to prepare SEM cross sections to within 500 nm of a feature, this is a time consuming process in which the sample has to be repeatedly imaged (either using an optical microscope or an SEM) to ensure the feature is not polished through.

FIB prepared site-specific cross sections have been used to investigate a wide range of material problems such as interfacial debonding in FeAl based composites [14], the grain structure around surface cracks in Grade 448

Focused Ion Beam Systems: Basics and Applications, ed. N. Yao.
Published by Cambridge University Press. © Cambridge University Press 2007.

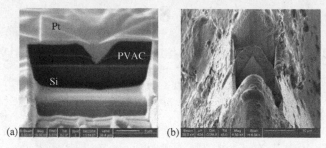

Figure 8.1 SE images of an FIB prepared lamella through (a) a nanoindent into a 4 μm thick polymer film on silicon and (b) the top edge of a drill bit.

(X-56) pipe line steel [15], tungsten contacts to carbon nanotubes [16], and Cu interconnects [17].

Other factors that have contributed to the rapid growth in use of FIBs for sample preparation are:

1. TEM lamellae can be prepared with near parallel sidewalls enabling quantitative compositional analysis to be performed over the entire area of the lamellae. In contrast, dimpling followed by BIB milling or electro-polishing produces tapered specimens. Also, as the FIB is perpendicular to the surface of the sample this limits the amount of preferential milling enabling lamellae and cross sections to be prepared from samples composed of materials with vastly different sputter rates. For example FIBs have been used to prepare TEM specimens of co-block polymers on silicon [18], polymer solar cells on glass [19], and teeth [20,21] (overcoming problems due to differences in hardness between the enamel and dentine). Figure 8.1(a) shows a secondary electron (SE) image of an FIB prepared cross section through the center of a nanoindent into two 2 μm thick poly (vinyl acetate) (PVAC) polymer films (separated by a 100 nm thick gold streamer layer) which were spun cast onto silicon and cross-sectioned to investigate the deformation and mechanics of multilayer nanoimprint lithography.
2. TEM lamellae can be prepared from volumes as small as 20 μm^3 and samples with irregular geometries, removing the need to embed the samples into a supporting matrix. For example, lamellae have been prepared from heterogeneous catalysts that were 30–40 μm in diameter [22], fragments of meteorite samples [23], and powders [24]. Figure 8.1(b) shows an FIB prepared lamella (before it is lifted out) milled into the top edge of a drill bit.
3. The preparation time can be as short as two hours.

This chapter discusses the use of FIBs for preparing specimens for physical and chemical analysis. The first sections focus on the techniques used to prepare TEM lamellae and emphasis is placed on the steps used to enhance the success yields. After discussing the relative advantages of the different techniques, the preparation of plan-view lamellae and SEM cross sections are outlined. This part of the chapter concludes with a discussion of sequential FIB cross-sectioning for three-dimensional (3D) reconstruction and the preparation of specimens for other analysis techniques such as AP field ion microscopy.

Unfortunately, numerous detrimental artefacts can arise when using FIBs for the preparation of TEM lamellae and cross sections, such as damage to the sidewalls. These artefacts and the methods used to minimize them are discussed in detail in the second part of the chapter.

8.2 Preparation of TEM cross sections

Three FIB based methods are routinely used for the preparation of TEM lamellae: the ex-situ lift-out technique [25,26], the H-bar technique [27,28], and the in-situ slice technique [29,30]. The latter technique can be viewed as a hybrid of the H-bar and ex-situ lift-out techniques combining their relative advantages (i.e., no tilt limitations in the TEM and a high success rate). Currently (circa 2005), of the three techniques the in-situ slice technique is in vogue while the H-bar technique is being used less frequently. As each technique has its own relative advantages all three are discussed in detail.

8.2.1 The H-bar technique

The H-bar technique was the first FIB based method developed to prepare TEM lamellae. It involves preparing a slice of material about 50 μm wide and 2.5 mm long which contains the ROI. This slice is mounted onto a U-shaped TEM grid (a slot grid cut in half) and trenches are milled on either side of the ROI to make a pathway for the electron beam in the TEM. The name arises because the trenches and lamella create an "H" shape.

The first step in the H-bar technique involves preparing a slice from the sample. These have been prepared by dicing with a diamond saw and using mechanical wheel, tripod, and hand polishing [31–35]. The choice of which method to use is dependent on both the sample and the equipment available. Dicing with a high precision diamond band saw, which is suitable for wafers and glass substrates, enables slices as narrow as 20 μm to be prepared in 10 to 20 minutes. (The narrower the slice the shorter the time required to FIB mill the trenches.) To aid handling and mounting of the slices onto the U-shaped TEM grids, T-shaped slices can be made by dicing four cuts into the sample (Figure 8.2(a)). Tripod polishing can be used to prepare wedges that narrow to a few micrometers at the apex [35]. An advantage of using such narrow slices is it considerably reduces the required FIB milling time enabling a number of

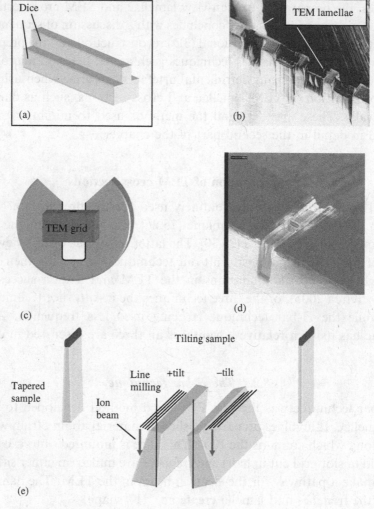

Figure 8.2 (a) T-shaped slice prepared by dicing four cuts, (b) an SE image of a tripod polished slice into which numerous TEM lamellae have been milled, (c) schematic of a slice mounted on a U-shaped TEM grid, (d) SE image of a finished H-bar sample ready for TEM, and (e) schematic of the tilting of a sample into the ion beam to produce a lamella with parallel sidewalls.

lamellae to be prepared along its length relatively quickly (Figure 8.2(b)). The most widely used method to prepare a slice is hand polishing and as this can be readily combined with the other methods and only requires polishing pastes or sheets, it is outlined in detail below.

The first step in preparing a slice by hand polishing consists of identifying the ROI under an optical microscope. If it is too small to be observed, FIB milling or Pt deposition can be used to render it visible. An area of approximately 2.5 mm^2 surrounding the ROI is then cleaved, cut, or polished. This is then mounted, using wax, onto a microscope slide such that it protrudes over the slice's edge and is polished using diamond paste or Al$_2$O$_3$ sheets with a series of polishing grades (30 μm down to 0.5 μm). A good support for the Al$_2$O$_3$ sheets is a wet glass pane as this keeps them flat and prevents them slipping. During polishing the sample is repeatedly examined under an optical microscope and the polishing is stopped when the ROI is to within 30–50 μm of the edge of the slide. The slide is then heated on a hot plate to melt the wax and the sample is removed and remounted with the other side protruding over the edge of the glass slide. It is then polished again until the slice is 50–100 μm wide. Typically, this hand polishing takes about two to three hours but the exact time depends on the hardness of the material and the initial dimensions of the square. The slice is then removed from the slide (again by melting the wax), washed in acetone to remove any residual wax and fixed using conductive paint to a TEM slot grid which has been cut in half (Figure 8.2(c)). The sample is then mounted in a suitable clamp and placed into the FIB or FIB/SEM system chamber with the top of the slice perpendicular to the FIB.

In the FIB or FIB/SEM system, the ROI is identified and a 1.5 μm wide and 1–3 μm thick Pt strap is deposited over it using ion beam assisted metal deposition [36,37]. This planarizes the surface and prevents rounding and milling of the top of the sample during FIB milling. If the sample is not planarized, height steps can lead to striations in the face of a cross section (the "curtain effect") due to changes in the sputter yield [38,39]. (The required thickness of the Pt deposited is dependent on the ion beam profile and the topography of the surface.) If the deposited Pt strap obscures the ROI, lines for positioning the TEM lamella during the FIB milling can be milled outside the area where the Pt will be deposited.

After depositing the Pt strap, two trenches are milled at a distance of 1.5 μm from each side of the Pt out to the edges of the slice, using a large beam current (2 nA to 20 nA) and by scanning the ion beam in a serpentine pattern. The exact dimensions of the trenches (and the Pt strap) are dictated by time constraints, the size of the feature or ROI, and the angles of tilt or rotation required in the TEM (see Section 8.2.4).

Having milled the trenches the remaining material between them is then thinned to create an electron transparent lamella. For this step the magnitude of the ion beam current is reduced to 1000 pA or 500 pA (to decrease the spot size to give better control of the subsequent thinning/milling steps). In addition, the milling regime is changed. The ion beam is scanned in lines and is moved inwards in steps of half a beam diameter; a process frequently referred to as FIB polishing. This decreases the amount of material redeposited onto the sides of the lamella and reduces the probability of accidentally milling the lamella due to ion beam or stage drifts. As the sputter rate is higher when milling at the edge of a step than when milling into a flat surface the "depth" to which the lines are milled should be at least 30% less than the "depth" that the trenches were milled, if the same milling parameters are used (i.e., beam overlap, sputter yields, and dwell times). Discussions on the effect of different milling regimes on the sputter rates can be found in the papers by Ghandhi and Orloff [40] and Santamore *et al.* [41]. Furthermore, as a result of milling by the tails of the ion beam distribution and the increase in the sputter yields as the angle of incidence of the ion beam to a surface increases, the sidewalls of the remaining material are tapered (Figure 8.2(e)). (For a mathematical description of the evolution of this taper see Boxleitner [42].) If a lamella with parallel sidewalls is required, for example for quantitative chemical analysis (and throughout the rest of the chapter it is assumed that this is always the case), this taper can be removed by tilting the sample into the ion beam while it is being thinned using the line milling procedure (Figure 8.2(e)). The tilt angles used have ranged from 1° to 4°, depending on the material, the magnitude of the beam current, and the depth of the lamella being milled.

Once the remaining material is approximately 500 nm the magnitude of the ion beam is further reduced to 300 or 100 pA (again to produce a smaller spot size) and it is thinned, using the line milling and tilting procedure, until it is approximately 100 nm thick. Some materials are under intrinsic stress and as such may buckle during the final thinning to electron transparency. Walker [43] showed that a cut made at the side edge of such lamella can relieve the stress, resulting in the straightening of the lamella and enabling it to be further thinned to electron transparency.

During the final thinning, if a single ion beam system is being used the sample is repeatedly imaged and the real-time SE signal is monitored for any sudden increases in intensity to ensure that neither beam nor stage drifts cause the lamella to be milled. If an FIB/SEM system is being used the electron beam can be used to monitor and end the thinning once the thickness of the remaining Pt has been reduced to a few hundred nanometers. Techniques to thin the lamellae below 100 nm are discussed in Section 8.6.3. A finished sample, through some CMOS transistor gates, ready for TEM, is shown in Figure 8.2(d).

As FIB milling is relatively slow to remove large volumes of material it is not practical to use it to prepare lamellae of features that are hundreds of micrometers beneath the surface of a sample. To mill away a trench that is

100 μm deep, 50 μm wide, and 60 μm long would take over 8 hours when using a beam current of 20 nA. To overcome this limitation to enable lamellae to be prepared from crack tips located hundreds of micrometers beneath the surface of the sample both Deshais *et al.* [44] and Huang *et al.* [45] combined the H-bar technique with BIB milling. They both used conventional dicing and polishing to prepare the slices for the H-bar technique, but then used BIB milling to sputter the slice to decrease the distance of the crack tips from the surface (which also reduced the width of the slice) before the sample was placed in the FIB system. Mechanical polishing to the crack tips was not used as there were concerns that this might affect the integrity of the cracks.

8.2.2 Ex-situ lift-out

The ex-situ lift-out technique was first described by Overwijk *et al.* [46] and involves cutting a lamella free from the sample, lifting it out, and placing it onto a TEM support grid using a needle, micromanipulator, and optical microscope. From the mid 1990s until recently (circa 2005) this has been the main technique used.

In this technique, the whole of the sample (if it is of suitable dimensions, otherwise it has to be diced or polished) is placed in the FIB chamber. As with the H-bar technique, the ROI is identified and coated with a Pt strap. Staircase-shaped cuts, positioned 1.5 μm from either side of the Pt strap, are then milled using a large beam current (2 to 20 nA) (Figure 8.3(a)). Staircase-shaped cuts rather than rectangular ones are milled to reduce the milling time as in the subsequent steps the slice/lamella is only ever milled/viewed at 45°. The staircase cuts are typically milled 15–20 μm wide and 8–12 μm deep. If the dimensions are much smaller than this, it becomes difficult to lift out the lamella with the needle and micromanipulator.

The material between the staircase cuts is then thinned to approximately 300 nm by using the line milling and tilting procedure, outlined for the H-bar technique. The sample is next tilted to 45° and the base and side edges (one completely and the other only 80% along its length) of the slice/lamella are milled through (Figure 8.3(b)). The changes in intensity of the SE signal during the milling can be used to determine when the lamella has been cut through. It is very important that the base and sidewalls are completely cut through otherwise the lamella can not be lifted out from the FIB cuts. To ensure this the box used to cut through the base should be positioned at least 1 μm above the base of the slice and the depth to which it is milled should be

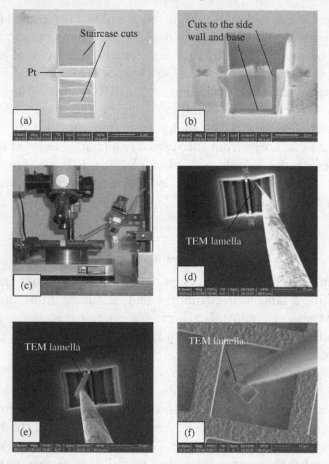

Figure 8.3 SE images of (a) staircase cuts milled on either side of the Pt, (b) sample tilted to 45° whose sidewalls and base have been cut, (c) an ex-situ micromanipulator and optical microscope, (d, e) a needle being pushed against a lamella, and (f) a lamella being placed onto a carbon support membrane.

at least twice the thickness of the slice. (As the box is milled at 45° redeposition onto its sidewalls will occur which reduces the sputter rate.) In a single beam system if there is any uncertainty as to whether the slice/lamella has been cut through the sample should be rotated by 180° and the opposite side of the slice/lamella inspected for the cuts. If using an FIB/SEM system it is not necessary to rotate the sample. Once the FIB has cut through the slice/lamella it will mill a U-shaped cut into the opposite staircase which can be seen by electron beam imaging as this is perpendicular to the surface of the sample (see Figure 8.5(c)).

The sample is then tilted back to zero degrees, and again the tilting and line milling procedure and a beam current of 300 pA or 100 pA is used to further thin the lamella to 100 nm. The remaining material at the holding sidewall is then cut to free the lamella from the sample. Again, if a single ion beam system is being used the intensity of the SE signal can be used to determine when the lamella has been cut free from the sample. Usually, the lamella falls into the path of the ion beam resulting in a sudden increase in the intensity of the SE signal. Once cut free, in a single beam system care must be taken not to image the sample at too high a magnification (>5000 times) in order to prevent further ion implantation and damage to the lamella. If an FIB/SEM system is being used electron beam imaging can be used to check that the lamella has been cut free from the sample. This advantage and other advantages of using an FIB/SEM system for the preparation of TEM lamellae are discussed in detail in Section 8.5.

The next step involves lifting the "free" lamella away from the FIB cuts and placing it onto a carbon coated TEM grid. As this can be as time consuming as preparing the lamella and more difficult to do it is described in detail.

The sample is removed from the FIB chamber and placed under an optical microscope equipped with long working distance lenses (typically 13 mm or 18 mm), to enable a micromanipulator to be used beneath them (Figure 8.3(c)). (An invaluable aid when teaching the lift-out step is to have a TV camera attached to the microscope.)

Although both glass and metal needles have been used for lifting out the lamellae the majority of researchers use glass needles as it is easier to get the lamellae to attach to these. Ebel *et al.* [47] investigated using glass and metal needles for lifting out GaAs lamellae. They reported the highest success rate when using floating glass needles and for the metal needles it was necessary to bias them to get the lamellae to adhere.

Glass needles can be made by melting the central part of 1 mm diameter glass rods and pulling the two ends at a controlled rate. Metal needles can be prepared by a modification of the electro-polishing methods used for making scanning tunneling microscopy tips [48]. A recipe that produces suitable tungsten needles consists of dipping 50 μm diameter wire in 2 M KOH for 15 minutes while applying 20 volts and then for 6 minutes while applying 6 volts. Dipping [49] the needles into the electrolyte produces needles with long shanks which enables their tips to reach to the base of the FIB cuts.

Using a low magnification (50×), the FIB cuts are placed at the center of the optical axis and the needle is positioned about 5 mm above them. The optical magnification is then increased to 500 times and the needle tip brought into focus. To prevent crashing the needle into the sample as it is lowered, the microscope is over-focused and the needle lowered and brought into focus. This process is repeated until the needle's tip is positioned above the FIB cuts. The sample is then rotated to align the lamella parallel to the needle's shaft. To illustrate the lift-out steps SE images, from when the technique has been performed within an FIB/SEM system, are used (this alternative approach is discussed at the end of this section) as these are more informative than the equivalent optical images. The needle is further lowered and repeatedly swept through the FIB cuts and pushed against the lamella until the latter sticks to it through electrostatic forces (Figures 8.3(d) and (e)). Once the lamella is attached to the needle, the needle is raised away from the sample and the sample is replaced with a carbon coated TEM grid. The needle and lamella are then lowered (again by focusing on the needle tip, over-focusing, and lowering the needle into focus) and then repeatedly swept across the surface of the carbon membrane until the lamella sticks to the carbon. Generally, the adhesion between the lamella and carbon membrane is greater than that between the needle and lamella. If necessary, the needle is then used to push the lamella flat onto the carbon (Figure 8.3(f)). The adhesion between the lamella and carbon is sufficiently strong that only rarely does a TEM lamella fall off during transportation to or mounting in the TEM. As well as continuous carbon membranes, holey and lacy carbon films can be used [50,51]. However, adhesion of the lamella to these supports is not as good (due to the smaller contact area) and typically 1 in 30 samples may be lost while transferring the grids to the TEM.

The success rate for the lift-out step ranges from 60% to 90%. Failure can occur when the lamella is being lifted out from the FIB cuts or as it is being placed onto the supporting membrane. During the lifting out step, the lamella may "jump" up the shaft of the needle, away from the tip, such that it cannot be brushed against the carbon membrane. Several approaches have been used to reposition the lamella at the needle's tip. These include (i) using a second micromanipulator, (ii) brushing the needle against the rim of the TEM grid, and (iii) tapping the micromanipulator so that the lamella jumps off the needle and onto the support membrane. Of these three methods, tapping the micromanipulator is probably the quickest and easiest one to do.

The lift-out can also fail if the carbon membrane is accidentally punctured and ripped by the tip of the needle as the lamella is being placed onto it (Figure 8.4(a)). This can result in (i) the carbon support membrane rolling around and encasing the lamella, (ii) the lamella on the ripped carbon membrane being at a large angle to the electron beam in the TEM such that it cannot be tilted onto a crystal axis (Figure 8.4(b)), and (iii) the lamella falling through the resulting hole and being lost. To prevent these problems TEM

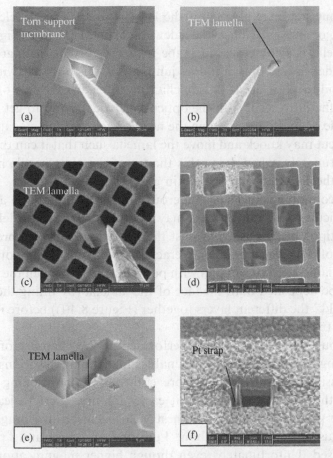

Figure 8.4 SE images of (a) a carbon support membrane punctured by the tip of the needle, (b) a lamella sat on a torn membrane, (c) a Cu grid without a support membrane onto which a TEM lamella has been placed and (d) the same lamella after being pushed over a hole, (e) a lamella that has moved against the sidewall of an FIB cut, and (f) a slice with a Pt strap that has been used to "weld" the different layers.

grids that have 10 µm sized holes and no support membranes can be used [52]. Also, as the needle's tip can pass through the grid holes, this makes it easier to get the lamella off the needle. The absence of the carbon support membrane is also advantageous for high-resolution electron microscopy (HREM). Once the lamella is on this type of grid (Figure 8.4(c)) it is possible to manoeuvre it over a grid hole (Figure 8.4(d)) by pushing it with the tip of the needle. As with the carbon coated TEM grids adhesion of the lamellae to these grids is very good. Typically, only about 1 in 40 lamella is lost during transport to and mounting in the TEM.

The lift-out step may also fail if the lamella jumps (due to electrostatics) away from, or moves against, the sidewalls of the FIB cuts (Figure 8.4(e)). As the lamella is symmetrical in shape it can readily be distinguished from any dirt or debris and thus, if it does jump away, it is possible, with patience, to find it and to lift it up. If the lamella moves against the side of the cuts it becomes difficult to see it with the optical microscope and to get it to attach to the needle. Repeatedly swinging the needle back and forth against the side of the FIB cut may knock and move the lamella such that it can then be lifted out. Unfortunately, when doing this the tip of the glass needle may shatter showering the FIB cuts and lamella in glass fragments.

Samples containing cracks or numerous layers that are poorly adhered can also be difficult to lift out as these may fracture or the layers delaminate as the needle tip is being pushed against them. Huang *et al.* [45] prevented the fracturing of lamellae containing cracks by leaving a supporting frame around the lamellae. For samples with poor adhesion between the layers, a Pt strap may be deposited onto the face of the lamella (by tilting the sample to 45°) to "weld" the different layers together (Figure 8.4(f)) before it is thinned to electron transparency.

The lift-out step has also been performed in the chamber of FIB/SEM systems using SE imaging [53]. This enables higher magnifications to be used than is possible optically giving more control over positioning the needle relative to the lamella and making it easier for new users to learn. In this approach it was found that the highest yield occurred if a tungsten needle with a radius of curvature of about 1 μm and an accelerating potential of 5 kV was used. Unfortunately, even though higher magnifications are possible the success yield of this approach (70%) is comparable to that of the ex-situ lift-out method. This, when coupled with the fact that the FIB system is being used for this step, means there is limited value in using this approach relative to the ex-situ method or the in-situ slice method which is discussed in the next section.

8.2.3 The in-situ slice technique

This technique can be viewed as a hybrid of the ex-situ lift-out and H-bar techniques. It was first reported by Ohnishi *et al.* [29] but for many reasons (such as patent issues and availability of suitable in-situ micromanipulators) it has taken many years to become accepted within the FIB community and it is only recently (circa 2005) that it has become the main FIB based technique being used. The chief reason why this method has superseded the ex-situ lift-out technique is because its success yield can be close to 100%. This high

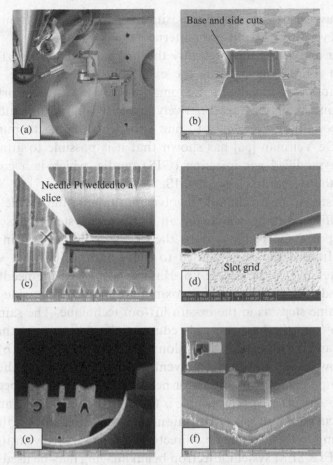

Figure 8.5 (a) Image of in-situ micromanipulator mounted on the stage, (b) SE image of a 500 nm thick slice that has had its base and sidewalls cut through, (c) SE image of a needle "welded" to a slice, (d) SE image of a slice being attached to a TEM slot grid, (e) SE image of a TEM grid with both tabs and V-slots, and (f) SE images of thinned lamellae mounted on a V-shape groove and a tab (insert) on a TEM grid.

success yield is because Pt "welding" is used to fix a slice of the sample onto the needle's tip as against relying on electrostatic forces.

The in-situ micromanipulator is an integral part of this technique. These have been mounted in the gas injector ports, on the stage, or on the sidewalls of the chamber (Figure 8.5(a)). An advantage of mounting the micromanipulator independently of the stage is that the sample can be moved to different locations while the tip of the needle remains fixed on the electro-optic axis. In addition to manipulating TEM lamellae in-situ micromanipulators in SEMs have also been used for positioning multi-wall carbon nanotubes onto cantilever tips [54] and pre-patterned electrodes [55].

From the mid 1990s until recently the main technique used for the preparation of TEM lamellae was the ex-situ lift-out technique, and many groups do not possess an in-situ micromanipulator as such. Unfortunately, as these micromanipulators are relatively expensive it is difficult for many groups, especially in academia, to find sufficient funding to purchase one. However, De Veirman [56] has shown that it is possible to attach a metal needle to a modified gas injection (GIS) needle, which is a cost effective solution if the system has a spare GIS.

The first step in this technique involves positioning the needle's tip at the stage eucentric position, which for an FIB/SEM system is also the beam coincident point. Once this is done the needle is not moved in the x or y directions. Instead the stage is used to move and position the sample and TEM grid relative to the needle tip. Having positioned the needle it is then retracted and a slice of material approximately 500–1000 nm wide is prepared using the same steps as in the ex-situ lift-out technique. The sample is then tilted by 45° and the base and side edges of the slice are cut through (one completely and the other only 80% along its length) (Figure 8.5(b)). Next the sample is lowered by 100 μm (to prevent accidentally crashing the needle tip into the sample) and the manipulator needle and gas insertion needle for the Pt "welding" are inserted. The sample is then raised, and if using an FIB/SEM system the cut end of the slice is positioned 100–200 nm beneath the tip of the needle. Owing to the two imaging directions this is relatively straightforward to do in an FIB/SEM system. Electron beam imaging may be used to position the edge of the slice beneath the tip of the needle while ion beam imaging can be used for the z-positioning. However, if using a single beam system and the sample is perpendicular to the ion imaging direction it is difficult, due to the large depth of field, to position the slice's edge a few hundred nanometers beneath the needle. If the needle and slice are too far apart (>300 nm) the deposited Pt will not be able to bridge/"weld" them together. Therefore, in a single beam system it is better to push the slice against the tip of the needle. Contact between the two is indicated by a change in the image contrast, movement of the needle tip, or the touch alarm being triggered.

The tip of the needle is then "welded" to the slice using in-situ Pt deposition (Figure 8.5(c)). The size of the Pt square deposited is 1 μm^2 and the deposition time is about 30 seconds when using a beam current of 50 pA. Care has to be taken not to use too large a beam current when depositing the Pt (i.e., >100 pA) in order not to locally deplete the Pt precursor gas. In this technique the two most critical steps are that the base of the slice is completely cut through and the needle is Pt welded to the slice's top edge. To ensure that the base is cut through the points highlighted in the ex-situ lift-out technique should be followed. To make sure that the needle is being "welded"/attached to the slice during the Pt deposition the grounding current can be monitored. As the two become "welded" (i.e., electrically connected) there will be a sharp change in the magnitude of the grounding current. After the Pt has been deposited another method to check that the two are "welded" is to gently tap the plinth of the system. If the two are fixed together the needle tip will not be able to vibrate.

Once the needle and slice are welded together the material holding the slice to the sample is cut. The plinth is then again gently tapped to check if the slice is completely free from the sample; if it is the needle and slice should vibrate together. If the slice is free the stage is then lowered and replaced by a TEM slot grid (which has been cut in half) and the slice is Pt welded to the top edge of the slot grid (Figure 8.5(d)) in essentially the reverse process of the lift-out procedure. Once the slice is fixed to the grid the needle is cut from it and retracted. The final step involves milling the slice to electron transparency using the tilting and line milling procedure while being careful not to mill away the Pt fixing "weld." In addition to slot grids, TEM grids with tabs and V-shaped grooves (Figure 8.5(e)) can also be used. An advantage of using a grid with tabs is that as the edges of the slice are straight (as against the base which has been cut at 45°) it is easier to Pt weld this to the side of a tab and there is no material beneath the slice to be sputtered and redeposited onto the lamella as it is being thinned. Figure 8.5(f) shows SE images of thinned slices Pt welded to a V-shaped groove and tab (insert).

8.2.4 Comparison of the different techniques

The choice of technique to use is dictated by numerous factors such as the secondary equipment available, the type of information required, the geometry of the sample, the amount of material, how unique a defect or sample is, the turn around time, and also to some extent the familiarity of a user with a given technique. Probably, the most decisive factors affecting the choice are the yield, how unique a defect is, and the turn around time. In the semiconductor industry the down time of a production line is very costly, while within academia the large numbers of users on a system may limit the available access time. In the following section the relative advantages of the different techniques are discussed and these should ideally form the basis for selecting which technique to use.

The success yields of the H-bar and in-situ slice techniques are similar, with both approaching 100%. Unfortunately, for reasons discussed in the previous sections the yield of the ex-situ lift-out technique can be as low as 60% depending on the experience of the user. Thus if a defect or sample is unique the H-bar or in-situ slice techniques should be used.

Although the in-situ slice technique is currently in vogue because of its high success yield, its main drawback is that the FIB or FIB/SEM system is used for the lift-out step, which typically takes 30–60 minutes per sample. Also, after the lift-out step the slice still has to be thinned to electron transparency. Thus to prepare three TEM lamellae with this technique typically takes 3–6 hours in addition to the time spent (typically overnight) milling the slices. In contrast, the FIB milling in the H-bar and ex-situ lift-out techniques can be automated to produce up to ten specimens overnight that are approximately 150–200 nm thick and as such require only 5–10 minutes per sample to thin them to 100 nm.

If there are many users on a system making lamellae, another important consideration is how many sessions are required before a user becomes proficient at a technique. Typically it takes three or four 4-hour sessions to become competent at the H-bar technique and seven or eight 4-hour sessions to become experienced with the other two techniques.

In the ex-situ lift-out and in-situ slice techniques, the pre-FIB sample preparation time is minimal (the samples can generally be placed directly into the system), whilst in the H-bar technique the preparation of the slice can take 2–3 hours if it is being prepared by hand polishing. In contrast, after FIB milling the H-bar's samples can be placed immediately into the TEM whilst the lifting out of the lamella and placing it onto a TEM support membrane in the ex-situ technique can take between 10 and 30 minutes each.

The primary drawback of the H-bar samples is that their geometry limits the angles of tilt or rotation during TEM analysis; the base or sidewalls of the trenches may block the path of the electron beam. For example, if a trench is 50 μm long and 10 μm deep, the maximum angle of tilt, if the electron beam is to pass through the center of the lamella, is only 6°. For ex-situ lift-out lamellae and slice samples the maximum angle of tilt is determined by the TEM holder. The trenches in the H-bar samples may also affect energy dispersive X-ray (EDS) analysis [57,58]. Electrons scattered from the lamella can irradiate the sidewalls and generate additional X-rays. The intensity of these additional X-ray signals can be reduced by 90% by preparing the lamella near the side edge of the slice [59], i.e., U-shaped samples.

Another advantage of the ex-situ lift-out and in-situ slice techniques over the H-bar technique is that the surrounding sample remains intact, which enables numerous lamellae to be prepared in close proximity to one another.

Another advantage of the in-situ slice and H-bar samples is that they can be returned to the system for further thinning if they have been prepared too thick. In a single-beam system this has also been used to prepare site-specific TEM lamellae of subsurface features in DRAM samples [60] and to cross section through the center of 1 μm diameter stratified Fe particles [61] embedded in a resin. For the subsurface features in the DRAM samples, the TEM lamellae were prepared 500 nm thick and the position of the defect within the lamellae was determined using HV SEM and energy filtered TEM, while for the Fe particles, the lamellae were initially prepared several micrometers thick and the diameter of the particles was determined using STEM. If an FIB/SEM system is being used then electron beam imaging can be used to end the FIB milling.

As will be discussed in Section 8.6.3, using high-energy FIB beams (30 keV) can result in damage being generated at the sidewalls of a lamella. To reduce the amount of damage post-FIB BIB milling using a low energy (100–500 eV) argon beam can be used. The geometries of the ex-situ and in-situ samples are both suitable for BIB milling whereas the geometry of the H-bar samples is such that they are not suitable. If the lamella is at a shallow angle to the BIB direction then the base and sidewalls of the trenches will be at a large angle (resulting in a higher sputter rate), and thus material can be sputtered from these and onto the lamella.

Ideally, if there is access to all the required secondary equipment (i.e., high-precision saws, micromanipulators) then the technique used should be chosen based on the required compositional and physical information, the geometry of the sample, and how unique the sample is. For example, for semiconductor devices on a silicon wafer from which only critical dimension measurements are required, if a high-precision diamond saw is available then the H-bar technique is the most appropriate to use. However, if the samples to be analyzed are unique and access time to an instrument is not limited, then because of its high yield the in-situ slice technique is the best method to use. If access time is limited and the samples are not unique then the ex-situ method should be used as it can prepare numerous specimens overnight, and if in the worse case scenario only 50% of the lamellae are successfully lifted out this is still around 25% more than can be prepared in the equivalent time using the in-situ lift-out technique.

8.3 Preparation of plan-view lamellae and SEM specimens

In addition to preparing cross section lamellae, FIBs are also used to prepare plan-view lamellae. One of the first methods was reported by Young *et al.* [62] who milled from the underside of the sample. Shortcomings with this

 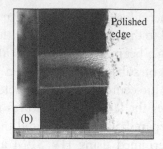

Figure 8.6 SE images of plan-view lamellae prepared using (a) the lift-out and (b) the H-bar techniques.

approach include making site-specific lamellae and stopping at the layers of interest.

Methods based on both the trench and ex-situ lift-out techniques have been used to prepare plan-view lamellae: the sample is cleaved or polished and mounted end-on in the FIB system. Figures 8.6(a) and (b) show SE images of a plan-view lamellae prepared using the ex-situ lift-out and H-bar techniques, respectively. In the approach based on the in-situ lift-out technique, it has been reported that the cut free lamella invariably falls into the FIB cut box and that during the lift-out step the needle should be swung towards the staircase cut so that the lamella is knocked into the FIB cuts [63]. These methods have been used to prepare plan-view lamellae of tungsten plugs [64], $CeO_2/Gd_2Zr_2O_7$ multi-layers [65], silicon nanowires [66], and to prepare numerous plan-view specimens at different heights in an yttria stabilized zirconia layer to study the change in texture through the layers [67].

Unfortunately, it is difficult to make site-specific plan-view lamellae using methods based on the H-bar and ex-situ lift-out techniques. The fracture position of the cleave cannot be controlled to better than 1 mm and polishing close to the ROI to reduce the amount of FIB milling can be time consuming. Site-specific plan-view lamellae can be prepared using the in-situ slice technique [29] by mounting the TEM grid at 90° to the sample and thus the slice. Other advantages of using the in-situ lift-out method are that the surrounding material is conserved, numerous specimens can be prepared in close proximity to one another, and the success yield can be near to 100%.

For some defects and materials it is advantageous to image them in both plan view and cross section. Methods to do this based on both the ex-situ and in-situ lift-out techniques have been reported by Langford *et al.* [68] and Minowa *et al.* [69]. Langford *et al.* used TEM grids without any support membrane (see Figure 8.4(c)) which enabled the lamella to be lifted off after imaging in the TEM and to be placed onto a piece of silicon where it could

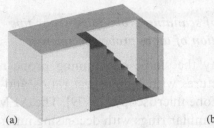 

(a)   (b)

Figure 8.7 (a) A schematic of an FIB cross section for SEM and (b) an SE image of a cross section through some metal interconnects.

then be cross-sectioned. Minowa *et al.* [69] Pt welded the needle to the plan-view lamella and then cut this out and placed it onto the top edge of another slice sample where it was then cross-sectioned.

The steps outlined previously for preparing the lift-out lamella can also be used to prepare cross sections, i.e., a Pt strap is deposited, a staircase-shaped cut is milled, and then the line milling procedure is used to mill up to the ROI. Figure 8.7(a) shows a schematic of an FIB cross section and Figure 8.7(b) shows an SE image of a cross section through a series of metal interconnects. Cross sections have been used to investigate diffractive optical elements [70], grain sizes [71], and the vertical structure of nanoporous aluminum oxide membranes [72], and to prepare samples for other analytical techniques such as electron backscatter diffraction analysis (EBSD) and Auger [12]. For example, Prasad *et al.* [73] prepared cross sections through shallow wear scars in electroformed Ni and used EBSD to study wear induced microstructural changes.

8.4 FIB sample preparation for other analytical techniques
8.4.1 Electron holography

FIBs have also been used to prepare site-specific lamellae for electron holo-graphy. In electron holography the region for analysis has to be adjacent to the vacuum therefore it is not possible to deposit a Pt protective layer over the ROI. To prevent ion milling the sample and the curtain effect occurring (which affects the interpretation of the data), both Dunin-Borkowski *et al.* [74] and Scharz *et al.* [75] prepared lamellae for electron beam holography by using the silicon beneath the ROI as a planarizing protective strap. Newcomb *et al.* milled from the back of the silicon wafer to create a free-standing cantilever and milled the lamella into this, while Mcarthy *et al.* used the in-situ slice technique to mount a slice upside down on the TEM grid.

8.4.2 Modification of scanning probe microscopy tips and the fabrication of atom probe specimens

FIBs have been used to modify the shape of scanning probe microscopy (SPM) tips for measuring structures with high aspect ratios and to deposit metal layers for electric force probe microscopy [76–79]. The AFM tips were modified by milling a series of annular rings with decreasing inner diameters down to a few hundred nanometers. The sharpening of the AFM tips to a radius of curvature of 5–20 nm is due to the milling by the tails of the ion beam distribution.

FIB based techniques have also been used to prepare specimens for AP analysis [80,81] from planar samples. AP field-ion microscopy (APFIM) is an atomic resolution technique, which requires a needle-shaped specimen, with an apex radius of curvature of 10–100 nm. Prior to using FIB milling and micromanipulation, the metal layers had to be deposited onto pre-sharpened tips thus preventing direct correlation with macroscopic properties. AP specimens have been prepared using methods based on both the ex-situ [82] and in-situ lift-out [83] techniques. In the approach based on the ex-situ technique the metal layers were grown onto Si wafers that had been patterned into pillars using reactive ion etching. The base of the pillars was then mechanically broken and the ex-situ micromanipulator was used to pick up a pillar and to mount it onto an electro-polished needle. The pillars were then shaped and sharpened by milling a series of annular rings, with the tails of the ion beam distribution again being used to sharpen the tips to 10–100 nm. The trimming and shaping of the pillar is considerably easier in an FIB/SEM system where real-time imaging can be used. In a method based on the in-situ technique a $2 \, \mu m^2$ pillar has been FIB milled from the substrate and this has been lifted out and Pt welded to the electro-polished needle. If numerous pillars are being prepared the repeated milling and lifting out of the pillars from the substrate is a time consuming step. The total time to prepare many pillars can be reduced by milling a 20–30 μm slice, fixing its end onto the electro-polished needle, and cutting the pillar from this slice and then repeating this step. Figure 8.8(a) shows an SE image of an FIB prepared slice that contains numerous pillars along its length as one of the pillars is being Pt welded to a needle, and Figure 8.8(b) after a pillar has been welded and the electro-polished needle lowered.

8.4.3 Three-dimensional reconstruction

The capability to make site-specific cross sections is also exploited for 3D reconstruction [84]. In this process, sequential slices (like cutting a loaf of

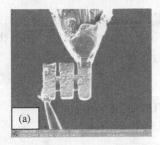 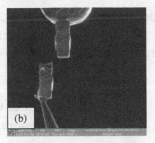

Figure 8.8 SE images of (a) a slice, into which numerous pillars have been FIB milled, being mounted onto an electro-polished needle, and (b) having cut a pillar from the slice and lowered the needle.

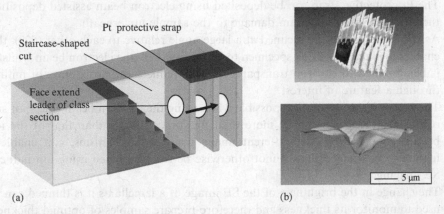

Figure 8.9 (a) Schematic of sequential FIB slicing for 3D reconstruction and a set of slices taken through the sample shown in Figure 8.7(b), and (b) the reconstructed surface of a nanoindented PVAC film.

bread) are milled through a feature or ROI, and the information from each cross section is then used for the 3D reconstruction. Figure 8.9(a) shows a schematic of the serial sectioning and a set of slices taken through the metal interconnects shown in Figure 8.7(b). Figure 8.9(b) shows a reconstruction of the surface of the nanoindent into the 4 μm thick PVAC layer shown in Figure 8.1(a).

FIB sequential slicing and 3D reconstruction have been used to investigate the shape of FIB milled vias in photonic structures [85], to examine Cu-Al multi-layers which had been deformed by nanoindentation [86], and have been combined with both Auger and SIMS for chemical analysis of bonding wires in an integrated circuit [87]. (3D chemical analysis can also be performed using image depth profiling in SIMS in which the beam is rastered perpendicular to the surface of the sample. However, the roughening of the base of the crater due to preferential milling, redeposition, and different sputter rates limits the z (depth) resolution [88].)

8.5 Advantages of using an FIB/SEM system for the preparation of TEM lamellae

There are numerous advantages of using FIB/SEM systems for preparing TEM lamellae relative to single beam FIB systems [89,90] and these include:

1. Navigation to a ROI can be performed using the electron beam, thus preventing any ion damage to the surrounding regions of the sample.
2. The Pt protective strap can be deposited using electron beam assisted deposition thereby preventing ion beam damage to the sample underneath.
3. As the two columns are inclined at a large angle relative to each other (52°), this enables the face of a TEM specimen to be imaged with the electron beam whilst it is being milled to electron transparency. This reduces the probability of milling through a feature of interest.
4. The higher image resolution possible when using the electron beam (the spot size of the electron beam can be more than five times smaller than that of the ion beam), coupled with the different image contrast mechanisms, can enable a feature to be imaged that cannot otherwise be resolved when using ion induced SE imaging.
5. The change in the brightness of the SE image of a lamella as it is thinned can be used to monitor its thickness and therefore prepare samples of optimal thickness for TEM analysis.
6. The two imaging directions enable the relative heights of the needle to the sample and the TEM grid to be readily determined, which can be difficult to do in a single beam system, owing to the large depth of field.
7. Analysis techniques such as EDS or EBSD may also be performed.

8.6 Artefacts resulting from the use of FIBs

Although FIBs are invaluable tools to prepare TEM lamellae, there are several detrimental effects associated with their use. Typically, the incident energies of the ion beams range between 30 and 50 keV. Therefore, during milling and ion imaging, Ga will be implanted into the top and sidewalls of the specimens. The implanted Ga may, through collision cascades, create vacancies and interstitials which, if of sufficient density, can result in the formation of dislocations or an amorphous layer. Figure 8.10(a) shows schematically the damage to a TEM lamella and Figure 8.10(b) is a BF TEM

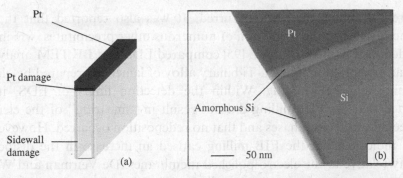

Figure 8.10 (a) Schematic of FIB damage to a lamella and (b) a BF TEM image of the FIB induced damage at the sidewalls of an FIB prepared lamella.

image in which the amorphous layer at the side of a lamella, created when it was cut free from the sample, is visible as a bright band. The increase in point defect density and the rise in local temperature may also result in intermixing at interfaces and diffusion giving rise to changes in the local composition and structure of a sample. Furthermore, preferential sputtering, redeposition, and segregation effects may also occur [91], which, depending on the material, may result in a phase transformation [92].

Numerous studies of the artefacts associated with using FIBs to prepare TEM lamellae have been made and these can be divided into two categories: those in which a comparison is made between lamellae prepared by FIB milling and other techniques and those that measured the depth of damage at the sidewalls and investigated methods to decrease it. Generally, in the FIB literature, the depth of the damage refers to the thickness of the amorphous layer, i.e., the region where the depth of the damage exceeds the critical point defect density [93]. However, it should be noted that the damage and its effect on the properties of a material may penetrate considerably further than this into the samples.

8.6.1 Comparisons of lamellae prepared by FIBs and other techniques

Ma [94] compared the microstructure of Inconel 783 lamellae prepared by jet polishing, BIB, and FIB milling. The precipitation of spherical Ni_3Al was only clearly observed in the lamella prepared using jet polishing. In the BIB prepared TEM lamella only part of the microstructure seen in the electro-polished lamella was visible and preferential sputtering of some of the spherical precipitates had occurred. For the FIB prepared lamella the morphology of the different phases and even the grain boundaries were not visible; however, preferential milling had not occurred. It was also reported that the FIB milling resulted in the formation of numerous other precipitates, which were not identified. Hutchison *et al.* [95] compared EDS and BF TEM analysis of cellular precipitation in a Cu-Ti binary alloy of lamellae prepared by electro-polishing and FIB milling. Within the detection limits of EDS, it was reported that the FIB milling did not result in "smearing" of the elements between the different phases and that no redeposition occurred. However, BF TEM indicated that the FIB milling caused an increase in the dislocation density relative to the electro-polished membranes. De Veirman and Weaver [96] made a comparative study of preparing lamellae of semiconductor samples coated with photoresist using tripod polishing, BIB, and FIB milling. They reported that FIB milling produced the best samples in terms of homogeneity and thinness while the BIB milling preferentially, sputtered away the photoresist and the tripod polishing caused the photoresist to be torn away. -

These comparative studies indicate some of the types of artefacts that can occur when using FIB milling and that these need to be considered when interpreting the chemical and structural data obtained from FIB prepared lamellae.

8.6.2 Damage to the top surface due to ion beam assisted metal deposition

During the FIB preparation of a lamella, metal is deposited over the ROI using ion beam assisted deposition, in order to planarize the surface of the sample to prevent the "curtain effect" and to reduce the amount of rounding/milling of the top of the lamella. (In an FIB/SEM system the Pt may also be deposited using the electron beam and this is discussed later in the section.) As the ion beam assisted deposition is an energetic process it can result in damage, Ga implantation, and intermixing of the depositing species into the top of the sample. Figure 8.11 shows an EELS map of a Pt strap deposited on Si using a 30 keV ion beam. The damage from the ion beam breakdown of the precursor has been studied by numerous authors [97–99]. Walker and Broom [98] investigated the resulting depth of damage for Si, In P, and GaAs and reported depths of damage of 30, 20, and 20 nm, respectively. Detailed Auger and PEELS/EDS analysis of Pt layers deposited on TiN and W have shown that the interface consists of two bands, a brighter band that is carbon rich and a darker band that is an intermixed zone with the substrate [63]. Kempshall *et al.* [99] studied the effects of reducing the incident energy to 5 keV and using e-beam induced deposition of Pt in an FIB/SEM system.

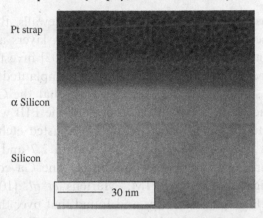

Figure 8.11 EELS image of a Pt strap on silicon.

They reported that reducing the ion beam energy from 30 to 5 keV, decreased the depth of damage in Si from 50 nm to 13 nm and that no damage beneath the e-beam deposited Pt layer could be observed. The e-beam deposition rate is dependent on the incident energy of the electrons, i.e., the number of SE generated, to break down the precursor gas and thus can be increased by a factor of ten by reducing the electron beam energy from 30 keV to 5 keV. However, if an electron beam of 400 pA at 5 keV is used the deposition rate is still ten times slower than that when using a 300 pA, 30 keV ion beam. Therefore, to deposit the Pt strap the first 100 nm can be deposited using e-beam deposition and the rest completed using ion beam deposition [89].

The simplest method to prevent ion beam damage to the surface of a sample if using a single beam system is to coat the surface with a 100 nm thick protective film prior to placing it into the FIB system. A range of materials have been used for this purpose, including Au [100], plasma-polymerized layers [101], and amorphous silicon [102]. A disadvantage of this is that as the whole sample is coated it may prevent its use for subsequent processing and device fabrication.

8.6.3 Ga ion implantation and sidewall damage

During the milling of a TEM lamella, steady state sputtering conditions will be reached in which the rate of Ga implantation balances the rate of Ga sputtering. Ishita *et al.* [102] modeled ion retention during steady state sputtering and reported a good correlation between their model and experimental results. They used EDS to determine the Ga concentration in a lamella prepared using a 30 keV ion beam. On the assumption that the implanted Ga was within the first 10 nm of the sidewalls, the authors calculated the Ga concentration in silicon and tungsten layers as 4 and 9 at.%, respectively. Langford and Petford-Long *et al.* [103] investigated the effect of different ion beam milling conditions on the implanted steady state Ga concentration in silicon lamellae. They reported that the Ga concentration decreased by a factor of 2.5 when the energy of the FIB was reduced from 30 to 5 keV. Lamellae milled using iodine gas assisted etching (GAE) with 30 and 5 keV ion beams were reported to contain 2.7 and 3 times less Ga than those prepared using a 30 keV ion beam. The Ga concentration has also been investigated using APFIM. Martens *et al.* [104] analyzed the distribution in FIB prepared AP tips and found that over the first 20 nm the Ga concentration was 30 at.%. Ferryman *et al.* [105] investigated the Ga distribution in silicon lamellae using XPS and reported that the Ga concentration varied over the lamella, and attributed this to the possible taper of the lamella. Ga ions implanted into the sidewalls of a lamella have been reported to coalesce in some materials to form Ga rich regions. Tanaka *et al.* [106] used a TEM that had been modified to incorporate an FIB to show that in GaAs the implanted Ga precipitated as liquid or amorphous particles of ~200 nm diameter because of the selective etching of As, while in Si, SiO_2, and SiC samples it precipitated as small dots with a diameter of 1–10 nm.

The depth of the damage at the sidewalls of FIB prepared TEM lamellae has been investigated for a wide range of materials [107–110]. To do this cross sections of the lamellae and cross sections of the ends of milled trenches have been made. Kato *et al.* [109] and Sutton *et al.* [110] investigated the effect of different beam currents on the depth of the damage in Si, GaAs, and InP. Both authors reported that over the range of 100 to 500 pA the depth of damage was independent of the magnitude of the beam current.

In addition to using TEM, scanning probe microscopy (SPM) has also been used to assess FIB induced artefacts. Huey *et al.* [111] and Basnor *et al.* [112] used AFM (atomic force microscopy) to characterize the swelling in silicon when implanting with low doses of Ga. They both reported that for doses above 10^{13} and below 10^{15} Ga/cm^2, swelling of the implanted region occurred. Brezna *et al.* [113] used scanning capacitance microscopy (SCM) to measure the FIB damage beneath FIB milled trenches in silicon. They reported that the electrical properties of the silicon were affected by distances up to 620 nm beneath the FIB milled regions.

Typically FIB prepared TEM lamellae are of the order of 100 nm thick and this when coupled with the damage at the sidewalls may affect HREM. Therefore, many methods have been used to reduce both the thickness of

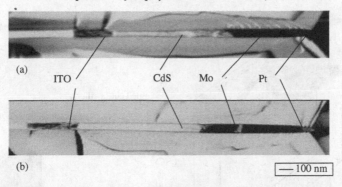

Figure 8.12 BF TEM images of a CdS solar cell which has been prepared using (a) LE FIB milling and (b) standard FIB milling.

the lamellae and the damage at the sidewalls. The most common practice is to reduce the energy of the FIB [114] (to typically between 1 and 5 keV) and to raster this over the sidewalls of the lamellae as they are tilted to between ± 5 and 10°. (The line milling procedure can not be used due to the increase in the spot size and the decrease in the signal to noise ratio.) Unfortunately, as the lamella is not rotated and because of the large incident angles preferential sputtering may occur during this low energy (LE) milling [115,116]. The amount will depend on the sputter yields of the different materials, the angle of tilt used, and how much material is sputtered away from the sidewalls. In Figure 8.12(a) a BF TEM image of a cross section of a CdS solar cell lamella which was LE FIB milled using a 50 pA, 5 keV beam at 10° for 2 minutes on each side shows that preferential milling of the different layers has occurred. (These conditions were determined to sputter approximately 70 nm from either side of the lamella.) For comparison, Figure 8.12(b) shows the equivalent BF TEM cross section of a lamella prepared using a 100 pA beam at 30 kV and a tilt of 1.5° during the line milling thinning.

The suitability of FIB prepared lamellae for HREM has also been improved by making angle cuts across the top edge of the lamella [117] (Figure 8.13(a)). For silicon lamellae the apex of the resulting taper will be completely amorphous but as the distance from the apex is increased, an optimum region in terms of the thickness of the crystalline region and damaged layer will occur for HREM. The angle cuts may also be made using an LE FIB beam to further improve the suitability of the lamellae for HREM. Figure 8.13(b) shows a BF TEM image of an angled cut lamella. Tapered regions in the vertical direction may also be prepared by tilting the lamellae to large angles [69,118] (4° and above) during the line milling

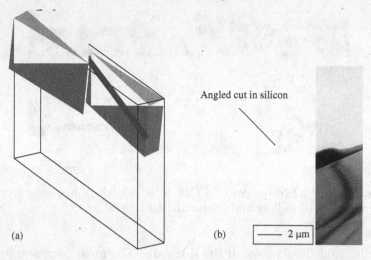

Angled cut in silicon

(a)   (b)   ——— 2 μm

Figure 8.13 (a) Schematic of FIB angle milling to produce a tapered specimen and (b) BF TEM of a lamella after an angle cut.

procedure. Furthermore, milling at large angles is also reported to reduce the depth of the damage at the sidewalls [119].

GAE involves the exposure of an area to a reactive gas during FIB milling. It is reported to increase the etching rate, reduce the redeposition of material, and decrease the number of crystal defects and amount of Ga incorporated into a sample [120]. Sugimoto *et al.* [121] used chlorine GAE to reduce the damage in their GaAs samples and Yamaguchi and Nishikawa [122] used iodine GAE to improve InP cross section specimens for HREM. Yamaguchi and Nishikawa [122] reported that GAE reduced the thickness of the damage layer from 31 nm to 2–5 nm and prevented the formation of some FIB artefacts such as the occurrence of micro-crystals. Langford and Petford-Long [103] and Kato *et al.* [109] investigated the use of iodine GAE on silicon cross sections and although it increased the sputter yield, both reported that it did not reduce the thickness of the amorphous layer.

BIB milling, which enables lower incident energies to be used than is currently possible in FIB systems (if a floating gun is used the energies can be as low as a few hundred electronvolts), has been used to reduce the FIB damage from Si p-n junctions [123] and steel specimens [124]. Kato *et al.* [109] reported that BIB milling of FIB prepared silicon lamellae, using a 4 keV beam at 4°, reduced the depth of the damage at the sidewalls of FIB milled pillars from 30 nm to 8 nm. When post-BIB milling FIB prepared lamellae the mounting of the sample in the miller is important to prevent redeposition

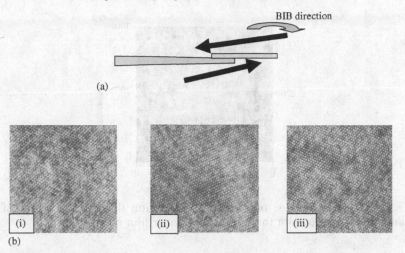

Figure 8.14 (a) Mounting of a sample for LE BIB milling to remove the FIB damage and (b) lattice images of InP lamellae prepared using (i) standard milling, (ii) GAE, and (iii) LE FIB milling.

onto the lamellae. Figure 8.14(a) illustrates the ideal mounting of an in-situ lift-out or an ex-situ lift-out sample for BIB milling on a 200–300 nm thick support membrane such that the top and bottom sides of the lamella can be milled simultaneously while it is being rotated. All the edges and surfaces close to the lamella should be at a shallow angle of incidence relative to the incident beam to prevent sputtering and redeposition onto the lamella. Figure 8.14(b) shows HREM lattice images of InP lamellae prepared using (i) the standard milling conditions, (ii) iodine GAE, and (iii) LE (5 keV) FIB milling, viewed along the <110> zone axis. Comparing the lattice images indicates the benefits of using GAE and post-BIB milling. The relative improvement of the lattice image of the GAE lamella will partly be due to the decrease in the thickness of the damage at its sidewalls, while the relative improvement of the lattice image of the LE FIB milled lamella will be due to both a decrease in its thickness and the depth of the damage at the sidewalls. Cross sections of these three lamellae showed that the thickness of the amorphous layers were 15 nm for the lamellae prepared using the standard FIB milling conditions and 3–4 nm for the lamellae prepared using the GAE and LE FIB milling.

Another method [125] that has been used so that ex-situ lift-out prepared lamellae can be BIB milled has consisted of depositing Pt straps from the corners of the lamella to the edges of the support grid such that these straps

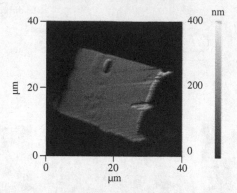

Figure 8.15 AFM image of an FIB cross section through the center of a nano-indent into a 4 µm thick PVAC polymer film on silicon.

support the lamella once the carbon support membrane has been milled away during the BIB milling.

8.6.4 Redeposition and preferential milling

In addition to the damage generated at the sidewalls, two other FIB artefacts that may occur are redeposition onto the lamellae and preferential milling [115,126–128]. The amount of redeposition will be dependent on the geometry of the FIB cuts and the milling regime used to thin the lamella. Cairney and Muroe *et al.* [128] reported that sequential line milling reduces, but does not completely eliminate, redeposition and that LE milling does not affect the amount of redeposition at the sidewalls of lamellae prepared without using the line milling procedure.

Preferential milling of samples composed of materials with different sputter rates will inevitably occur and the amount will be dependent on the differences in the sputter yields and the structure and geometry of the samples. Figure 8.12(b) shows a BF TEM image of a cross section through an FIB lamella of a CdS solar cell prepared using standard FIB milling conditions, which shows that preferential milling of the different layers has occurred. The line milling step was performed using a beam current of 100 pA and at a tilt angle of 1.5°. Figure 8.15 shows an AFM image of an FIB cross section through a nanoindentation into a 4 µm thick PVAC spun onto silicon. The cross section was milled using a 30 kV, 100 pA beam current without any tilting during the line milling procedure. The image shows that relative to the silicon the PVAC has been recessed in places by up to 200 nm. In addition to affecting quantitative chemical analysis, preferential milling will also place a limitation on the minimum thickness that a specimen can be prepared.

8.7 Summary

In this chapter the different FIB techniques used for preparing specimens for physical and chemical analysis have been discussed. The diverse range of examples given illustrates the versatility of the different techniques and the considerable impact that they have had and are having on material/device analysis. It can only be expected that the range of novel uses and applications will continue to rapidly increase.

References

[1] R. M. Langford, S. O'Reilly and I. Mewn. *Mater. Res. Soc. Symp. Proc.*, **739** (2002), 65.

[2] Y. Katoh, M. Ando and A. Kohyama. *Nucl. Mater.*, **323** (2003), 251–62.

[3] D. R. G. Mitchell, D. J. Attard, G. A. Collins and K. T. Short. *Coatings Tech.*, **165** (2003), 107–18.

[4] H. Saka, K. Kuroda, M. H. Hong *et al. ICEM*, **13** (1994), 1009.

[5] D. M. Longo, J. M. Howe and W. C. Johnson. *UltraMicroscopy*, **80** (1999), 69.

[6] A. J. Smith, P. Munroe and T. Tran. *J. Mater. Sci.*, **36** (2001), 3519.

[7] C. Burmer, S. Gorlich and S. Pauthner. *Microelectronics Reliability*, **38** (1998), 987.

[8] G. Mardingly. *Inst. Phys. Conf. Ser.*, **164** (1999), 575.

[9] S. Tsuji, K. Tsujimoto and H. Iwama. *IBM J. Res. Develop.*, **42** (1998), 509.

[10] J. C. Reiner, P. Gasser and U. Sennhauser. *Microelectronics*, **42** (2002), 1753.

[11] S. Morris, S. Tatti, E. Black, N. Dickson *et al. Proc. 17th Int. Symp. Testing and Failure Analysis*, (1991), 471.

[12] D. Verlleij. *Microelectronics Reliability*, **38** (1998), 869.

[13] D. J. Larson, A. K. Petford-Long, A. Cerezo and G. W. D. Smith. *Acta Mater.*, **47** (1999), 4019.

[14] J. M. Cairney, P. R. Muroe and J. H. Schneibel. *Scripta Mater.*, **42** (2000), 472.

[15] R. Pantel, G. Auvert and G. Mascarin. *Microelectronic Eng.*, **37/38** (1997), 49.

[16] B. Wei, K. -R. Philipp, U. Bader *et al. UltraMicrosc.*, **85** (2000), 93.

[17] E. Zschech, E. Langer, H. -J. Engelmann and K. Dittmar. *Mater. Sci. Semicond. Process.*, **5** (2003), 457.

[18] H. White, Y. Pu, M. Rafailovich, R. Sokolov, A. H. King *et al. Polymer*, **42** (2001), 1613.

[19] J. Loos, K. J. Jeroen, F. Morrissey and R. A. J. Janssen. *Polymer*, **43** (2000), 7493.

[20] K. Hosi, S. Ejiri, W. Probst *et al. J. Microsc.*, **201** (2001), 44.

[21] H. Engqvist, J. -E. Schultz-walz, J. Loof *et al. Biomaterials*, (2003).

[22] R. Haswell, D. W. McComb and W. Smith. *J. Microsc.*, **211** (2003), 161.

[23] P. J. Heaney, E. P. Vicenzi, L. A. Giannuzzi and K. J. T. Livi. *Am. Mineral.*, **85** (2001) 1094.

[24] B. I. Prenitzer, L. A. Giannuzzi, K. Newman *et al. Mater. Trans. A*, **29A** (1998), 2399.

[25] F. Shaapur, T. Stark, T. Woodward and R. J. Graham. *Mater. Res. Soc. Proc.*, **480** (1997), 173.

[26] D. P. Basile, R. Boylan, B. Baker, K. Hayes and D. Soza, *Mater. Res. Soc.* **254** (1992), 23.

[27] E. Kirk, D. Williams and H. Ahmed. *Inst. Phys. Conf. Ser.*, **100** (1989), 501.

[28] K. Park. *Mater. Res. Soc. Proc.*, **199** (1990), 271.

[29] T. Ohnishi, H. Koike, T. Ishitani *et al. Proc. 25th Int. Symp. Testing and Failure Analysis*, **449** (1999).

[30] B. W. Kempsall, B. W. Schwarzx and L. A. Giannuzzi. *ICEM*, **249** (Durban, South Africa, 2002).

[31] F. Altmann and D. Katzer. *Thin Solid Films*, **344** (1999), 609.

[32] D. M. Schraub and R. S. Rai. *Prog. Cryst. Growth Charact. Mater.*, **36** (1999), 99.

[33] R. M. Anderson and S. D. Walck. *Mater. Res. Soc. Proc.*, **480** (1997), 187.

[34] FEI service notes.

[35] S. J. Klepies, J. P. Benedict and R. M. Anderson. *Mater. Res. Soc.*, **179** (1988).

[36] L. R. Harriott. *J. Vac. Sci. Technol. B*, **11** (1993), 2200.

[37] R. J. Young, J. R. A. Cleaver and H. Ahmed. *J. Vac. Sci. Technol. B*, **11** (1993), 234.

[38] J. Szot, R. Hornsey, T. Ohnishi and S. Minagawa. *J. Vac. Sci. Tech. B*, **10** (1992), 575.

[39] T. Ishitani and T. Yaguchi. *Microsc. Res. Tech.*, **35** (1996), 320.

[40] A. Gandhi and J. Orloff. *J. Vac. Sci. Technol. B*, **8** (1990), 1814.

[41] D. Santamore, K. Edinger, J. Orloff and J. Melngailis. *J. Vac. Sci. Tech. B*, **15** (1997), 2346.

[42] W. Boxleitner, G. Hobler, V. Kluppel and H. Cerva. *Nucl. Instrum. Meth. Phys. Res. B*, **175** (2001), 102.

[43] J. F. Walker. *Inst. Phys. Ser.*, **157** (1997), 469.

[44] G. Deshais and S. B. Newcomb. *Eurem 12*, **573** (Brno, Czech Republic, 2000).

[45] Y. Z. Huang, S. P. Lozano-Perez, R. M. Langford, J. M. Titchmarsh and M. L. Jemkins. *J. Microsc.*, **207** (2002), 129.

[46] M. H. Overwijk, F. C. van den Heuvel and C. W. T. Bulle-Lieuwma. *J. Vac. Sci. Technol. B*, **11** (1993), 2021.

[47] J. Ebel, C. Bozada, T. E. Schlesinger, C. Cerny *et al. IEEE 36th Annual International Reliability Physics Symposium*, (Reno, Nevada, 1996).

[48] A. I. Oliva, A. Romero, J. L. Pena, E. Anguiano and M. Aguilar. *Rev. Sci. Instrum.*, **67** (1996), 1917.

[49] K. S. K. Kim, S. C. Lim, I. B. Lee *et al. Rev. Sci. Instrum.*, **74** (2003), 9.

[50] F. A. Stevie, T. C. Shane, P. M. Kahora *et al. Proc. Symp. Applied Surface Analysis*, **23** (Burlington, MA, USA, 1995) 61.

[51] L. R. Herlinger, S. Chevacharoenkul and D. C. Erin. *Proc. ISTFA '96*, **199** (1996).

[52] R. M. Langford and Petford-Long. *J. Vac. Sci. Tech. A*, **19** (2001), 982.

[53] R. M. Langford and C. Clincton. *Micron*, **35**:7 (2004), 607.

[54] A. Hall, W. G. Matthews, R. Superfine, M. R. Flavo and S. Wasburn *Appl. Phys. Lett.*, **82**(15) (2003), 2506.

[55] P. Avouris, T. Hertel, R. Martel *et al. Appl. Surf. Sci.* **141** (1999), 201.

[56] A. E. M. De Veirman. *Mater. Sci. Eng.*, **B00** (2003), 1.

[57] D. M. Longo, J. M. Howe and W. C. Johnson. *UltraMicroscopy*, **80** (1999), 69.

[58] Y. Yabuuchi, T. Okano, S. Tametou, M. Arai and T. Kouzaki. *Eurem 12*, **565** (Brno, Czech Republic, 2000).

[59] M. Saito, T. Aoyama, T. Hashimoto and S. Isakozawa. *Jpn. J. Appl. Phys.*, **37** (1998), 355.

[60] T. Ishitani, Y. Taniguchi, S. Isakozawa *et al. J. Vac. Sci. Technol. B*, **16** (1998), 1907.

[61] T. Yaguchi, T. Kammino, H. Kobayashi, H. Koike, K. Tohji, K. Nakatsuka and R. Urao. *Eurem 12*, **569** (Brno Czech Republic, 2000).

[62] R. J. Young, E. C. G. Kirk, D. A. Williams and H. Ahmed. *Mater. Res. Soc. Proc.*, **199** (1990), 205.

[63] R. M. Langford, Y. Huang, S. Lozano-Perez, J. M. Titchmarsch and A. K. Petford-Long. *J. Vac. Sci. Technol.*, **B19** (2001) 755.

[64] H. Bender. *Inst. Phys. Conf.*, **164** (1999), 595.

[65] T. Kato, T. Murago, Y. Iijima *et al. J. Electron. Microsc.*, **53** (2004), 501.

[66] T. Tsutsumi, E. Suzuki, K. Ishii *et al. J. Vac. Sci. Technol. B*, **17** (1999), 1897.

[67] S. B. Newcomb and W. A. J. Quinton. *Eurem 12*, **575** (Brno, Czech Republic 2000).

[68] R. M. Langford, D. Ozkaya, D. Zhou, A. K. Petford-Long and C. Stanley. Proc. *ICEM*, **251** (Durban, 2002).

[69] K. Minowa, K. Takeda, S. Tomimatsu and K. Umemura. *J. Crys. Growth*, **210**, (2000), 16.

[70] C. Dix, P. F. McKee, A. R. Thurlow *et al. J. Vac. Sci. Technol. B*, **12** (1994), 3708.

[71] J. M. Cainey, P. R. Munroe and M. Schneibel. *Scripta Mater.*, **42** (2000), 473.

[72] A. Heilmann, F. Altmann, D. Katzer *et al. Appl. Surf. Sci.*, **144** (1999), 628.

[73] S. V. Prasad, J. R. Michael and T. R. Christenson. *Scripta Mater.*, **48** (2003), 255.

[74] R. E. Dunin-Borkowski, S. B. Newcomb, M. R. McCartney, C. A. Ross and M. Farhoud. *Electron Microscopy and Analysis, Institute of Physics Conference Series*, **168** (2001), 485.

[75] S. M. Scharz, B. W. Kempshall, L. A. Giannuzzi and M. R. McCartney, *Microsc and Microanal.*, **9** (suppl. 2) (2003), 116.

[76] H. Ximen and P. E. Russell. *UltraMicroscopy*, **42** (1996), 1526.

[77] A. Olbrich, B. Ebersberger, C. Bolt *et al. J. Vac. Sci. Technol. B*, **17** (1999), 1570.

[78] C. Menozzi, C. G. Gazzadi, A. Alessandrini and P. Facci, *UltraMicroscopy*, **104**:3–4 (2005), 220.

[79] K. Akiyama, T. A. Eguchi, Y. Fujikawa *et al. Rev. Sci. Instr.*, **76**:3 (2005), 33705.

[80] D. J. Larson, A. K. Petford-Long, A. Cerezo *et al. Appl. Phys. Lett.*, **73** (1998), 8.

[81] D. J. Larson, D. T. Foord, A. K. Petford-Long, A. Cerezo and G. W. S. Smith. *Nanotechnology*, **10** (1999), 45.

[82] D. J. Larson, B. D. Wissman, R. L. Martens *et al. Microsc. Microanal.* **7** (2001) 24.

[83] M. K. Miller, K. F. Russell, G. B. Thompson. *Ultra Microscopy*, **102**(4) (2005), 287.

[84] H. Z. Wu, S. G. Roberts, G. Mobus and B. J. Inkson. *Acta Mater.*, **51** (2003), 149.

[85] R. M. Langford, G. Dale, P. J. Hopkins, P. J. S. Ewen and A. K. Petford-Long. *J. Micromech. Microeng.*, **12** (2002).

[86] B. J. Inkson, T. Steer, G. Mobus and T. Wagner. *J. Microsc.*, (2002).

[87] T. Sakamoto, Z. Cheng, M. Takahashi, M. Owari and Y. Nihei. *Jpn. J. Appl. Phys.*, **37** (1998), 2051.

[88] H. Ximen, R. K. Defreez, J. Orloff, J. Elliott *et al. J. Vac. Sci. Technol. B*, **8** (1990), 1361.

[89] R. Kruger. *Micron*, **221** (1999), 30.

[90] P. Gnauck and P. Hoffrogge. *Proc. SPIE: Reliability, Testing and Characterization of MEMS/MOEMS II*, **4980** (2003), 106.

[91] D. J. Barber. *UltraMicroscopy*, **52** (1993), 101.

[92] F. V. Nolfi, ed. *Phase Transformation During Irradiation* (London: Applied Science Publishers, 1983).

[93] J. S. Williams. *Mater. Sci. Eng. A*, **253** (1998), 8.

[94] L. Ma. *Micron*, **35**:4 (2004), 273.

[95] C. R. Hutchinson, R. E. Hackenberg and G. J. Shiflet. *Ultramicrocopy*, **94** (2003), 37.

[96] A. De Veirman and L. Weaver. *Micron*, **30** (1999), 213.

[97] S. Rubanov and P. R. Munroe. *Mater. Lett.* **4142** (2002), 1.

[98] J. F. Walker and R. F. Broom. *Inst. Phys. Conf. Ser.*, **157** (1997), 473.

[99] B. W. Kempshall, L. A. Giannuzzi, B. I. Prenitzer, F. A. Stevie and S. X. Da. *J. Vac. Sci. Technol. B*, **20** (2002), 286.

[100] N. I. Kato, N. Miura and N. Tsutsui. *J. Vac. Sci. A*, **16** (1998), 1127.

[101] N. Miura, K. Tsujimato, R. Kanehara, N. Tsutsui and S. Tsuji. *Proc. 22nd Int. Symp. Testing and Failure Analysis* (1995), 353.

[102] T. Ishitani, H. Koike, T. Yaguchi and T. Kamino. *J. Vac. Sci. Technol. B*, **1** (1998), 1907.

[103] R. M. Langford and A. K. Petford-Long. *Proc. 12th European Microscopy Congress*, Vol. II (2000), 557.

[104] R. L. Martens, D. J. Larson, T. F. Kelly, A. Cerezo, P. H. Clifton and N. Tabat. *Micros. Microanal.*, **6**, (suppl. 2) (2000), 522.

[105] C. A. Ferryman, J. E. Fulghum, L. A. Giannuzzi and F. Stevie. *Surf. Interface Anal.*, **33** (2002), 907.

[106] M. Tanaka, K. Furuya and T. Saito. *Jpn. J. Appl. Phys.*, **37** (1998), 7010.

[107] J. P. McCaffrey, M. W. Phaneuf and L. D. Madsen. *UltraMicroscopy*, **87** (2001), 97.

[108] H. Bender and P. Russel. *Inst. Phys. Conf. Ser.*, **157** (1997), 465.

[109] N. I. Kato, N. Miura and N. Tsutsui. *J. Vac. Sci. Technol. A*, **16** (1998), 1127.

[110] D. Sutton, S. M. Parle and S. B. Newcomb. *Inst. Phys. Conf. Ser.*, **168** (2002), 377.

[111] B. D. Huey and R. M. Langford. *Nanotechnology*, **14** (2003), 409.

[112] B. Basnor, A. Lugstein, H. Wanzenbeck *et al. J. Vac. Sci. Technol. B*, **21** (2003), 209.

[113] W. Brezna, H. Wanzenbock, A. Lugstein *et al. Physica E*, **19** (2003), 178.

[114] R. J. Young, P. D. Carleson, T. Hunt and J. Walker. *Proc. From the 24th Int. Symp. Testing and Failure Analysis*, (1998).

[115] R. M. Langford, D. Ozkaya, C. Kaufman *et al. Electron Microsc. Anal. Inst. Phys. Conf. Ser.* **168** (2001), 437.

[116] R. M. Langford, D. Ozkaya, B. Huey, A. K. Petford-Long. *Microsc. Semicond. Mater. Inst. Phys. Conf. Ser.*, **169** (2001), 511.

[117] Z. Wang, I. K. Takeharu, T. Hirayama *et al. J. Vac. Sci. Technol. B*, **21** (2003), 2155.

[118] A. Yamaguchi, M. Shibata and T. Hasshinga. *J. Vac. Sci. Technol.*, *B*, **11** (1993), 2016.

[119] A. Leslie, K. L. Pey, K. S. Sim, M. T. F. Beh and G. P. Goh. *Proc. 21st Int. Symp. Testing and Failure Analysis*, **353** (1995).

[120] P. E. Russell, T. J. Stark, D. P. Griffis, J. R. Phillips and K. F. Jarausch. *J. Vac. Sci. Technol. B*, **16** (1998), 2494.

[121] Y. Sugimoto, M. Taneya, H. Hidaka and K. Akita. *J. Appl. Phys.*, **68** (1990), 2392.

[122] A. Yamaguchi and T. Nishikawa. *J. Vac. Sci. Technol. B*, **13** (1998), 962.

[123] D. Cooper, A. C. Twitchett, R. E. Dunin-Borkowski *et al. Proc. of 13th European Microscopy Congress*, Vol. II (2004), 783.

[124] C. L. Collins, M. MacKemzie, A. J. Craven *et al. Proc. 13th European Microscopy Congress*, Vol. II (2004), 775.

[125] Z. Huang. *J. Microsc.*, **215** (2004) 219.

[126] S. Rajsiri, B. W. Kempshall, S. M. Schwarz and L. A. Giannuzzi. *Proc. Microsc. Microanal.*, **99** (2002), 50.

[127] T. L. Matteson, S. W. Schwarz, E. C. Houge, B. W. Kempshall and L. A. Giannuzzi. *J. Electron Mater.*, **31** (2002), 31.

[128] J. M. Cairney and P. R. Munroe. *Micron*, **34** (2003), 97.

9

In-situ sample manipulation and imaging

T. KAMINO AND T. YAGUCHI
Hitachi Science Systems

T. OHNISHI AND T. ISHITANI
Hitachi High-technologies Corporation

9.1 Process of TEM sample preparation

9.1.1 Pre-thinning

Since a focused Ga ion beam with a maximum diameter at the sub-micrometer level is used in the FIB technique, the sputtering speed is not so high as in other ion milling techniques [1,2]. To reduce the FIB milling time, pre-thinning of the sample using a polisher or cutter is required. There is no definite sample target thickness to be achieved by pre-thinning, but a typical thickness used for FIB milling is in the range of 50 to 100 μm in most cases. Preparing thinner samples promises shorter FIB milling times but we should be careful not to cause any structural and chemical changes in the sample at this stage [3]. Note that the thinner samples may become more stressed during the pre-thinning process than the thicker ones. There are various techniques and tools to be used for the pre-thinning process. A micro-cleaving system is widely used for pre-thinning of the devices with single crystal Si or GaAs substrates. The system allows cleaving of the sample including the circuit of interest quite easily and precisely and the cleaved sample may be directly FIB milled or first sliced to 10 to 20 μm using a fine blade prior to the FIB milling. In the case of polycrystalline materials such as metals, ceramics, and their composites, mechanical-polishing or dicing can be done. Relatively soft materials such as polymers, rubbers, and papers can be sliced with a razor blade. No special tool is required for these materials. Nonconductive materials such as glasses, polymer, and ceramics should be metal coated prior to FIB milling. Carbon or metals such as Pt,

Focused Ion Beam Systems: Basics and Applications, ed. N. Yao.
Published by Cambridge University Press. © Cambridge University Press 2007.

Au, or Pd are used for the coating. A coating thickness of 20–30 nm is sufficient.

9.1.2 Mount on a sample support

The pre-thinned sample is then sectioned into a plate about 1–2 mm wide and 2.5–3 mm long and directly mounted onto the FIB sample holder. If the sample is smaller than that, it should be sectioned into small plates and glued to a sample support made from a semicircular metal disk [4–5]. Conductive paste such as carbon paste or silver paste can be used to glue the sample to the support disk. Mo, Cu, and Al plates with a thickness of 30–50 μm are available as the support material. The choice of the materials should be made from the analytical point of view because the system peak from the support material may cause trouble in elemental analysis using the EDX system. In the case of particles or fibers, the sample may be embedded in epoxy. The embedded sample is sectioned into the foil with a thickness of 100 nm or less, and then glued to an edge of the support. Don't forget to coat the whole glued sample with evaporated carbon or metal layers. This is very important to avoid any instability of the ion beam position caused by the charged sample during the FIB milling.

9.1.3 Surface treatment

In TEM sample preparation, having a smooth surface is very important to get enough uniformity in the thickness. Always remember that the smoother the surface the better the uniformity in the thickness of the sample. Therefore, it is strongly recommended to keep the sample surface smooth before FIB milling. Methods such as polishing or slicing can be applied to make the sample surface smooth but be careful to avoid introducing any mechanical damage with this process. It is recommended that if you want to observe the cross-sectional structure of the top surface, omit the mechanical polish or metal deposition, but instead add a carbon coating using a vacuum evaporator. The recommended thickness of the carbon layer is 20–40 nm. FIB assisted metal deposition will follow as the final surface treatment. Deposited metal smoothes the rough surface of the sample. If you don't see any sharp protuberance on the surface in an SIM image after the metal deposition, it means the deposition is sufficient to obtain a uniformly thin sample. A deposition thickness of 0.5–2 μm is usually sufficient.

9.1.4 Thinning

Generally, thinning of the sample is carried out in three steps. First, we use a large beam to mill the sample as quickly as possible (rough milling). When the sample is thinned to a thickness of 2–5 µm, the ion beam current is decreased to reduce the sample damage caused by the FIB milling (medium milling). Finally, a small beam is used to finish the thin sample to minimize ion beam irradiation damage (final milling). When a Ga ion beam with an acceleration voltage of 30 kV is employed, the average ion beam diameters/ion beam currents for rough, medium, and final millings are 60 nm/10 nA, 100 nm/2 nA, and 40 nm/30 pA, respectively. It is usual for samples for TEM observation to be thinned to a thickness of 50 nm to 0.2 µm but the best thickness depends on what you want from the sample. A thickness of less than 0.1 µm is needed for high-resolution TEM image observation and 0.1–0.3 µm for EDX or EELS analysis. Note that the required sample preparation time is proportional to the size of the milling area and the sample damage is also nearly proportional to the total milling time. Don't forget to minimize unnecessary damage introduced to the sample. Also don't observe the cross-sectional SIM image during preparation. If you do, the sample may be damaged and what you observe in the TEM may not be the original one but the damaged one. Ga ion beams can easily change or damage the original structure of the sample.

9.2 Sample damage

Since FIB involves bombarding a sample with energetic Ga ions, the sample gets the ion implantation and the damage caused by the bombardment [6–7]. The Ga implantation may reduce the accuracy of the quantitative elemental analysis and irradiation damage may confuse the structural analysis. Ga concentration in the sample and structural change caused by the irradiation damage depend on the substances to be milled and the ion milling condition.

At the final stage of the FIB milling, the cross section is approximated to be in a steady state, where as many implanted atoms are removed by sputtering as are replenished by implanted Ga. This can be expressed as

$$S_{Ga} = Y_{imp}, \tag{9.1}$$

and

$$\frac{S_{Ga}}{S_M} = R_{Ga-M}\left(\frac{C_{Ga}}{C_M}\right), \tag{9.2}$$

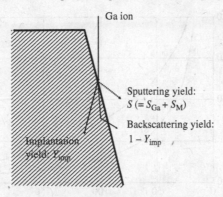

Figure 9.1 Basic parameters of Ga FIB cross-sectioning of a sample M, sputtering yield S $(=S_{Ga}+S_M)$, implantation yield (Y_{imp}), and backscattering yield $(1-Y_{imp})$.

where S_{Ga} and S_M are the partial sputtering yields of components Ga and M, respectively; S is the total sputtering yield $(=S_{Ga}+S_M)$; Y_{imp} is the implantation yield; C_{Ga} and C_M $(=1-C_{Ga})$ are the atomic concentrations of their components near the target surface; and R_{Ga-M} is the enhancement factor in the Ga preferential sputtering for the host matrix M (i.e., the ratio of the probabilities of a Ga atom near the surface to be sputtered to that of an M atom. The R_{Ga-M} value may not be unity because of differences in surface binding energies, sputter escape depths, and energy transfers within the collision cascade.

From (9.1) and (9.2), C_{Ga} is derived as

$$C_{Ga} = \frac{1}{R_{Ga-M}\left(S/Y_{imp} - 1\right) + 1}. \tag{9.3}$$

At $R_{Ga-M}=1$ it simply yields to

$$C_{Ga} = \frac{Y_{imp}}{S}. \tag{9.4}$$

Figure 9.1 shows the basic parameters of Ga FIB cross-sectioning of the sample M, sputtering yield S $(=S_{Ga}+S_M)$, implantation yield (Y_{imp}), and backscattering yield $(1-Y_{imp})$.

In the FIB prepared TEM sample, two trenches are milled, leaving a narrow strip of about 100 nm. The specimen thickness is usually larger than the transverse depth of incident Ga ions, typically about 20 nm for Si and 10 nm for W when a high voltage of 30 kV is applied to accelerate the Ga ion

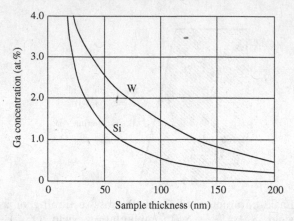

Figure 9.2 Calculated Ga concentration in a 100 nm thick W and Si sample milled at 30 kV as a function of sample thickness.

beam. The average Ga concentration ($C_{t,\text{av}}$) in a sample of thickness t is expressed as

$$C_{t,\text{av}} = \frac{2C_t}{2C_t + N}, \tag{9.5}$$

and

$$\frac{2C_t}{tN} \quad \text{when } C_t \ll \frac{tN}{2}, \tag{9.6}$$

where N is the atomic density of the sample, and C_t is the Ga concentration in the implanted area.

9.2.1 Implantation of Ga ions in the thinned sample

Figure 9.2 shows the calculated Ga concentration as a function of sample thickness in W and Si samples cross-sectioned at 30 kV. In the 100 nm thick W and Si samples, Ga concentrations of 1.5% and 0.5% are expected. Since the implanted Ga ions in the cross-sectioned sample are concentrated near the surface [8–10], the Ga concentration decreases with an increase in sample thickness. In fact, the Ga concentrations in 200 nm thick W and Si samples are 0.5% and 0.2%, respectively. If you want to minimize the relative intensity of the Ga peak in EDX analysis, do not thin your samples too much.

The concentration of Ga ions in the samples was measured experimentally. Figure 9.3(a) shows an STEM image of a 100 nm thick W/Si multi-layer sample and Figures 9.3(b and c) shows the results of EDX analysis of the two layers. The sample was prepared at 30 kV. In the figures, besides the W-M

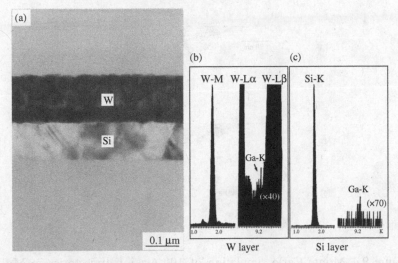

Figure 9.3 STEM image of a 100 nm thick W/Si layer and EDX spectra from the layers. (b) W-M peak and 40 times enlarged Ga-K peak, (c) Si-K peak and 70 times enlarged Ga-K peak.

and Si-K peaks, expanded Ga-K peaks are shown. Ga concentrations in the W and Si layers calculated from the intensities of these peaks were 1.6% and 0.4%, respectively. The measured values are in good agreement with the theoretical value in this case. The depth of Ga implantation is proportional to the energy of the incident Ga ion. It suggests that if you have too high a Ga concentration in the sample, one of the effective remedies against this problem is to use a low energy ion beam for final milling.

Monte Carlo simulation of 100 Ga ion trajectories in Mg for an acceleration energy of 40 kV (a), 30 kV (b), and 10 kV (c) are shown in Figure 9.4. The depth of area with high Ga concentration for the samples milled at 40 kV, 30 kV, and 10 kV are roughly 20 nm, 15 nm, and 10 nm, respectively. A practical example is shown in Figure 9.5. The sample is Mg-Al alloy and final milling was carried out at 40 kV and 10 kV. Mg alloy is one of the most sensitive materials to Ga ion beam irradiation, so the ion beam employed in the milling was relatively small and the milling time was longer. The calculated Ga concentrations for 40 kV and 10 kV are 4% and 1.3%, respectively.

9.2.2 Structural changes caused by the Ga ion beam irradiation

We should not forget that Ga ion beam irradiation can substantially alter the structure [11] and composition of the surfaces of a sample through displacement damage. The structural changes of the surfaces depend on the milling conditions and the nature or composition of the material. Details will

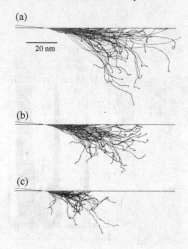

Figure 9.4 Monte Carlo simulation of 100 Ga ion trajectories in Mg at (a) 40 kV, (b) 30 kV, and (c) 10 kV at a glancing angle of 5°.

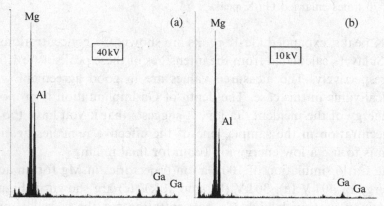

Figure 9.5 EDX analysis of a 100 nm thick TEM sample of Mg-Al alloy prepared by Ga FIB milling at (a) 40 kV and (b) 10 kV.

not be given here but you should be aware that choosing the correct glancing angle is very important to minimize the damage caused to a sample. In general, the lower the glancing angle the less the damage; on the other hand, the lower the glancing angle the lower the milling rate. In practice, the glancing angle is set from both the damage and milling speed points of view. We usually set lower angles of around 2° for lighter elements and around 6° for heavier elements. An example of the damage caused by Ga ion beam irradiation and a result of the remedy is shown in Figure 9.6. The sample is an InP single crystal which was cross-sectioned at the acceleration voltages of 40 kV and 10 kV. Bright dots on the surface of the sample milled at 40 kV

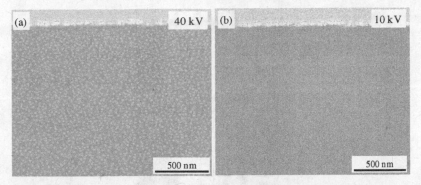

Figure 9.6 SEM images of InP cross-sectioned at (a) 40 kV and (b) 10 kV. Bright dots appearing on the sample cross-sectioned at 40 kV are the re-deposited In particles. No redeposition of In occurred at 10 kV.

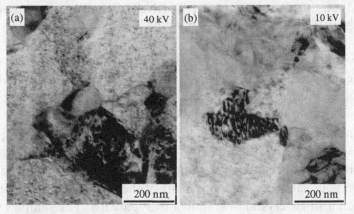

Figure 9.7 TEM images of Mg-Al alloy thinned at (a) 40 kV and (b) 10 kV. The dark dots in (a) are the point defects caused by Ga irradiation. No defects can be seen in (b).

(Figure 9.6(a)) are the particles of redeposited indium. On the sample cross-sectioned at 10 kV, however, no redeposited particles are observed (Figure 9.6(b)). The surface looks clean and smooth. Comparing the differences between these images it can be said that the employment of a low energy ion beam for final milling is quite effective in reducing redeposition.

In the case of metals, a different type of damage happens. A typical example of the damage to a metal sample caused by high energy Ga ion beam irradiation and its countermeasure are shown in Figure 9.7. The sample is Mg-Al alloy thinned at both 40 kV and 10 kV. In the TEM image of the alloy thinned at 40 kV (Figure 9.7(a)), many crystal lattice defects are observed as dark dots. In the TEM image of the sample thinned at 10 kV, however, no

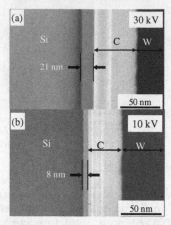

Figure 9.8 Cross-sectional dark-field STEM images of amorphous layers formed during FIB milling. 21 nm and 8 nm thick amorphous layers were formed at (a) 30 kV and (b) 10 kV, respectively. C and W in the photographs indicate carbon and tungsten deposition layers.

such defects are observed (Figure 9.7(b)). From the TEM image quality of the sample thinned at 10 kV, it is again obvious that the sample can be cleanly and smoothly thinned using a lower energy ion beam. These experimental results demonstrate well the importance of choosing a low energy ion beam for final milling in TEM sample preparation.

But these data does not tell us how deep samples will get damaged. When studying the surface region of a sample, it is worth knowing the depth of damage on your sample. A measurement of the thickness of the damage that occurred on a single crystal Si is shown in Figure 9.8. The samples were final milled at 30 kV and 10 kV, and the thickness of the amorphous layers formed as the result of ion beam irradiation damage are 21 nm and 8 nm, respectively. The results of the measurement reveal that the damage depth is almost proportional to the energy of the ion beam employed in the final milling. In case you still have a damage problem, even after reducing the ion beam energy to the practical limit of the FIB system, additional thinning by a conventional method, e.g., Ar ion milling, will work.

9.3 High-resolution TEM image observation of FIB prepared samples

Although a certain rate of damage occurs on the cross-sectioned surfaces of an FIB milled thin sample, high-resolution TEM image observation is possible in most cases if the sample is prepared properly. Figure 9.9 shows a high-resolution TEM image of a semiconductor device. The sample was

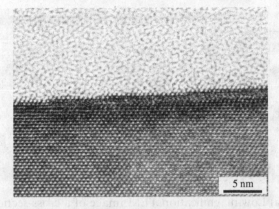

Figure 9.9 Cross-sectional high-resolution TEM image of an Si substrate and silicon oxide layer of a semiconductor device. The sample was thinned at 10 kV to the thickness of 80 nm. The atomic array at the top of the substrate and the fine structure of the amorphous silicon oxide layer can be clearly observed.

Figure 9.10 High-resolution TEM image of a hydroxyapatite crystal of human tooth. Lattice fringes of (100) crystal plane with a separation distance of 0.817 nm can be clearly observed.

thinned to a thickness of 80 nm at 10 kV. Although it has a 16 nm thick amorphous layer on both surfaces, as demonstrated in the previous section, crystal lattice fringes obtained with the electron beam incident along the (110) plane of the silicon and individual Si atoms at the interface between the Si substrate and the silicon oxide above it are clearly observed. This kind of image quality, however, only became possible after the FIB milling technique at low energy, e.g., 10 kV, was routinely adopted for TEM sample preparation. Figure 9.10 shows a high-resolution TEM image of hydroxyapatite, which is one of the most sensitive materials to electron beam irradiation, and

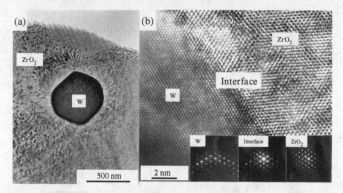

Figure 9.11 (a) Low-magnification TEM image of a cross-sectioned W fiber embedded in ZrO$_2$ and (b) a high-resolution TEM image of the interface with the insertion of electron diffraction patterns from the interface.

therefore considered to be difficult to thin using the FIB technique without causing damage. However, crystal lattice fringes of the (100) plane with a distance of 0.817 nm and other fine structures can be clearly observed.

From the image, it is evident that the quality of the high-resolution TEM image does not suffer from the damage that occurred during the FIB milling.

9.4 FIB capability in TEM sample preparation

Recent developments and improvements of the FIB techniques for TEM sample preparation have enabled us to observe fine structures of many composite materials [12–13]. Figure 9.11 shows a low-magnification cross-sectional TEM image of a tungsten (W) fiber embedded in zirconia (ZrO$_2$) (Figure 9.11(a)) and a high-resolution TEM image with a nano-probe electron diffraction pattern of the interface (Figure 9.11(b)). Since the material is a very hard metal and ceramic composite material, it was impossible to prepare an adequate TEM sample for the study of the interface between the two materials. As is shown here in the sample prepared by the FIB technique, however, we can observe the whole structure of the cross section of the W fiber and the interface structure of the W–ZrO$_2$ interface at an atomic resolution.

The capability of the FIB technique in practice is greater than expected. Figure 9.12 shows a TEM image of a toner particle thinned by the FIB technique. The sample was coated with an FIB assisted W deposition layer prior to the thinning process and thinned by FIB milling at an acceleration voltage of 30 kV. Although the material has a very low melting point and was therefore considered impossible to thin using the FIB technique, the original

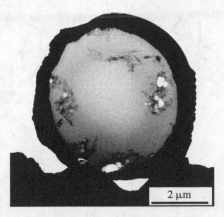

Figure 9.12 Bright field STEM image of a toner particle thinned at 40 kV. The toner is covered with an FIB assisted W deposition layer. The original shape of the toner particle and the inside structures are well preserved.

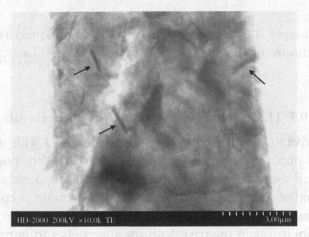

Figure 9.13 STEM image of a paraffin embedded human lung. A small piece of the sample was extracted and thinned at 30 kV.

shape of the particle and its internal structure are well preserved. So far the thin sectioning of this kind of material with a very low melting point has only been possible with a cryo-sectioning technique that requires special training.

Figure 9.13 shows a STEM image of a thin section of paraffin embedded human lung tissue which was prepared for characterization of inhaled mineral fibers. The sample was extracted from the paraffin block of tissue and thinned to a thickness of 5 μm at an acceleration voltage of 30 kV. Mineral fibers in the tissue (indicated by the arrows) can be clearly observed.

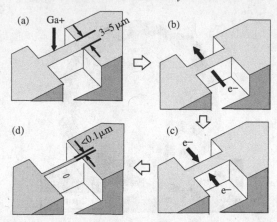

Figure 9.14 Procedure for TEM sample preparation from a specific site. (a) The sample is milled to a thickness of 3–5 µm. Then transferred to the STEM unit for (b) STEM and (c) SEM observation to find the site to be thinned. (d) Finally, transferred to the FIB system to make a <0.1 µm thick sample from the specific site.

From these images it can be said that the FIB technique has made TEM sample preparation of the materials of low-melting point much easier than before.

9.5 TEM sample preparation from a specific site

One of the advantages of the FIB technique compared with other conventional TEM sample preparation techniques is the excellent positional accuracy in setting the area to be thinned. However, because structures are becoming smaller and more complicated, TEM sample preparation from specific sites in state of the art materials such as semiconductor devices is no more easy than it was in the past. Various approaches to improve the positional accuracy are being developed and some have already been put to effective use. One of the most reliable systems used for this purpose is a dedicated FIB-STEM system. The system has a compatible sample holder so that sample preparation in the FIB system and its observation in the STEM are possible without reloading the sample [14].

Figure 9.14 shows the procedure for TEM sample preparation from a specific site using the FIB-STEM system. In this method, the sample is milled to a thickness of 3–5 µm and transferred to the STEM unit to observe STEM and SEM images. The presence of the site in the sample can be confirmed by the STEM observation and the position of the site can be assured by the SEM observation three dimensionally. The sample is then transferred back to

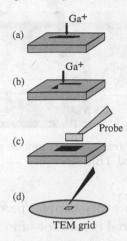

Figure 9.15 General flow chart of a lift-out technique. (a) Trench mill the surrounding area, (b) thin the site to be characterized, (c) lift the thinned sample out, (d) deposit the sample on a carbon coated TEM grid.

the FIB system to make a TEM sample exactly from the site to be characterized. In this method, we use STEM for the observation of interior structures so that we can minimize unwanted sample damage caused by Ga ion beam irradiation. This is an important advantage when the structural analysis of a specific site at atomic level resolution is required.

9.5.1 FIB lift-out technique

The FIB lift-out technique was developed for lifting out a thinned sample directly from a bulky sample [15–16]. The general flow chart of the lift-out technique is shown in Figure 9.15. First, the sample is inserted into the FIB system and the surrounding area of the site to be characterized is trench milled (Figure 9.15(a)) and the site is thinned to electron-transparent thickness (100 nm or less) (Figure 9.15(b)). The sample is then transferred to the stage of a light microscope and the thinned sample is lifted out using a glass rod (Figure 9.15(c)), and deposited onto a carbon coated TEM grid (Figure 9.15(d)). The technique may be applied to most of the substances used in industrial materials. Although the procedure is simple, the samples prepared by the technique have a clean surface and uniform thickness so that even atomic level high-resolution TEM images can be observed. Since the technique does not require any mechanical polishing or dicing prior to the FIB milling, the time required for the whole process is much shorter than that required for conventional FIB techniques.

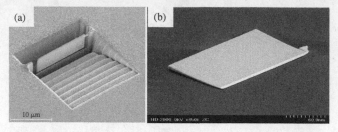

Figure 9.16 Trench milled area and a thinned sample (a) and the sample mounted on a carbon coated TEM grid (b).

This is an advantage when many samples need to be prepared in a short time. However, since a glass rod is used for lifting and mounting the sample, some experience is required to achieve a good success rate, and a better way to mount and align the sample is desired. Figure 9.16(a) shows a trench milled and cross-sectioned area, and the sample mounted on a carbon coated TEM grid is shown in Figure 9.16(b).

9.5.2 FIB micro-sampling technique

This technique has also been developed for TEM sample preparation of a specific site. The general flow chart of the FIB micro-sampling technique is shown in Figure 9.17. First, the area surrounding the site to be characterized is trench milled (Figure 9.17(a)). Second, cross section the site and pick up a sample piece (Figure 9.17(b)). Next the sample is mounted on a carrier (Figure 9.17(c)), and thinned using FIB milling (Figure 9.17(d)). All of the steps above are accomplished under vacuum in the FIB system. Since the sample is manipulated by an elaborate mechanical probe and bonded by FIB assisted deposition, the success rate is very high and therefore it is used as one of the standard TEM sampling techniques in a wide range of materials [17–18].

The technique has the following unique features:

1. No mechanical pre-thinning such as dicing or fracturing is necessary.
2. It is capable of both cross-sectional and plan-view TEM micro-sampling.
3. Plan-view TEM micro-sampling from identical areas in cross-sectional TEM view are possible, and vice versa.
4. It is well suited for TEM observation of magnetic materials because the sample is so small and the influence of the magnetic field from the sample is negligibly small.
5. Good results in quantitative EDX analysis can be expected because the sample is so small and any material may be used as the sample carrier.

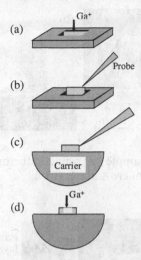

Figure 9.17 General flow chart of an FIB micro-sampling technique. (a) Trench mill the surrounding area, (b) cross-section and pick up using a mechanical probe, (c) mount on a carrier, (d) Thin the site to be characterized.

Figure 9.18(a) shows a micro-sample mounted on a carrier and Figure 9.18(b) shows a sample after thinning. A typical size of the micro-sample is 10–20 μm wide, 10–15 μm high, and 2–5 μm thick. The technique can be applied to a wide range of materials including semiconductor materials, metals, ceramics, polymers, and various kinds of composite materials.

Figure 9.19(a) shows a micro-sample of a coating of a beverage can and Figure 9.19(b) shows a magnified STEM image of a framed area in (a). The layers composed of an anti-bacterial coating, the printed layer and a protective layer of the can are thinned uniformly. Although the sample was extracted upward by a mechanical probe bonded to the surface of the micro-sample, no damage such as micro-cracking occurred at the interface between Al and coating. As is demonstrated here, the adhesive force of the FIB assisted metal deposition layer is strong enough to extract and carry the micro-sample, even if the surface is covered with a nonconductive and plastic polymer.

9.6 Summary and future prospects

Nanotechnology is based on well-designed and manufactured nanomaterials. In the characterization of the nanomaterials, the structural and compositional analyses of nanometer-scaled areas is essential. TEM is one of the

Figure 9.18 FIB micro-sample of semiconductor device mounted on a carrier (a) and a thinned micro-sample (b).

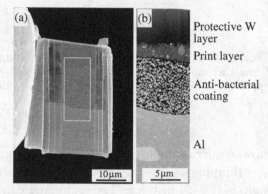

Figure 9.19 (a) SEM image of a micro-sample of a coating of a beverage can and (b) a dark field STEM image of the framed area in (a).

most used instruments in this characterization, and sample preparation for TEM observation is the first and most important thing to do. Among several TEM sample preparation methods in use, the FIB technique is distinguished as the best method to prepare a TEM sample from a specific site. In fact, the success rate in failure analysis of semiconductor devices using TEM improved significantly after the FIB technique was applied. In materials characterization, however, new results bring new requirements, and the requirements of the materials scientist are growing steadily more challenging. Actually, quite a few of the latest requirements coming from high-tech industry concern the capability of the single FIB technique, especially in positional accuracy. The combination of FIB and STEM systems, however, could improve the positional accuracy in TEM sample preparation drastically. In this method, electrons instead of Ga ions, are used for observation of the sample; consequent damage during searching the site or positioning the milling area is much less than that of conventional FIB technique.

In materials science, the importance of TEM investigation is constantly increasing and most investigations require site-specific sample preparation. The future will throw up many challenges in this world, and FIB will continue to be the best solution in the future as well.

References

[1] E. C. G. Kirk, D. A. Williams and H. Ahmed. *Microsc. Semicond. Mater. Sci., Inst. Phys. Conf. Ser.*, **100** (1989) 501–6.

[2] T. Ishitani, H. Tsuboi, T. Yaguchi and H. Koike. *J. Electron Microsc.*, **4** (1994), 322–6.

[3] A. R. Neureuther, C. Y. Liu and C. H. Ting. *J. Vac. Sci. Technol.*, **16** (1979), 1767–71.

[4] Y. Kitano, Y. Fujikawa, T. Kamino, *et al. J. Electron Microsc.*, **44** (1995), 410–13.

[5] Y. Kitano, Y. Fujikawa, T. Kamino, *et al. J. Electron Microsc.*, **44** (1995), 376–83.

[6] T. Ishitani and T. Ohnishi. *J. Vac. Sci. Technol. A*, **9** (1991), 3084–9.

[7] D. W. Susnitzky and K. D. Johnson. *Proc. Microsc. Microanal.*, **4** (1998), 636–7.

[8] J. Melngailis. *Electron Beam, X-ray and Ion-beam Submicrometer Lithographies for Manufacturing, Proc. SPIE*, **1465** (1991), 36–49.

[9] T. Ishitani. *Jpn. J. Appl. Phys.*, **34** (1995), 3303–6.

[10] T. Ishitani, H. Koike, T. Yaguchi and T. Kamino. *J. Vac. Sci. Technol. B*, **16** (1998), 1907–13.

[11] W. Hauffe. *Top. Appl. Phys.*, **64** (1991), 305–38.

[12] H. Saka, T. Kato, M. H. Hong, *et al. Proc. Galvatech.*, (1995).

[13] H. Saka. *The 10th Int. Symp. on Advanced Materials*, (Tsukuba, Japan, 2003), 57–8.

[14] T. Ishitani, Y. Taniguchi, S. Isakozawa, *et al. J. Vac. Sci. Technol. B*, **16** (1998), 2532–7.

[15] F. A. Stevie, R. B. Irwin, T. L. Shofner *et al. Characterization and Metrology: International Conference* (1998), 868–72.

[16] L. A. Giannuzzi, J. L. Drown, S. R. Brown, R. B. Irwin and F. A. Stevie. *Microsc. Res. and Tech.*, **41** (1998), 285–90.

[17] T. Yaguchi, T. Kamino, M. Sasaki, G. Barbezat and R. Urao. *Microsc. Microanal.*, **6** (2000), 218–23.

[18] T. Ohnishi, H. Koike, T. Ishitani, *et al. Proc. 25th Int. Symp. Testing and Failure Analysis*, (1999), 449–53.

10

Micro-machining and mask repair

MARK UTLAUT
University of Portland

10.1 Introduction

Transistors are, by far, the largest number of artificial objects made by human technology. This device has caused and fueled the "Third Wave" of history. As integration and miniaturization of devices continue to progress, auxiliary techniques have been invented to aid in their development and failure analysis. Focused ion beams (FIB) is one of those techniques without which such progress would have been very slow or even impossible. The primary use of FIB is in micro-machining, which is the programmed, controllable removal or addition of material for fabrication, analysis, or repair on a sample at the sub-micrometer scale. As modern small-scale fabrication and repair of structures progresses, FIB has become an essential tool for work at the micro- and nano-scales. An FIB system is capable of being used as a combined milling/deposition machine and as a scanning ion microscope (with different modes of contrast generation), so that as the milling/deposition work proceeds, it can be inspected. In addition, the marriage of an FIB with an SEM (scanning electron microscope) or an AFM (atomic force microscope), allows a natural integration of techniques so that metrology can be performed and the work can be viewed using different imaging modes. The term "milling," which is the sputtering of material, is borrowed in analogy with larger conventional machine tools such as lathes which remove material by cutting. FIBs are the micro- and nano-level lathes and milling machines used in modern small-scale technology [1,2,3]. As the feature sizes of structures become smaller, FIB technology is being pushed to the physical limits of operation. Some of these limits have been addressed in other chapters in this book, and we reiterate some of them as applied to micro-machining.

Focused Ion Beam Systems: Basics and Applications, ed. N. Yao.
Published by Cambridge University Press. © Cambridge University Press 2007.

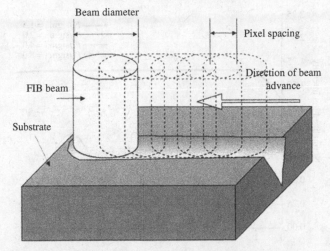

Figure 10.1 The ion beam is moved sequentially from one site to another, with the possibility of overlapping positions. (Courtesy of Ampere Tseng [4].)

10.2 Scanning strategy

The basic structure and operation of an FIB system has been covered in Chapter 2. In a typical FIB system, the beam is sequentially placed at adjacent positions to form an area of quadrilateral or more general shape. The "beam size," typically determined by the beam current and system optical properties, and the "beam overlap," typically an adjustable parameter, both influence the quality of the shape and quality of the milled region (Figure 10.1).

Most FIB systems employ digital scanning where the beam is stepped along pixels to fill out a programmed shape. When imaging, the beam is rastered across the sample and information generated in the sample in the form of secondary electrons or ions are synchronously collected to form an image. When milling or depositing material, the beam is rastered through the pattern shape for a predetermined time. In order to mill a smooth pattern, the ion dose delivered to the sample must be uniform. Since the ion beam has a quasi-Gaussian profile, in order to achieve uniform dose there must be sufficient overlap of the pixel spacing. The effect of beam overlap can be seen in Figure 10.2, where the ion dose uniformity is shown as a function of spatial distance for different amounts of beam overlap. Assuming the beam shape to be a Gaussian distribution, with standard deviation σ, one performs a convolution of the beam shape with the pixel spacing, p_s to calculate the ion dose spatial variation. As can be seen in the figure, a uniform dose is achieved when $p_s/\sigma \sim 1.5$, although for some applications a $p_s/\sigma \sim 2$ might be acceptable. In most FIB machines the beam overlap is an adjustable parameter

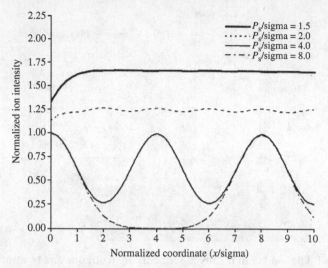

Figure 10.2 Convolution of the beam shape (assumed to be Gaussian) with different pixel spacings (beam overlap). For a fixed dwell time per pixel, the curves represent the spatial variation of the ion dose. For the case $p_s/\sigma = 1.5$, corresponding to a beam overlap of 64%. (Courtesy of Ampere Tseng [4].)

that can be chosen as a "percentage of beam size" overlap. For a Gaussian, the "beam diameter" $d_b = 2.35\sigma$. For $p_s/\sigma = 1.5$ this is equivalent to $p_s/d_b = 0.64$, or 64% overlap. If the pixel spacing does not have sufficient overlap, the resulting milled volume will have a bottom with divots, which for most applications is unacceptable [2,3,4].

10.3 Sputtering

The sputtering process occurs when incident ions onto the sample transfer sufficient momentum to surface and near-surface atoms to allow them to escape through collisional cascades [3,5]. The number of sputtered atoms per incident ion is called the sputter yield, Y. For ions incident perpendicular to the surface, and without the introduction of reactive gases, such as Cl, I, or XeF_2, typically $0 < Y < 5$ for energies of ions normally used ($\sim 35\,\text{keV}$), but for some materials it can be several times higher. If gases are introduced Y can increase by as much as 20×.

Often it is useful to express the sputter yield in terms of the volume of material removed per quantity of incident charge (cubic micrometers, μm^3, per nanocoulomb, nC). Typical values of sputter yields for 25 keV Ga^+ incident ions lie in the range 0.05–$0.7\,\mu m^3/\text{nC}$. The advantage of using these units is the ease afforded in calculating the removal rate (volume removed

Table 10.1 *Sputter yields for selected elements.*

Element	Density (g/cm^3)	Sputter yield $(\mu m^3/nC)$	Sputter yield (atoms/ion)
C (diamond)	3.57	0.18	2.73
Al	2.7	0.30	2.89
Si	2.33	0.27	2.08
Ti	4.5	0.37	3.35
Cr	7.19	0.09	1.20
Zn	7.13	0.34	3.57
Ge	5.32	0.22	1.55
Se	4.81	0.43	2.52
Mo	10.2	0.12	1.32
Ag	10.5	0.42	0.92
Sn	5.76	0.25	1.17
W	19.25	0.12	1.22
Pt	21.47	0.23	2.44

per time) by forming the product of incident beam current and the sputter yield. For example, the sputter yield of Si is $0.27\,\mu m^3/nC$ so that if one were using a beam current of 100 pA, the sputter rate would be $0.027\,\mu m^3/s$. If a 10 µm square hole 3 µm deep were to be milled, then it would take roughly 3 hours to remove the $300\,\mu m^3$. If, in this example, one were to use a beam current of 10 nA, the milling time would be a much more reasonable 1.85 min. Table 10.1 gives some measured sputter rates and yields for common materials of interest [6]. The sputter yield Y is calculated from the sputter rate Y_r by the relationship

$$Y = 96.4\frac{\rho}{m}Y_r,$$

where m is the mass (in amu) and ρ is the density of the sample.

Numerous attempts have been made to calculate sputter yields from basic theory. One model frequently used is that of Sigmund [7]. According to this model, for ion energies E_0, the total sputtering yield (integrated over a solid angle) is given by:

$$Y_{tot}(E_0) = \frac{4.2\times10^{14}\alpha S_n}{U_s} \quad [\text{target atoms/primary ion}],$$

where

$$\alpha = 0.15 + 0.13\frac{M_{target}}{M_{ion}},$$

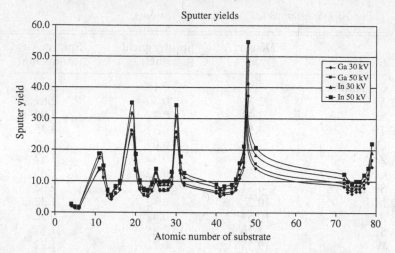

Figure 10.3 From the Sigmund model, calculated sputter yields for 30 keV and 50 keV Ga and In impinging on solid mono-elemental substrates [2].

and

$$S_n = 8.462 s_n(\varepsilon) \frac{M_{ion}}{M_{ion} + M_{target}} \frac{Z_{ion}Z_{target}}{(Z_{ion}^{2/3} + Z_{target}^{2/3})^{1/2}},$$

and U_s is the surface potential. The "reduced energy" ε and the "reduced nuclear cross section" s_n are given by

$$\varepsilon = \frac{aM_{sample}E}{Z_{ion}Z_{sample}e^2(M_{ion} + M_{sample})} \text{ [E in ergs]},$$

and

$$S_n(\varepsilon) = \frac{0.5\ln(1+\varepsilon)}{\varepsilon + 0.14\varepsilon^{0.42}},$$

where a is the Thomas–Fermi radius (in angstroms), and $a = 0.468$ $(Z_{ion}^{2/3} + Z_{sample}^{2/3})^{-1/2}$. Extensive tables [2] can be found for these quantities, and an illustration of the dependence of sputtering on the incident energy and sample composition is shown in Figure 10.3.

The dependence of sputter yield on incident energy is shown in one example in Figure 10.4, where it can be seen that there is an increase in yield with increasing energy that levels out somewhere near 40 keV [3,4]. This is due to the deeper penetration depth of the ions, where the collisional cascades are generated too deep to transfer sufficient energy to atoms to escape from the surface.

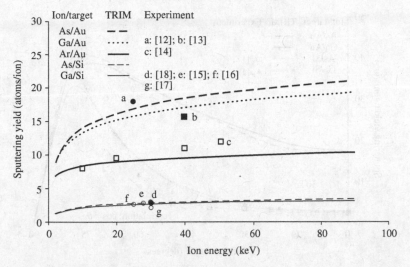

Figure 10.4 Theoretical curves and experimental data points of sputtering yield dependence on the energy of the incident ion for several ion/target combinations. (Courtesy of Ampere Tseng [4].)

The sputtering yield is also dependent upon the incident angle of the ion beam to the surface. In the Sigmund model the angular sputter yield dependence is

$$Y(\Theta) = Y(0)\cos^{-f}(\Theta),$$

where f is a constant depending on the mass ratio M_{sample}/M_{ion}. Simulations of this effect and experimental data are shown in Figure 10.5. This dependency can be used to advantage when milling shapes near edges [2].

10.4 Redeposition

One deleterious complication to milling is the redeposition of sputtered material into the volume that is being milled [3,4]. In conventional large-scale machining, the swarf is carried away by liquid or air streams. In FIB milling, the redeposition is very dependent on the milling strategy used. It has been shown that for the same total dose incident onto a sample, that repeated scans to mill a rectangular area cause less redeposition than a single slow pass. In the case of a single slow scan, redeposited material is not removed, while for many repeated scans some fraction of the redeposition is milled away by the primary beam. Figure 10.6 shows this effect.

These results can be understood partially by a simple geometrical analysis. Consider milling a groove in one dimension of width d and height h, as

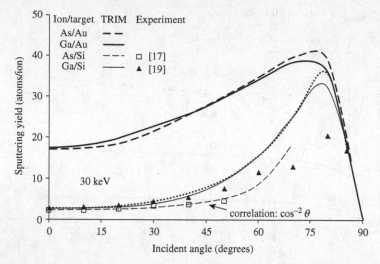

Figure 10.5 Angular dependence of sputtering yield of As and Ga ions onto Si and Au. (Courtesy of Ampere Tseng [4].)

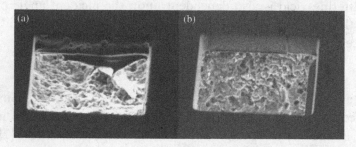

Figure 10.6 SEM photos of $5\,\mu m \times 5\,\mu m$ milled boxes in Si by $30\,keV\ Ga^{+}$ for the same total dose in each case. Box (a) was milled with a single pass of the ion beam, box (b) was milled with 50 passes. Both milled boxes are viewed at $45°$.

shown in cross section in Figure 10.7. Assuming that the sputtered efflux has a cosine distribution, and falls off as $1/r$, and that the sticking coefficient is unity for the F_0 atoms/length emitted, then it can be shown that the flux density of sputtered material from the bottom redeposited onto a sidewall at a height h is

$$F(h) = \frac{F_0}{2} \int \frac{\cos\varphi\cos\theta}{r}\,\mathrm{d}x = \frac{F_0}{2}h\int_{0}^{d} \frac{x\,\mathrm{d}x}{r^3},$$

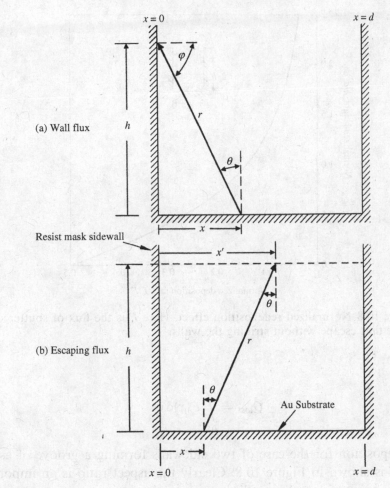

Figure 10.7 Geometry used for calculation of redeposition effects.

and the flux density of atoms that escape is

$$F(x') = \frac{F_0}{2} \int \frac{\cos^2\theta}{r} \, dx = \frac{F_0 h^2}{2} \int_0^d \frac{dx}{r^3},$$

where x' is the distance from the sidewall. The total flux onto a sidewall and the escaping flux are given by

$$Q_{wall} = \int_0^h F(h) \, dh,$$

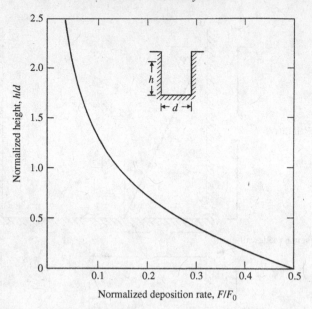

Figure 10.8 Normalized redeposition effect. Here F is the flux of sputtered atoms that escape without striking the walls.

and

$$Q_{lost} = \int\limits_0^h F(x)\mathrm{d}x.$$

The redeposition for the case of two sidewalls forming a groove of aspect ratio h/d is shown in Figure 10.8. Clearly the aspect ratio is an important parameter to consider when milling a volume.

10.5 Channeling

Another effect that can cause complications when micro-machining is due to channeling [3,5]. As discussed in Chapter 3, channeling occurs in crystalline materials where ions can penetrate deep into low index directions, interacting so deep that sputtering cannot happen. In polycrystalline materials, this means that low index crystals will not sputter as rapidly as others, resulting in differential milling rates in mono-elemental samples. Figure 10.9 shows the result of milling into polycrystalline Cu with $30\,\mathrm{keV}$ Ga^+.

The uneven bottom in the milled volume is due to differential sputtering rates of the microcrystals, and not redeposition.

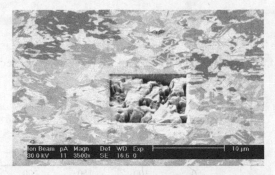

Figure 10.9 A 10 µm square milled in Cu. The uneven bottom is due to the differential milling rate of the different microcrystals.

Figure 10.10 The same region of Cu milled in Figure 10.9, but with the introduction of a gas, which uniformly clears the milled area by completely disturbing the effects of channeling.

There are remedies for reducing or nearly completely eliminating channeling. One method is to introduce a gas near the sample surface. The channeling axes are very sensitive to incident ion direction, and the introduction of sufficient scattering centers on the surface can deflect incident ions enough to eliminate channeling, thus reducing differential milling. This may happen at the expense of reduced effective milling rate, but can yield good results (Figure 10.10).

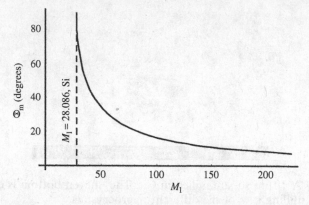

Figure 10.11 Plot of $\Theta_{\max}$ for ions heavier than Si impinging onto Si. There is less lateral spread for heavier ions, resulting in less damage. The same basic trends occur for any substrate material, where ions heavier than the substrate will form similar curves.

10.6 Damage considerations

The inherent structure of the sample is altered as incident ions, after losing their energy, come to rest in the sample after colliding with several target atoms. Damage to the structure appears as atoms are dislocated, and the incident ions are implanted. A crystalline structure can become amorphous. Referring to Figures 3.3 and 3.4 in Chapter 3, the range and straggle of incident ions depends not only on their initial energy and angle of incidence, but also on the type of incident ion. Ions that lose their energy in a shorter distance will produce less damage at greater depths. From an analysis [8] of the dynamics of scattering, it can be shown that there is a maximum scattering angle of the incident ion, $\Theta_{\max}$, given by

$$\Theta_{\max} = \cos^{-1}\sqrt{1 - \frac{M_{\text{sample}}^2}{M_{\text{ion}}^2}}, \quad 0 \le \Theta_{\max} \le \pi/2,$$

which, for example, for incident ion masses greater than 28 impinging into Si, is shown in Figure 10.11. This analysis leads to the conclusion that, in order to minimize the lateral spread of collisions, heavy ions should be used. Monte Carlo simulations agree with this conclusion, as shown in the data of Table 10.2 [8].

Verification of these theoretical suggestions were carried out by Jamison [8]. By preparing TEM samples with 30 keV Ga^+ and In^+, the amorphous layer produced atop crystalline Si could be measured directly in the TEM. The results of this study are shown in Figure 10.12.

Table 10.2 *Simulated damage depths layers in Si for various energies of two different incident ions, Ga$^+$ and In$^+$ [8].*

Ion energy (keV)	Ga damage layer (A)	In damage layer (A)	Reduction (%)
5	78	57	27
10	126	88	30
20	199	133	33
30	262	171	35
40	307	195	36
50	364	240	34

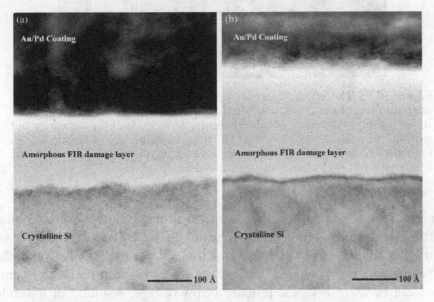

Figure 10.12 TEM micrographs from a study by Jamison [8] comparing the damage layer thicknesses of different impinging ions. In (a) 30 keV In$^+$ impinged on Si, while in (b) 30 keV Ga$^+$ was used. These results agree with the simulations in Figure 10.2, clearly showing the thickness of the damaged crystalline structure is greater for Ga than In. The Au/Pd coating was added as a preservation layer in making the TEM samples.

10.7 Charging of the sample

Samples are not perfect conductors, and many samples have insulating material on them which can accumulate excess charge on their surface as the primary ion beam is scanned across them. This excess charge buildup is generally positive, due to the positive ionic contribution from the primary

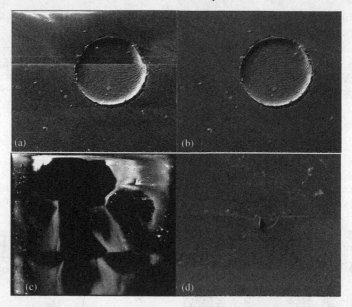

Figure 10.13 The effects of charging on two different samples and the neutralization of the effect. The top images are of an insulating membrane. Image (a) shows the effect of mild charging, while (b) is of the same area but with charge neutralization present. The bottom images are of plastic. Image (c) shows considerable charging, while (d) was collected under the same conditions, but with charge neutralization used.

beam, and the release of negative charge from the sample by the secondary electrons generated by the primary beam. In general, the primary beam contributes about a third to a half of the total charge buildup. As the surface of the sample acquires charge (positive), a voltage is built up on the sample which can be as high as the primary beam voltage, resulting in unwanted deflections of the primary beam and secondary electrons. In the worst case, the primary ions normally impinging onto the sample can be entirely deflected from the sample. Even in cases where the charging is not so severe, the primary beam is deflected from its intended landing point, resulting in a failure to cause the desired effects to happen where intended. Charging manifests itself as complete or partial loss of imaging capability (the image can completely disappear due to the secondary electrons being forced back to the sample) or in the image jumping or showing regions of flaring brightness [3]. In the case of milling or deposition of material, charging makes that process happen at the wrong position (primary beam deflection) or completely fails due to stopping the beam. Figure 10.13 shows examples of slight and considerable charging on samples. In the first case (Figure 10.13(a)), the charging effect shows up as the bright area at the top left of the image, and the horizontal line a third of the way down the image is where the primary beam "glitched" due to charge buildup. The lower micrograph (Figure 10.13(c)) shows how catastrophic the effect can be. There is complete loss of control of where the primary beam impinges the sample, and the dark areas are where the image information being collected from the secondary electrons is extinguished due to the electrons being attracted back to the sample instead of being collected by the imaging detector. In Figures 10.13(b) and (d) the charge neutralization system has been employed to reduce the charging effect to satisfactory levels.

There are several techniques to minimize or completely eliminate sample charging. In some situations the sample can be pre-coated with a thin conductive layer (C, Pt, or Au for example) giving the unwanted charge a path to ground. While this can be used for some laboratory investigations, and some failure analysis situations, it is not in general possible to radically so alter a sample. In those cases where such a coating can be used, it is typically applied with a thin film evaporator to a thickness of 0.1 nm, which if necessary can be removed by RIE.

Another technique that works in many situations is to image a charging sample by collecting secondary ions instead of secondary electrons. By placing a negative bias on the front of the detector, the secondary electrons from the sample are returned to the sample surface, while positive secondary ions are collected in the detector. Unfortunately, the secondary ion yields are several orders of magnitude less than secondary electron yields for most samples, so that reduced image quality may become a serious issue.

An adjunct technique used with secondary ion imaging, to reduce charging effects, is to flood the surface with low energy electrons from an electron source placed near the sample surface. By using low energy (10–100 eV) electrons to flood the sample surface, sites that are positively charged on the sample attract the electrons until they are neutralized. The disadvantage of this method is that secondary ion imaging must be used with the concomitant loss of information for forming an image. One technique that can overcome this situation is to multiplex the imaging detector and the electron gun providing the neutralizing electrons. Typically, a control grid on the electron gun is biased so as to provide electrons to the sample surface, while the detector is biased not to collect electrons. The electron gun is then biased to eliminate the electron current, the detector is enabled to collect electrons, the primary ion beam is unblanked before scanning on the sample, and a line of image data is collected. The primary ion beam is then blanked, and the process is repeated. In a typical situation requiring charge neutralization, this technique only adds about 1% to the total image collection time.

In some situations a useful charge neutralization technique is to introduce water vapor onto the sample surface simultaneously while imaging or milling. At the correct flux of water vapor, the excess surface charge can be bled from the area undergoing modification or imaging. A user must introduce water vapor with care, as some samples which contain hydrocarbons will have their milling rate substantially increased selectively for the hydrocarbon material.

10.8 Introduction of gases for deposition and milling enhancement

The use of auxiliary gases introduced into the sample chamber are covered extensively in Chapter 4 of this volume. We point out here that the introduction of gases can substantially and selectively increase the milling rate of the sample [1,2,3]. This effectively reduces the work time and can be used in many situations to differentially mill regions with a variety of materials present. As pointed out in the previous section, the use of a gas can reduce or eliminate the effects of charging.

One example of gas assisted milling is shown in Figure 10.14, where this is shown to emphasize the ability of the right gas to perform milling selectively. Gases have been found that can selectively mill insulators rather than metals (as shown in the figure) or that can selectively mill metals rather than insulators. One might argue that the majority of future advancements in the use of FIB will be centered around the development of ion beam–gas chemistry interactions.

10.9 Milling artefacts

Due to the physics of the interaction of the milling beam with the sample, there are artefacts that are caused by differential milling rates and geometric effects. These artefacts initially were seen when making early attempts at preparing sample cross sections, but can affect the work done while performing a variety of other applications.

"Waterfalls" (Figure 10.15) are the vertical stripes in the surface of a face that has been exposed by milling adjacent material away. This effect arises from surface topography and large differences in sputter rate for the materials present. This effect can be highly accentuated if gases are introduced which are selective between the materials that are present.

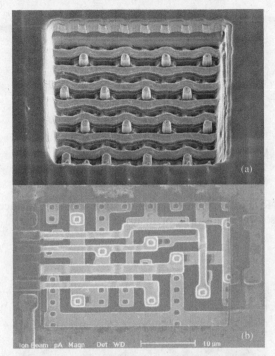

Figure 10.14 Two images showing the great selectivity of gas assisted etching. Both regions were milled with Ga^+ while XeF_2 was introduced. Only the metal structures remain after the milling.

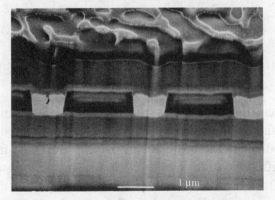

Figure 10.15 "Waterfalls" (the vertical stripes in the face) are caused by topographic and differential sputter rates.

"Ridges" (Figure 10.16) are the horizontal stripes that are caused by either the primary beam drifting during a final "polishing" step or insufficient milling time to clean the face. When large beam currents are used to reduce the necessary time to remove material to expose a face, the beam "tails" do not

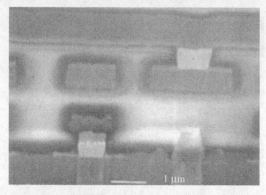

Figure 10.16 "Ridges" are the horizontal stripes on the surface of the face, which are caused by beam drift or insufficient milling time.

Figure 10.17 The small black droplets in this TEM sample are probably Ga.

allow a sharply cut face to emerge from the mill. If it essential to view the structure of the face, a low current (the beam "tails" are reduced) polishing step must be used where the beam is run parallel along the face to effect the polish. Beam drift, which removes the beam from the face, and insufficient milling time are the culprits.

Small drops (Figure 10.17) sometimes form on the face of a cross section, and are almost certainly Ga which has congregated via diffusion. The material in the primary beam (Ga) has to end up somewhere, and in some materials into which it has been implanted, the diffusion rate is sufficiently large to allow the Ga to "pool" into droplets. Sometimes they can be sputtered away, and in some cases they can be removed chemically (via a dilute solution of HCl and alcohol to promote wetting).

10.10 Beam placement accuracy – metrology and milling

Often overlooked when considering the use of machining tools at the micro- and nano-level is the ability to accurately place the FIB beam onto the sample where the user wants it. A pristine ion beam placed incorrectly, in many cases, is more unwanted than a beam of moderate profile accurately placed. Milling in the wrong place can be a waste of time in a good case and a total disaster in the worst case. Potential data can be irretrievably lost or a microstructure can be destroyed. There are several contributions to the beam placement error that must be understood and minimized in order to achieve the necessary performance. These effects can be divided into two major categories, one of which involves the engineering of the FIB system, while the other involves the limits imposed by nature. Some potential contributions to beam placement accuracy are as follows:

1. Engineering considerations that can be minimized:
   - scan magnification mis-calibration;
   - scan noise (AC electronic noise);
   - scan–sample stage mis-alignment;
   - sample stage mechanical jitter;
   - ion optics instability – ion source instability either electrical or mechanical;
2. Issues involving nature with hard limits:
   - low signal/noise images;
   - sample charging;
   - sample damage.

The uncorrelated contributions would have their effects added in quadrature. The effects which involve mis-calibration, electronic noise, and mechanical vibration, can in principle be lowered to levels acceptable for work at the nanometer scale. Because the ion beam is charged and sputters material from the sample onto which it is scanned, the last three contributions are physical limitations that nature places on the operation of the system. The optimization of image acquisition for the purpose of metrology causes an increase in sample charging and damage. Likewise, the optimization of reduced sample charging and damage, yields images with insufficient signal/noise to be able to perform the necessary metrology. In the case of the repair of photo-lithographic masks, the requirements for beam placement accuracy at present are to be able to place the beam to within 15–25 nm of the desired landing point depending on the mask type. With good engineering practices, this is achievable, but due to the damage imparted to the sample by the ion beam, requires the marriage of FIB with SEM and AFM technologies.

10.11 Micro-machining applied to the repair of masks

Most modern microelectronics is produced using optical lithography, where light is used to expose photosensitive resist coatings. Almost all the steps in these complex production processes are preceeded by a lithography step where, for example, contact metalization or implantation doping is defined. This patterning is achieved by the use of photomasks, which selectively block and transmit light to the resist, or through the introduction of an attenuated phase shift or a strong phase shift, thus transferring the pattern contained in the mask to the resist. As feature sizes in microelectronics have shrunk, several different kinds of masks have been developed. In the past, the mask structure was chrome-on-glass (COG), also known as a binary mask, where the chrome was used as the blocking medium. These are generally produced by electron beam lithography, followed by chrome etch, resist removal, and subsequent inspection for defects. There has been considerable work done in developing phase shift masks to extend the resolution, contrast, and depth of focus of optical lithographic tools beyond what is achievable with binary chrome masks. In addition, techniques to bring about the desired patterns have been developed called optical proximity correction.

Once the mask has been made, defects must be located and cataloged. Since a single printable defect on a mask can bring the yield for a semi-conductor device to zero, the defects must be removed, and fall into two broad classes: clear defects and opaque defects. Clear defects are where absorber material has not been placed properly, leaving holes through which the light can pass, and the repair consists of depositing opaque material to fill in the hole. Opaque defects are where excess absorber material blocks the transmission of light, and the repair consists of removing the excess material. There is a third kind of defect in alternative phase shift masks where the substrate material has the wrong thickness to cause the correct phase shift. Because of their ability both to mill away material and to deposit new material at the sub-micrometer level, FIBs are natural candidates for use in the repair of defective masks. As feature sizes have shrunk, new variations of masks are under development, which include phase shifting masks and the use of optical proximity correction techniques. While FIB is a candidate to repair the various types of masks, fundamental limitations of FIB technology may force FIBs to be used in conjunction with other techniques such as electron

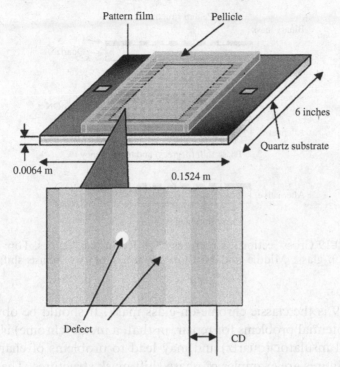

Figure 10.18 Generic mask structure showing a typical size and component parts, as well as two types of defects.

beams (EB) and atomic force microscopy (AFM). The integration of an FIB and SEM has been covered in detail in Chapter 5. All of the facets of FIB technology capability and limitation potentially converge in the repair of masks.

10.12 Mask structure

The basic structure of a photomask is shown in Figures 10.18 and 10.19. The generic opaque and clear defects are shown, as well as cross sections of generic types of masks [9,10]. The substrate of the mask, called the reticle, contains the pattern to be transferred to the wafer, and is in general quartz or some variation of quartz. The pellicle is a thin membrane or plate that is placed between the mask and the projection optics in order to protect the mask from contamination. Because pellicles can be made extremely thin and uniform, when inserted into the optical path, they provide necessary protection, but do not introduce image degradation. The top cross section in

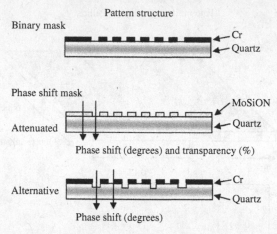

Figure 10.19 Cross-sectional structures of different mask types. Top: Binary chrome-on-glass. Middle and bottom: examples of two phase shift mask designs.

Figure 10.19 is the classic chrome-on-glass mask. It should be obvious that there are potential problems for repair, in that, a metal (chrome) is on top of a very good insulator (quartz) and may lead to problems of charging. The lower two figures are examples of phase shift mask structures. The MoSiON (molybdenium silicide) is the phase shifting layer, which allows approximately 10% transmission through the film while imparting a 180° phase shift. An alternative phase shift mask structure is accomplished by removing precise amounts of the substrate material to achieve the desired phase shift.

Once a mask is made it must be inspected for defects, the position and type of defect must be cataloged and the data made available to the machine that will effect repair. Since the price of mask-making is so high, development costs have to be minimized, so that the number of development cycles should be reduced and shortened in length. Efforts to this aim can be realized if the repair system has sufficient "smarts" to recognize, prescribe, and effect repairs. This requires not only a robust optical system with sufficient resolution in metrology to allow milling and deposition placement accuracy and quality, but must also possess sophisticated image recognition and pattern generation and overlay capabilities. Thus, the repair machine must be highly sophisticated in both hardware and software.

The system has to be calibrated and the mask must be aligned so that navigation to the defect sites can be carried out. Once a site is within the field of view the final alignment must be made to place the beam properly to perform the repair. One of the serious problems encountered during this process is charging on the sample. With so much insulating material present,

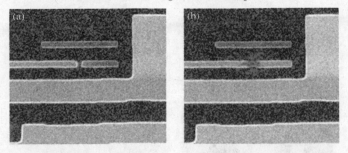

Figure 10.20 Images of a clear defect that was repaired by depositing a carbon patch. The broken line in (a) was patched together as seen in (b). (Courtesy of FEI Co.)

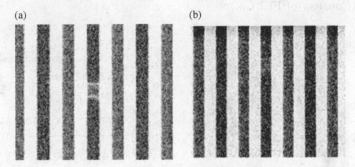

Figure 10.21 Images of an opaque defect bridging two lines (a) that was repaired by milling it away with the Ga ion beam (b). (Courtesy of FEI Co.)

in lieu of charge neutralization controlled navigation, milling, and deposition would be impossible. In general, a flood electron gun provides sufficient negative charge to cause neutralization.

Examples of repaired defects in a mask are shown in Figures 10.20 and 10.21. In the case of clear defects, it is fairly easy to deposit carbon thick enough to fix the defect. A variety of proprietary materials are available. The resultant opacity is sufficient to make the mask useable As the repair systems become more sophisticated, there must exist graphical user interfaces capable of allowing the operator to use software that will allow a copy of a known good area of the mask onto a defect site, so that multiple repairs at different sites can be performed consecutively easily. This requires the system to have the ability to generate an image of the good site with sufficient signal/noise and then align it and overlay it onto the defect site, and if necessary modify the image appropiately if needed, and proceed with the repair. The system should also be able to determine what geometries are defective, and whether

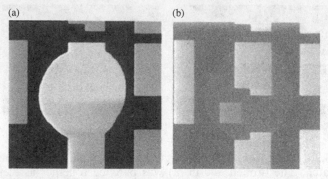

Figure 10.22 An example of a defect corrected by overlaying a good site image onto the defect and automatically determining how to fix the defect. Image (a) shows the original defect, while image (b) is the same area after the repair. (Courtesy of FEI Co.)

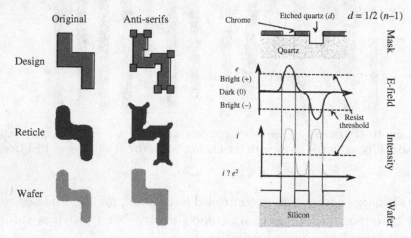

Figure 10.23 An example of the intent and nature of optical proximity correction (OPC). The purpose is to suitably modify the design so that it will be replicated into the wafer faithfully by correcting for both optical and process distortions.

the defect is clear or opaque. An example of such an operation is shown in Figure 10.22, where a defect (which was buried) was automatically corrected by overlaying an image of a good site onto the defect site.

On masks that employ optical proximity correction (OPC) [9–11], where there are complex and intricate shapes, this technique is especially useful as shown in Figures 10.23 and 10.24. OPC applies systematic changes to photomask geometries to compensate for nonlinear distortions caused by optical diffraction resist process effects. These distortions include line-width

Figure 10.24 (a) An FIB image of a defective OPC pattern (0.2 µm defect). (b) The same area after the repair. (Courtesy of FEI Co.)

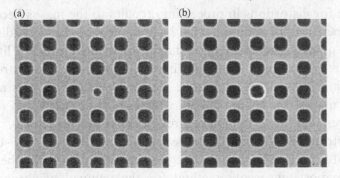

Figure 10.25 Repair of an opaque defect in a MoSiON phase shift mask. In (a) there is an obscured hole that must be opened up. This flaw is shown corrected in (b). (Courtesy of FEI Co.)

variations dependent on pattern density, and on line-end shortening which can break connections to contacts. Other distortions which can be reduced are diffusion and loading effects during resist and etch processing of the mask. OPC makes changes to the IC layout that anticipate the distortions. To compensate for line-end shortening for example, the line is extended using a hammerhead shape that results in a line in the resist that is much closer to the original intended layout. To compensate for corner rounding, serif shapes are added or subtracted from corners to produce corners in the silicon closer to the ideal layout As masks are designed to attain the highest resolution possible from the steppers which use them, one of the resolution enhancement techniques is the embedded phase shifter mask, of which the most common type is the MoSiON mask. A demonstration of the ability to mill selectively the mask absorber instead of the substrate is shown in Figure 10.25, where a contact hole in a phase shift MoSiON mask is obscured. The gas used in this process shown in Figure 10.25 is proprietary, and demonstrates the kind of high selectivity necessary in order to minimize the effect of the incident ion beam on the clear parts of the mask. One of the concerns in using Ga ion beams to remove opaque mask material is that implantation of Ga into the quartz substrate can cause staining in the clear areas which modifies the optical properties of the quartz. There are known post processing techniques to remove the staining, but they add time and complexity to the repair.

10.13 Future FIB mask repair machines

As feature sizes become smaller and masks become more complex in order to compensate for distortions in processing, repairs to the masks become more difficult to perform. In addition, what have in the past been two-dimensional repairs are quickly becoming three-dimensional in nature. As lithography heads below the printing of 0.13 μm lines to the 65 nm node, strong phase shifting techniques will be imperative. One such technique uses alternating phase shift masks (see Figure 10.19), which are very costly and difficult to manufacture, and have complex three-dimensional structures. Since the quartz substrate has varying thickness, repairs of the bumps in the quartz become problematic. Clearly a process not involving Ga would be welcome. However, a most efficaceous method for the removal of chrome at high resolution is found in the use of the Ga beam. To resolve this, a suitable combination of three techniques (FIB, SEM, and AFM) appears to be a good solution. The dual-beam machines discussed in Chapter 5 offer the possibility of suitable mask repair solutions into the future. The SEM is capable of high resolution imaging without implanting baryons into the sample, and under suitable conditions can compensate for sample surface charging and also can initiate deposition of materials if gases are introduced. Figure 10.26 demonstrates the capability of an "environmental" SEM to form pristine images in cases where normal SEMs exhibit serious charging. Thus the SEM can be an adjunct partner to the repair process, both as a navigation tool and a means of performing gas assisted operations.

In addition to the association of SEM technology with FIB technology, there also is another nanoscale metrology tool available for integration: the AFM. The AFM is capable of profiling a surface at the nanometer scale, so that the three-dimensional profiles of defects can be accurately mapped and measured, and the data used to arrive at an optimal repair strategy. The AFM can also be used to scrape away and remove material, either substrate or absorber. Figure 10.27 shows the possibility of such technology.

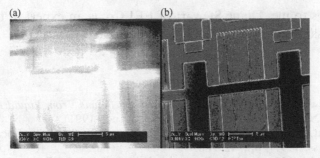

Figure 10.26 Two images of the same area on a photomask. (a) A normal high vacuum SEM image exhibiting serious charging of the sample, rendering the image useless. (b) An ESEM (environmental SEM) image of the same sample taken at elevated pressures, showing that it is possible to completely remove the sample charging problem. (Courtesy of FEI Co.)

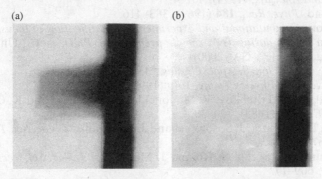

Figure 10.27 The removal of material with an AFM. The defect in (a) has been removed by the AFM, as shown in (b). (Courtesy of FEI Co.)

10.14 Conclusion

FIB technology, by utilizing the interactions of ions with matter, is capable of performing micro-machining and deposition of material at the micro- and nano-scales. FIB has been invaluable in the development and production of microelectronics through failure analysis and circuit modification, but also in repairing the masks that generate the devices in Si. The use of FIB will continue, and mask repair is a potential future area of contribution for FIB, but fundamental physical limitations may require that it be integrated with SEM and AFM technology to be able to be useful in the future.

Acknowledgements

It is always a pleasure to thank those who share their knowledge and skills freely. Too many people to list here have taken their time to show and explain things to me. Diane Stewart of FEI Co. is an expert in the repair of masks, and she gave me invaluable assistance here.

References

[1] P. D. Prewet and G. L. R. Mair. *Focused Ion Beams From Liquid Metal Ion Sources* (New York: John Wiley and Sons, Inc., 1991).

[2] M. Utlaut. *Handbook of Charged Particle Optics*, ed. J. Orloff (Boca Raton, FL: CRC Press, 1997), pp. 429–88.

[3] J. Orloff, M. Utlaut and L. Swanson. *High Resolution Focused Ion Beams: FIB and its Applications* (New York: Kluwer Academic/Plenum Publishers, 2003).

[4] A. A. Tseng. *J. Micromech. Microeng.*, **14** (2004), R15–R34.

[5] A. Benninghoven, F. G. Rudenauer and H. W. Werner. *Secondary Ion Mass Spectrometry* (New York: John Wiley and Sons, 1987).

[6] A. Leslie. *5th European Symposium on Reliability of Electron Devices, Failure Analysis and Analysis*, (1994), 401–4.

[7] P. Sigmund. *Phys. Rev.*, **184** (1969), 383–416.

[8] R. Jamison. *Computational and experimental quantification of focused ion beam damage in silicon during TEM sample preparation, Ph.D. thesis*, University of California, Berkeley, CA (2000).

[9] M. D. Levenson. *Jpn. J. Appl. Physics.*, **33** (1994), 6765–73.

[10] M. D. Levenson. *Physics Today*, July (1993), 28–36.

[11] J. Randall, A. V. Tritchkov, R. M. Jonckheere, P. Jaener and K. G. Ronse. *Proc. SPIE*, **3334** (1988), 124–30.

[12] X. Xu, A. D. D. Ratta, J. Sosonkina and J. Melngailis. *J. Vac. Sci. Technol. B*, **10** (1992), 2670–80.

[13] P. G. Blauner, Y. Butt, J. S. Ro and J. Melngailis. *J. Vac. Sci. Technol. B*, **77** (1989), 609–17.

[14] O. Almen and G. Bruce. *Nucl. Instrum. Methods*, **11** (1961), 257–78.

[15] L. Frey, C. Lehrer and H. Ryssel. *Appl. Phys. A*, **76** (2003), 1017–23.

[16] J. G. Pellerin, G. M. Shedd, D. P. Griffs and P. E. Russell. *J. Vac. Sci. Technol. B*, **7** (1989), 1810–12.

[17] D. Santamore, K. Edinger, J. Orloff and J. Melngailis. *J. Vac. Sci. Technol. B*, **15** (1997), 2346–9.

[18] H. Yamaguchi. *J. Physique Coll.* C6 **48** (Suppl. 11), C6165–C6170.

[19] C. Lehrer, L. Frey, S. Peterson and H. Ryssel. *J. Vac. Sci. Technol. B*, **19** (2001), 2533–8.

11

Three-dimensional visualization of nanostructured materials using focused ion beam tomography

DERREN DUNN

IBM Microelectronics

ALAN J. KUBIS AND ROBERT HULL

University of Virginia

11.1 Introduction

One of the fundamental goals of materials science is to establish structure property relationships. Historically, structure property relationships in materials have been established with great success using imaging and spectroscopic methods that measure signals that vary in one and two dimensions. For example, X-ray and neutron diffraction techniques are typically used to measure average physical and crystallographic properties of relatively large volumes of material. While different modes exist for these techniques such as spot modes and surface reflectance modes, they effectively average over volumes of materials properties and produce information that varies in at most two-dimensions. Auger electron spectroscopy (AES) and secondary ion mass spectroscopy (SIMS), while capable of high spatial resolution in surface mapping modes and depth profiling modes, are traditionally used to produce maps and profiles in one and two dimensions.

High-resolution imaging techniques, such as scanning and transmission electron microscopy, generally produce two-dimensional real space intensity maps of surfaces or an image that has been averaged through a sample thickness, as is the case in conventional transmission electron microscopy.

Scanning probe microscopy techniques, such as atomic force microscopy and scanning tunneling microscopy, may be regarded as providing a measure of surface variation in three dimensions. In particular, both techniques are capable of measuring in-plane variations in position with near atomic scale

Focused Ion Beam Systems: Basics and Applications, ed. N. Yao.
Published by Cambridge University Press. © Cambridge University Press 2007.

lateral resolution as well as sub-atomic scale measurements parallel to a local surface normal. These scanning probe techniques are limited to surface measurements, however, and in general cannot directly measure sub-surface information.

In an effort to investigate structural relationships in three dimensions, several two-dimensional characterization techniques have been adapted to incorporate tomographic reconstruction principles to reconstruct object volumes and chemical maps in three dimensions. There are primarily two methods employed to generate three-dimensional reconstructions of materials, methods that reconstruct three-dimensional objects from projections and direct sectioning methods. Projection methods make use of images and chemical maps collected from an object at many different angles. These images and maps are then transformed using Radon transforms and back-plane projections to reconstruct volumes and form a three-dimensional representation of an object. One of the most common probes for tomographic reconstructions that use projection methods are X-rays. Copley *et al.* [1] give an overview of this technique and a comparison of various instrumental variables. X-ray tomography records X-ray attenuation as a function of angle to produce a two-dimensional image. A series of these images can be obtained at various positions and then a three-dimensional volume is reconstructed using projection methods.

A new X-ray tomography technique, cone-beam X-ray tomography, is currently being investigated for medical applications [2]. The advantage of this technique is that an array detector is implemented which has a higher effective photon collection efficiency so that there is no need for interpolation. There is an added complexity to this technique in that complex deconvolution algorithms are needed and scattered photons contribute to the noise. Whether this technique will find applications in materials analysis is yet to be seen. In some cases elemental distributions in a material can be inferred [3], but since attenuation is dependent on the density of the material and the linear attenuation coefficient of elemental constituents, care should be taken in interpreting data.

Many different types of X-ray sources can be used to perform X-ray tomography but to achieve high spatial resolutions, a synchrotron source is used to obtain a monochromic primary beam. With synchrotron radiation spatial resolution of up to 1 μm can be obtained. The major advantage of this technique over most other micro-tomographic techniques is that it is non-destructive. This allows a sample to be used for other testing or it allows the evolution of a material to be followed as a function of time and experimental conditions [4–9].

Transmission electron microscopy (TEM) imaging techniques have also been adapted to produce tomographic reconstructions in several modes. Bright-field imaging has been used in tilt series modes to collect through thickness projections of objects as a function of angle. Back projection and Radon transform methods are then applied to these tilt series images to reconstruction volume representations of an object. For example, this method has been used successfully to reconstruct vias used in wiring structures in integrated circuits [10]. In particular, reconstructions using BF images were used to investigate the roughness and step coverage of liner materials in via structures, which is a critical area of investigation in product failure analysis for the semiconductor industry.

Recently, another application of TEM to reconstruct three-dimensional objects has been realized using electron holography. In electron holography, a bi-prism is used in conjunction with field-emission electron sources to measure both the amplitude and phase in transmitted images. As part of holographic image processing, phase information is used to produce three-dimensional reconstructions of an object. Electron holography has been used successfully to elucidate the three-dimensional structure of ultra-fine particles used in catalysis [11].

Direct section methods are typically done in two modes, atom-by-atom methods and serial sectioning. Atom-by-atom methods typically employ field ion microscopy (FIM) [12–14]. In FIM, atomic scale reconstructions of three-dimensional elemental positions are performed with the aid of time of flight mass spectrometry (TOFMS) and a position sensitive detector. A sample is formed into a probe with a tip radius of less than 50 nm and a taper angle of less than 10°. Monolayers of atoms are field desorbed in a pulse mode and spatially and mass analyzed in the TOFMS. This information is then reconstructed into a three-dimensional volume with close to atomic resolution. The field of view of this technique is limited to less than a cubic micrometer, and requires specialized sample fabrication into very specialized geometries. Recent advances in sources, detectors, and sample preparation methods have eased constraints for this technique and have shown considerable promise for ultra-high-resolution characterization of materials with 0.2 nm resolution [15–17].

Serial sectioning methods have found wide application and involve the slicing of an object at regular increments and the collection of images and chemical maps from each slice. Volume reconstructions from these images and chemical maps are then created by interpolating image and chemical signals between slices to form a three-dimensional representation.

Scanning electron microscopy (SEM) and transmission electron microscopy (TEM) have both been widely used in conjunction with serial sectioning to produce three-dimensional tomographic reconstructions [18–20]. In these studies, SEM was used to look at metal precipitates while TEM was used for looking at cellular structures and organelle. In the case of SEM the sample needs to be removed from the microscope and polished between images. This allows slices to be taken at 200–300 nm depth increments. Another way SEM has been used is by looking at backscattered electrons as a function of primary electron energy [21]. Backscattered electrons come from increasing depths as the primary beam energy is increased. Noise for a particular depth image can be reduced by energy filtering the backscattered electrons. In this case the method is nondestructive. In the TEM method serial sections of 1–2 μm in thickness were taken from the structure of interest and then analyzed slice by slice in the microscope [20].

Another technique that can be used for serial section tomographic reconstructions with high spatial resolution is scanning force microscopy (SFM). Proof of concept for the SFM was done on a block copolymer sample. Sectioning of this sample was accomplished using plasma etch processes. An advantage of this technique is that changes in topography due to preferential etching can be followed with the SFM so that planarization is not needed. Lateral resolution for this experiment was found to be as high as 10 nm [22].

A number of ways have been found to use primary ion beams for reconstructing three-dimensional volumes from both shape and elemental information obtained by serial sectioning. A nondestructive method using particle induced X-ray emission (PIXE) and scanning transmission ion microscopy (STIM) has been developed using focused ion beams (FIB) [23,24]. An ion beam, usually a proton beam, is rastered over the surface of a sample and the emitted X-ray intensity is recorded as a function of position. Since X-ray attenuation is a nonlinear process, local X-ray attenuation factors are calculated from STIM density reconstructions to account for differences in X-ray production cross sections [25]. Through accurate modeling of the experiment and advanced algorithms for concatenating experimental output, accurate three-dimensional representations can be obtained with spatial resolutions of around 0.5 μm [26]. This technique has been used to study both biological samples [25] as well as metallurgical samples [27].

Secondary ion mass spectrometry (SIMS) can be used in a mode that is equivalent to serial sectioning in order to measure three-dimensional elemental information from a solid sample [28]. A primary ion beam is used to sputter material from a surface. Ions produced in this way are extracted into a mass filter and their yield as a function of mass is recorded. When an ion microscope is used as a detector, a specified mass is collected and imaged on a detector that can measure ion intensity as a function of position [29]. Since the primary beam is sputtering the surface away, images can be collected at given time intervals corresponding to depths in the sample [30]. Slices can then be reconstructed into a three-dimensional volume with a lateral resolution of ~2 µm and a vertical resolution of ~20 nm [31–34]. One drawback of this method is that sputtering efficiency varies dramatically with both composition and crystallographic orientation [28] making topographical changes as the sample is sputtered an increasing problem with increased depth. This can be compensated for through use of atomic force microscopy data or by milling the surface perpendicular to the imaging plane, but this involves removing the sample from the ion microscope and can lead to contamination problems [35].

These techniques leave an important part of dimensional space inaccessible, corresponding to length scales of tens of nanometers to tens of micrometers. As the field of nanotechnology advances, these length scales become increasingly critical. Further, the ability to engineer on the nanoscale (i.e., on length scales of the order of tens to hundreds of atomic dimensions) implies control (and therefore measurement) of properties in three dimensions. It is to this important regime that focused ion beam tomography is ideally suited.

The essence of focused ion beam tomography is the serial acquisition of images at different depths in a structure during sputtering along the surface normal. These images may be created by secondary electrons, by secondary ions, or by mass-filtered secondary ions to create chemical maps. The resolution of these images is determined by a combination of the primary ion beam diameter at the sample surface, secondary electron and ion escape depth, and implant straggle of primary ions in the sample. For a state-of-the-art primary ion beam diameter of 10 nm, and a beam energy of 30 keV, this translates into a practical spatial resolution of the order of 20–30 nm for most inorganic materials, both parallel and perpendicular to the sample surface normal, as will be described later in this chapter. Sets of images recorded at different depths from the original sample surface are then aligned and concatenated in a computer using appropriate interpolation algorithms between slices. This forms a three-dimensional representation of the sample which can contain more than 10^7 individual volume elements (voxels), each of which are separately addressable. Three-dimensional structural and chemical relationships between components in the structure may thus readily be investigated.

A major issue in the generation of such three-dimensional reconstructions is signal to noise. The secondary electron yield (number of secondary electrons emitted per incident ion) is relatively high, generally greater than one for conducting or semi-conducting materials. Thus, even at 10 nm resolution there are high numbers of secondary electrons to detect. In contrast, secondary ion yields are many orders of magnitude lower, and this affects the signal to noise ratio in secondary ion mass spectroscopy (SIMS) elemental maps. Collection efficiencies and transmission through quadrupole mass filters typically are also relatively low. This means that chemical mapping using mass-filtered secondary ion species may often be signal limited. Compounding this issue is the fact that there is only one opportunity to detect a sputtered ion. This is in contrast to techniques such as electron energy loss spectroscopy in the transmission electron microscope, where low interaction cross sections between the primary electron and the appropriate chemically sensitive inelastic scattering event in the sample may be overcome by acquisition of spectra from literally billions of primary electrons incident on the same volume of sample. This "one shot" aspect of FIB-SIMS reconstructions means that efficient strategies for secondary ion detection are crucial. Despite these limitations, it is our experience that extremely useful chemical reconstructions may be obtained for inorganic materials at local atomic concentrations of the order of a few percent or greater, and at spatial resolutions of the order of tens of nanometers.

Focused ion beam microscopy (FIB) is clearly well suited for tomographic reconstructions using serial sectioning methods. Several groups have attempted to use the FIB for three-dimensional analyses but were limited by the capability of computers to handle the amount of data generated [36]. With the advent of fast microcomputers, images could be acquired at increasing depths and elemental distributions compared as a function of depth. This methodology has the same inherent topography problem encountered in the ion microscope but has better lateral resolution. Since the FIB is an ion mill and material can be removed with nanometer precision in-situ, serial sectioning can be performed. This is done by first milling a flat surface parallel to the incident ion beam, and then rotating the sample to image the newly formed surface [37–41]. This process is repeated to obtain images at various depths in the sample. These images can then be compared to investigate changes in composition. The microstructure of a multi-layer device as well as embedded particles have been examined in this manner [37,38]. In the case of the layered device a trench was milled in the surface of the sample at the position of the device of interest. The sample was then rotated 70° so that the newly cleaned surface could be imaged and ion mapped. Distributions of various elements were then compared. Since the surface being imaged is not normal to the primary ion beam, the image is a shortened projection of the actual surface. This can easily be compensated for in various commercial software packages.

FIB tomography has also been used to examine fracture and bonding surfaces [42–44]. In one case images were taken between serial milling and the morphology compared slice to slice. Since the new surfaces were easily recognized due to their smoothness, elemental mapping was not needed and image reconstruction was adequate. For the bond pad investigation elemental maps were desired for comparison to the images. To minimize the ambiguity due to surface morphology the sample was imbedded in molding resin prior to analysis. This minimized surface damage at deeper layers and allowed images to be more easily compared to elemental maps.

Another advancement in viewing three-dimensional volumes is in the area of computer aided tomographic reconstructions. Volume reconstructions for medical applications have fueled a good deal of research into interpolation routines as well as algorithms for noise reduction [45–49]. Many of these algorithms can be used for rendering volumes of FIB generated data [50]. Volumes have been reconstructed from both secondary electron images and ion maps so that the distribution of elements in semiconductor via structures could be compared. A more advanced algorithm using shape based interpolation was used to look at metal precipitates with irregular features. These reconstructions can be combined with animation routines so that the whole of a complex structure can be easily studied and visualized. While limits to resolution come from both beam interactions and redeposited material during milling [51], a lateral resolution of ~10 nm can be obtained.

11.2 Quantitative three-dimensional tomographic reconstructions using focused ion beams

Fundamentally, FIB tomography is a nanoscale application of classic serial sectioning that is capable of yielding quantitative three-dimensional insight into problems in materials science. An implicit assumption in FIB tomography is that the physical properties of a feature can be accurately reconstructed in three dimensions from discrete two-dimensional slices. Ideally, one might like to disassemble a feature of interest atom by atom using a continuous sputter process, sort atoms by species and position, and then display physical and chemical information graphically for analysis. Unfortunately, continuous sputter processes yield poor spatial resolution for most materials because differential sputtering and redeposition seriously degrade resolution and chemical sensitivity. Problems due to differential sputtering and redeposition can be overcome by discretely sectioning a three-dimensional feature of interest as a function of depth and then interpolating geometric and chemical information between sections. In this case, a section or slice is taken at discrete depths specifically chosen to avoid ambiguities and distortions due to continuous sputter processes.

For each slice, images are collected using either secondary electron or ion images. Chemical information is also collected for each slice using SIMS, energy dispersive X-ray analysis (EDX), Auger electron spectroscopy (AES), or any other spectroscopic technique capable of producing two-dimensional elemental maps. These slices are then concatenated and a sampled volume is reconstructed using interpolative algorithms. As such, focused ion beam tomography can be thought of as a two-step process, data collection and volume reconstruction. In the following sections, we explain an algorithm for focused ion beam tomography that uses secondary electron images and SIMS elemental maps. The reader should keep in mind that these are only two of many possible signals that can be used to generate tomographic recon-structions. The volume reconstruction methods discussed below are useful for a variety of imaging and spectroscopic signals.

11.2.1 Data collection

Data collection is the most crucial step in focused ion beam tomography because it determines the ultimate accuracy and resolution of three-dimen-sional reconstructions. A typical data collection algorithm used for focused ion beam tomography is as follows.

First, an area of interest is found in a sample and fiducial marks are cut into the periphery. These marks will subsequently be used to align slice data during volume reconstruction and are re-cut at several depths to avoid slice misalignment.

Next, secondary electron images and/or SIMS elemental maps are collected as a function of depth into a feature of interest. Each image and elemental map is divided into pixels, where pixel size is chosen to be approximately equal to the probe size. A probe of appropriate size is then scanned over the image area and held at each pixel for a user-selected dwell time, typically ranging from 20–40 μs for images and up to 4 ms for SIMS elemental maps. At each pixel, secondary electron and SIMS signals are collected and stored. The feature of interest is then sputtered down to a specified depth with the beam parallel to the data collection surface, and another set of images and elemental maps is collected. To a first approximation, slice thickness is determined by the amount of material removed by the beam when collecting secondary electron images and SIMS elemental maps. Secondary electron images and SIMS elemental maps are repeatedly collected as a function of depth until a volume of interest has been sampled.

There are several aspects of this procedure that warrant further explanation. First, it is very important that a feature of interest be sectioned with the beam parallel to the surface for data collection because this minimizes deviations from a planar section due to differential sputtering. If one chooses to remove material between slices using normal incidence sputtering, surface undulations due to differential sputtering will increase in magnitude and severely degrade depth resolution.

Second, it is important that fiducial marks used for alignment remain sharp and undistorted in each slice. Typically, at least three square fiducial marks are cut into the periphery of each slice so that both rotational and translational drift can be corrected.

Third, care must be taken to ensure that the depth of each slice is measured accurately. The most straightforward method for measuring the depth of slices is to measure sputtered depth, edge-on using secondary electron images. In particular, the depth from the original surface is measured with the beam parallel to the sputtered surface. Inherent in this method is the assumption that the image plane is flat and not inclined along the beam. A second method for determining slice depth is to measure the final sputtered depth using scanning tunneling microscopy or atomic force microscopy. The total depth is then divided by the number of slices to estimate the average spacing between slices.

Finally, it is important to choose a probe size that does not over or under sample image and elemental map signals in a particular slice. Since focused ion beam imaging and SIMS are inherently ion processes, one has to take account of the normal and lateral extents of signal generation. A typical FIB image or elemental map is divided into pixels with an image plane area determined by the digital-to-analog characteristics of beam scan coils. Since only those ions within 0.5 nm of the surface have enough energy to escape [52], a more important consideration is the lateral range of ions once they have impacted a sample surface. Quick estimates of lateral range for most materials can be made using a typical spreadsheet and approximations to Linhard, Scharff, and Schiott (LSS) theory [53]. Using this information, an appropriate probe size can be chosen to ensure that signals measured do in fact arise from within voxel bounds.

Table 11.1. *Estimated lateral ranges for 30 keV Ga ions.*

Material	Lateral range (nm)
Al	7
Ti	7
Si	3

The ultimate resolution of FIB tomographic reconstructions is directly determined by instrumental factors and physical aspects of sputter events. Since we are discussing a sputter based technique, both lateral and depth resolution must be considered.

Lateral resolution is controlled by two factors, the smallest ion probe achievable in a given FIB and the lateral range of probe ions in a feature of interest. The smallest achievable probe is determined by the FIB column contrast transfer function and Coulombic interactions of ions within the beam [54]. Using modern FIB instrumentation, it is possible to achieve ion probes 10 nm in diameter. High current ion probes smaller than 10 nm in diameter are susceptible to perturbations due to Coulombic interactions [54]. If dual-beam FIB systems are used to collect data, higher resolution data are achievable with modern secondary electron microscopy columns. Ultimate resolution of these columns is, however, still limited by the contrast transfer function of the particular column used but can be as high as 1 to 2 nm.

Since ion probes are used to sputter surfaces, the effect of ion–sample interaction has to be taken into account when considering the ultimate resolution of image and SIMS elemental maps. Because secondary electron images and SIMS elemental maps are incoherent, lateral resolution is directly proportional to effective probe size. As was mentioned earlier, the effective probe size is increased by the lateral range of probe ions in features of interest. To estimate the increase in probe size due to lateral straggle, one can use a variety of techniques. The simplest method is to use LSS theory. Table 11.1 shows estimates for 30 keV Ga ions in Al, Ti, and Si.

The effective probe size can then be estimated by adding the lateral range to probe ions in sample materials to the smallest achievable incident probe. Using the lateral ranges in Table 11.1 and a 10 nm incident probe, the ultimate lateral resolution of images and elemental maps from these materials is approximately 24 nm for Al and Ti and less for Si. As was mentioned earlier, lateral range is strongly dependent upon sample material and should be calculated on a case-by-case basis.

Depth resolution in FIB tomography is affected by differential sputtering and damage profiles produced as ions traverse sample materials. To minimize differential sputtering, it is important to section a feature of interest with the ion beam parallel to the surface for data collection. This minimizes differential sputtering during sectioning.

Damage profiles along the beam direction during data collection directly affect the minimum distance between slices for a feature of interest. Knock-on displacement events can distort sample features, so the minimum depth between slices is limited by depth of maximum damage for a given sample material. For 30 keV Ga ions, this depth ranges from a few nanometers to tens of nanometers depending upon sample material.

11.2.2 Volume reconstruction

Once data from a feature of interest have been collected, they are processed to reconstruct three-dimensional image and chemical information. There are several methods that can be used to reconstruct volumes from these data, but we will focus on linear intensity interpolation and shape based interpolative methods.

11.2.3 Linear intensity interpolation

In linear interpolative schemes, collected data are concatenated as a function of depth to produce a three-dimensional set of discretely sampled data. Voxels between slices are then interpolated using linear interpolation algorithms. In linear interpolative schemes, secondary electron intensities or SIMS intensities are measured at each voxel in a slice and then voxels are interpolated from slice to slice.

Linear interpolative schemes are useful for objects that can be generated by extrusion operations such as columnar grains and vias used in multilevel interconnect structures for integrated circuits. For example, Figure 11.1 shows a 3×4 array of vias from a structure in an integrated circuit. This volume reconstruction was done using secondary electron image data from eight slices through these vias, parallel to the top surface of the reconstructed volume.

Using the same interpolative methods, three-dimensional chemical maps can be reconstructed using linear interpolative schemes. Shown in Figure 11.2 is a set of Al vias reconstructed using linear interpolation. These vias are cut into SiO_2 interlayer dielectric, lined with Ti, and then filled with Al.

It is clear from Figure 11.2 that linear intensity interpolation is capable of accurately reconstructing via structures. However, as was mentioned

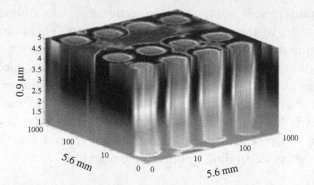

Figure 11.1 A secondary electron image volume reconstruction of a 3×4 array of vias in an integrated circuit macro. This reconstruction was done using eight slices that were cut parallel to the top surface of this volume. In this reconstruction, contrast is observed from differences in secondary electron yield due to different materials and crystallographic orientations.

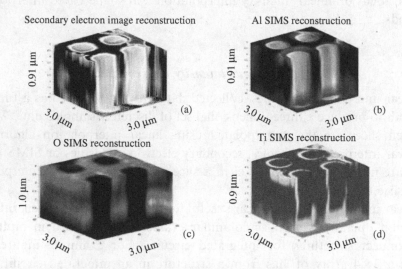

Figure 11.2 Image and SIMS chemical maps reconstructed from secondary electron images and SIMS elemental maps. (a) Secondary electron image reconstruction calculated using linear interpolation from five secondary electron images. (b) Al elemental reconstruction calculated using linear interpolation from five SIMS Al elemental maps. In this figure, light gray indicates the highest Al signal, while dark gray and black indicate low Al intensity. (c) SIMS elemental reconstruction using negative O ion elemental maps. In this reconstruction, gray represents highest O concentration, while black represents background. (d) Ti SIMS reconstruction using five Ti SIMS elemental maps. In this figure, light gray represents the highest Ti signal and dark gray represent background level signals [50].

previously, this type of interpolation does not accurately reconstruct features that cannot be generated by extrusion operations. If for instance, these vias contained voids, void edges would be blurred during the interpolation process. This would make it nearly impossible to make quantitative estimates of void volume, surface area, connectivity, or morphology.

11.2.4 Shape based interpolation

Shape based interpolation is a scheme developed for medical tomographic imaging that uses interfaces present in three-dimensional data sets to reconstruct complex features. Shape based interpolation is more accurate than intensity based reconstruction techniques because the shape of the object, rather than image intensity, is being interpolated, thus inaccuracies and edge blurring often observed in intensity interpolation are avoided [55,56]. Since FIB images and elemental maps consist primarily of spatially varying intensities, shape based interpolation can be readily adapted to three-dimensional tomographic reconstructions of geometric and chemical data from FIB microscopy data. Shape based volume reconstructions can be used to not only establish three-dimensional chemical and geometric relationships, but also to obtain quantitative information such as the sharpness of interfaces, surface area, volume, volume fraction, and connectivity of features of interest.

Shape based interpolation is a method in which the shortest distance of a voxel to the edges of a feature within a slice is calculated using either Euclidean or city-block distances [55] as is shown in Figure 11.3. If a voxel is inside the edges of a feature, this distance is entered into a voxel as a positive distance; if a voxel is outside the edges of a feature, its closest distance to feature edges is entered into a voxel as a negative distance. Voxels that fall on the edges of a feature, by definition, have zero distances. The determination of whether a voxel is inside or outside of a feature is done using standard inside-out tests [57].

Once each experimentally collected slice has been processed and edge distances recorded, distances for slices between recorded data slices are interpolated using bi-linear interpolation [55]. After interpolation has been completed, voxels that have negative edge distances are turned off (set to zero) and the resulting volume is represented by a volume of voxels that have zero or positive distances to feature edges.

The accuracy of tomographic reconstructions calculated using shape based interpolation is directly dependent upon the accuracy with which the

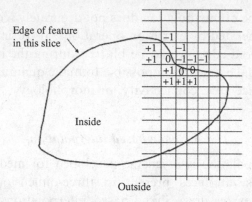

Figure 11.3 A schematic diagram of a typical slice of an object in which the distances to the edges have been calculated and placed in each pixel position. Pixels that land on an edge have 0 distance, pixels inside an edge have positive distances and those outside of object edges have negative distances to an edge.

positions of feature edges can be determined within experimentally measured slices. Typically, standard image processing and edge detection methods are applied to each experimentally measured slice prior to reconstruction. The particular methods used for each reconstruction require some investigator input to determine which features will be reconstructed and which methods are best suited to determining edge positions within each slice. An in depth discussion of image processing and edge detection methods are beyond the scope of this chapter, but the reader is encouraged to consult standard image processing text books to determine which methods are best for his or her particular data [58].

SIMS elemental maps can be processed using shape based interpolation methods similar to those described for image data. In some cases, however, SIMS elemental maps have a low signal to background ratio making it difficult to find edges using typical edge detection methods. In these cases, a combination of shape based interpolation methods and intensity interpolation are used to analyze these data. In particular, SIMS data are concatenated into a three-dimensional array and volume between slices is linearly interpolated to produce a three-dimensional map of SIMS intensities [50]. These data are then further processed to determine the chemical distribution of selected elements within or outside the edges of features of interest. Since the volumes reconstructed from image data have positive voxel values for voxels inside a feature and zero for those outside of a feature, these volumes can be used as masks to select SIMS intensities inside a given feature by logical AND operations. In addition, the chemical distribution outside of

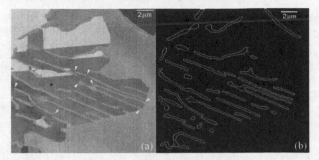

Figure 11.4 Typical SEM images and edges collected as part of a serial section experiment to investigate Cu-15In colonies. (a) An SEM image of a sample showing Cu-15In colonies. White arrows in this image point to examples of lamellae that make up a colony. (b) Edges determined from the SEM image shown in (a) [59].

selected features can be selected by NAND logical operations. Logical operations can then be used to investigate changes in chemistry across the interfaces of selected features.

As an example of shape based volume reconstructions, we now present volume reconstructions of the structure and chemistry of lamellar structures from Cu-15In colonies [59]. Shown in Figure 11.4 is a typical SEM image taken from a Cu-15In colony. These colonies consist of lamellar In rich regions, which are marked by arrows in Figure 11.4(a). Figure 11.4(b) shows an image that consists of edges for Cu-15In lamellae determined from the SEM image in Figure 11.4(a). These edges and others from consecutive slices through this colony were used to reconstruct a three-dimensional volume from this colony using shape based interpolation.

Figure 11.5 shows a montage of shape based volume reconstructions calculated from slices similar to those shown in Figure 11.4(b). In Figure 11.5(a), the entire set of Cu-15In lamellae shown in Figure 11.4 have been reconstructed. In this reconstruction, all positive distances and interfaces have been colored gray showing the spatial variation of lamellae within this colony structure. In Figures 11.5(b) and 11.5(c), a set of lamellae have been extracted from this colony for further investigation. It has long been thought that these colonies consist of disconnected plates. It is clear from this reconstruction, however, that these lamellae are connected in Figure 11.5(c). If this colony were viewed only from a two-dimensional projection of Figure 11.5(c), an investigator might erroneously assume that this colony was made up of disconnected plates.

There is a great deal of geometric and spatial information available in shape based volume reconstructions like those shown in Figure 11.5. From

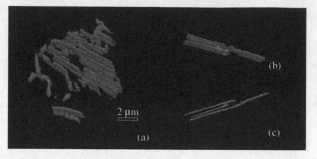

Figure 11.5 Reconstructed volumes from Cu-15In colonies. (a) A volume reconstruction calculated from SEM images using shape based interpolation. (b) A few select lamellae from the colony shown in (a). This set of lamellae appear to be connected in this view. (c) The same lamellae from (b) but at a different depth. At the depth shown in (c), these lamellae appear to separate as a function of depth. If a single slice is observed at the last slice for (c), these lamellae might be mistakenly reported as not connected [64].

these reconstructions, volume, volume fraction, and connectivity relationships are easily calculated. These reconstructions can also be used to help understand the spatial variation of chemical species measured by SIMS. Figure 11.6 shows a montage of reconstructed volumes and three-dimensional In SIMS elemental maps taken from cellular colonies. Figure 11.6(a) shows a shape based volume reconstruction of several cellular structures taken from the colony in Figure 11.5(a). This reconstruction shows connected lamellae that diverge near the top surface of the reconstruction.

Figure 11.6(b) is a three-dimensional In elemental map showing the spatial variation of In both in lamellae and the matrix. This chemical reconstruction was done using linear interpolation of the In signal between slices. If the elemental map shown in Figure 11.6(b) is operated on using a logical AND operation with the volume in Figure 11.6(a), the spatial variation of In within lamellae is obtained as shown in Figure 11.6(c). This operation shows not only the variation of In concentration within lamellae but also yields the position of interfaces between lamellae and the surrounding matrix. Using this position, one can extract data across a specific interface and investigate the chemical and spatial variation of interfaces.

Another example of the power of shape based reconstructions is their application to quantum dot structures. Recently, Kubis *et al.* [59] have applied these techniques to investigate growth characteristics of Ge/Si quantum dots (QD). One of the most important problems investigated in these systems is how dots are arranged in the matrix during growth. Figure 11.7 shows a shape based volume reconstruction from a Ge/Si quantum dot

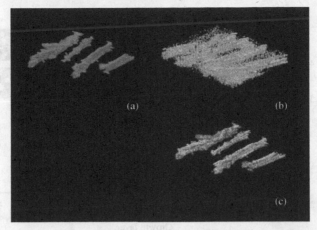

Figure 11.6 A montage of volume reconstructions from a Cu-15In colony. (a) A volume reconstruction calculated using shape based interpolation in In rich colony regions. (b) An SIMS volume reconstruction using positive In ions and direct interpolation of intensities. Dark and light gray represent different concentrations of In and black defines background. (c) A volume reconstruction that is the result of a logical AND operation of the shape based reconstruction shown in (a) with the In SIMS elemental map shown in (b). This reconstruction yields a clearer view of the variation of In within colony lamellae [64].

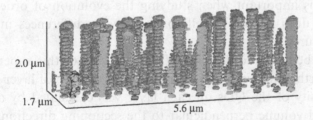

Figure 11.7 A shape based volume reconstruction of quantum dots in a Ge/Si quantum dot superlattice. In the near left-hand corner is an example of a QD column that has terminated during growth. Spatial resolution between the layers was intentionally decreased so that the evolution of the columns could more easily be observed.

superlattice. In the near left-hand corner as highlighted in the reconstruction is an example of a QD column that has terminated during growth of the superlattice and does not continue through the entire structure.

In many cases tomographic reconstructions have been used empirically to visually inspect spatial distributions of quantum dots. In the case of the QD superlattice shown in Figure 11.7, close examination shows that all of the QD

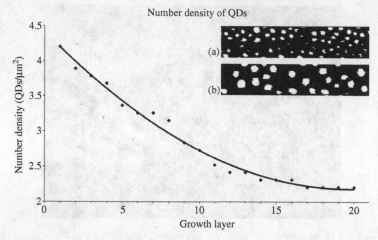

Figure 11.8 A plot of the number density of QDs as a function of growth layer in this superlattice structure. Inset (a) is a processed slice through the lowest layer and (b) shows a slice through the uppermost layer.

columns in the reconstruction do emanate from the first layer grown and in some cases are extinguished prior to the 20th layer. The termination of QD columns after several layers of growth was predicted by Tersoff *et al.* [60] and Liu *et al.* [61] for QD columns that were too closely spaced. While this observation is important when studying the evolution of ordering in these structures it does not lend itself to quantifying differences in comparably grown structures.

To gain a better understanding of the ordering of this superlattice, these data were further analyzed and compiled as a function of layer number. The number density of the QDs on each layer was calculated by analyzing the reconstructed volume perpendicular to the sectioning direction. Figure 11.8 shows a graph of the number density of QDs on each of the grown layers. The number of QDs on each successive layer decreases until a minimum is reached at the 17th layer. As can be seen in the inset of Figure 11.8 the lower layers consist of a high number density of small QDs while the upper layers have a lower density of larger QDs. As some columns disappear in Figure 11.7, QDs in other columns increase in size. This clearly shows one effect of the self-assembly process. This size distribution and volume relationships shown in Figures 11.7 and 11.8 would not have been easily accessible without shape based volume reconstructions.

Volume reconstructions using FIB microscopy and shape based interpolation can provide a wealth of critical information for problems in materials science and condensed matter physics. In problems involving solid state phase transformations and reaction kinetics, it is important to ascertain the connectivity of component phases and their respective volume fractions, surface to volume ratios, etc. Typically, this type of information is obtained through stereological analyses. One drawback to stereological analyses is that feature geometry must be assumed. Features are typically assumed to take on model shapes such as plates, cylinders, spheres, and oblate and prolate spheroids, and then methods of geometric probability are used to infer mean feature dimensions, volume fractions, surface areas, etc. By using FIB three-dimensional reconstruction techniques, one can measure these quantities directly without having to assume model feature shapes.

11.3 Summary and future work

Focused ion beam tomography is proving to be a very powerful technique for investigating three-dimensional relationships in a number of materials systems. Volume reconstructions for a wide array of materials can be calculated with lateral resolutions as high as 20 nm and depth resolutions of 10 nm. When combined with accurate reconstruction algorithms such as shape based interpolation, FIB tomography is capable of yielding quantitative information such as surface area, volume, perimeter, volume fraction, and connectivity. In addition, three-dimensional variations in chemistry can also be investigated if SIMS, AES, or EDX mapping techniques are incorporated into volume reconstructions. If image volume reconstructions are used as masks to process chemical reconstructions, it is possible to investigate three-dimensional distributions of chemical species within features of interest and their variation across interfaces.

In this chapter, we have shown examples from metallurgy and solid state physics in which FIB tomography has yielded new insight that would have been difficult to obtain using traditional two dimensional methods of characterization. FIB tomography will continue to be an important technique for objects with critical dimensions ranging from 10 nm to 10 μm and beyond. There are however, several advances in FIB microscopy systems that will undoubtedly increase the range of applicability of FIB tomography and may well extend its range of critical length scales below 10 nm. Perhaps the most important change can be seen in the maturity of dual-beam FIB micromachines. In dual-beam systems, an SEM column is included in addition to the ion column. Using this dual-beam configuration, a sample can be sectioned using the primary ion beam and imaged with the SEM column. In this type of FIB, lateral resolution is no longer confined by the beam characteristics and contrast transfer function of the primary ion column, thus the ultimate resolution of image slices can be increased to that of a current state of the art SEM. In addition, since sputtering is not occurring during image collection, less material is lost making it possible to achieve much finer depth resolution when collecting slices.

Outside of higher resolutions attainable with SEM imaging on dual-beam FIBs, most modern dual-beam systems allow for SEM imaging with electron accelerating voltages less than 2.0 keV. Low voltage imaging modes also extend the applicability of FIB tomography to organic systems and those materials systems that are easily damaged by electron radiation. In particular, low voltage SEM imaging modes make it possible to investigate organic dielectrics in the semiconductor industry and many polymer systems in other fields of materials science. Limitations for these materials systems will most probably be confined to ion effects such as differential sputtering and whether these materials can be sectioned with sufficient accuracy via sputter processes to allow for volume reconstruction.

In addition to dual-beam systems with high-resolution SEM columns, a great deal of work is currently being done on ion sources to attain smaller probe sizes. In particular, there are efforts to develop gas ion sources [62,63], which, in theory, should yield smaller virtual source sizes and lead to effective ion beam sizes of less than 10 nm. If successful, these sources could be mass-matched to materials of interest and allow for greater control of sputtering, and therefore much finer slice sampling during data collection. While these sources are unlikely to improve lateral resolutions for single beam FIB systems, they could make a great difference in the depth resolution of volume reconstructions done using dual-beam FIB systems. In addition, these sources might offer the possibility of increasing ionization yield for SIMS in focused ion beam systems, thereby increasing sensitivity and signal to noise characteristics of chemical element volume reconstructions. These sources are still at the development stage, however, and will probably not be incorporated into production FIB systems for some time.

In summary, techniques for FIB tomography using serial sectioning with a lateral resolution of approximately 20 nm and a depth resolution of 10 nm have been presented. This critical dimension scale is well within the realm of application for many nanotechnology applications. When coupled with chemically sensitive analytical techniques such as SIMS, AES, and EDX, structural and chemical relationships can be readily developed in three-dimensions from volume reconstructions for nanostructured materials systems.

Acknowledgements

The authors would like to acknowledge G. Shiflet (University of Virginia) and G. Gilmer (Bell Laboratories) for provision of samples used to generate tomographic reconstructions presented in this article. Funding from IBM, the Defense Advanced Research Projects Agency (Virtual Integrated Prototyping Program), and the National Science Foundation (Materials Research Science and Engineering Center) is gratefully acknowledged.

References

[1] D. C. Copley, J. W. Eberhard and G. A Mohr. *J. of Metals*, **46** (1994), 14–26.
[2] M. Defrise. *Comput. Med. Imag. Graph.*, **25** (2001), 113–16.
[3] J. W. Owens, L. G. Butler, C. Dupard-Julien and K. Garner. *Mater. Res. Bull.*, **36** (2001), 1595–602.
[4] L. Babout, E. Maire, J. Y. Buffiere and R. Fougeres. *Acta. Mater.*, **49** (2001), 2055–63.
[5] C. L. Lin and J. D. Miller. *J. Chem. Eng.*, **77** (2000), 79–86.
[6] D. Lu, M. Zhou, J. H. Dunsmuir and H. Thomann. *Mag. Res. Imaging*, **19** (2001), 443–8.
[7] C. F. Martin, C. Josserond, L. Salvo *et al. Scripta Mater.*, **42** (2000), 375–81.
[8] A. Guvenilir, T. M. Breunig, J. H. Kinney and S. R. Stock. *Acta. Mater.*, **45** (1997), 1977–87.
[9] W. Ludwig and D. Bellet. *Mater. Sci. Eng.*, **A281** (2000), 198–203.
[10] H. Stegmann, H. J. Engelmann and E. Zschech. *Microelectron. Eng.*, **65**, 171–83.
[11] A. Tonomura. *Electron Holography.* Springer Series in Optical Sciences, Vol. 70 (New York: Springer-Verlag), pp. 41–2.
[12] M. K. Miller. *Mater. Charac.*, **44** (2000), 11–27.
[13] B. Deconihout, C. Pareige, P. Pareige, D. Blavette and A. Menand. *Microsc. Microanal.*, **5** (1999), 39–47.
[14] D. Blavette, A. Bostel and J. M. Sarrau. *Nature*, **6428** (1993), 432–5.
[15] T. F. Kelly and A. A. Gribb. *Microscopy Today*, September/October (2003), 8–12.
[16] T. F. Kelly, P. P. Camus, D. J. Larson, L. M. Holzman and S. S. Bajikar. *Ultramicroscopy*, **60** (1996), 29.
[17] T. F. Kelly and D. J. Larson. *Mater. Charac.*, **24** (2000), 59.
[18] M. A. Mangan, P. D. Lauren and G. J. Shiflet. *J. Microsc.*, **188** (1997), 36–41.
[19] M. A. Mangan and G. J. Shiflet. *Scripta Mater.*, **37** (1997), 517–22.
[20] G. E. Soto, S. J. Young and M. H. Ellisman. *NeuroImage*, **1** (1994), 230–43.
[21] H. Neidrig and E. I. Rau. *Nucl. Instrum. Methods Phys. Res. B*, **142** (1998), 523–34.
[22] R. Magerle. *Phys. Rev. Lett.*, **85** (2000), 2749–52.
[23] D. N. Jamieson. *Nucl. Instrum. Methods Phys. Res. B*, **136/138** (1998), 1–13.
[24] K. G. Malmqvist. *Nucl. Instrum. Methods Phys. Res. B*, **104** (1995), 138–51.
[25] R. M. S. Schofield. *Nucl. Instrum. Methods Phys. Res. B*, **104**, (1995), 212–21.
[26] A. Sakellariou, M. Cholewa, A. Saint and G. J. F. Legge. *Nucl. Instrum. Methods Phys. Res. B*, **130** (1997), 253–8.

[27] Y. K. Ng, I. Orlic, S. C. Liew *et al. Nucl. Instrum. Methods Phys. Res. B*, **130** (1997), 109–12.

[28] A. Benninghoven, F. G. Rudenauer and H. W. Werner. *Secondary Ion Mass Spectrometry Basis Concepts, Instrumental Aspects, Applications and Trends* (New York: John Wiley and Sons, 1987).

[29] H. Hutter and M. Grasserbauer. *Mikrochim. Acta*, **107** (1992), 137–48.

[30] A. J. Patkin and G. H. Morrison. *Anal. Chem.*, **54** (1982), 2–5.

[31] N. S. McIntyre, R. D. Davidson, C. G. Weisener *et al. Surf. Interface Anal.*, **18** (1992), 601–3.

[32] S. F. Lu, G. R. Mount, N. S. McIntyre and A. Fenster. *Surf. Interface Anal.*, **21** (1994), 177–83.

[33] H. Hutter, K. Nowikow and K. Gammer. *Appl. Surf. Sci.*, **179** (2001), 161–6.

[34] K. Gammer, S. Musser and H. Hutter. *Appl. Surf. Sci.*, **179** (2001), 240–4.

[35] M. L. Wagter, A. H. Clarke, K. F. Taylor, P. A. W. van der Heide and N. S. McIntyre, *Surf. Interface Anal.*, **25** (1997), 788–9.

[36] W. Steiger, F. Rudenauer, H. Gnaser, P. Pollinger and H. Studnicka. *Mikrochim. Acta Supp.*, **10** (1983), 111–17.

[37] H. Satoh, M. Owari and Y. Nihei. *J. Vac. Sci. Technol. B*, **9** (1991), 2638–41.

[38] Y. Nihei. *J. Surf. Anal.*, **3** (1997), 178–84.

[39] Y. Nihei, B. Tomiyasu, T. Sakamoto and M. Owari. *J. Trace Microprobe Tech.*, **15** (1997), 593–9.

[40] B. Tomiyasu, I. Fukuju, I. Komatsubara, M. Owari and Y. Nihei. *Nucl. Instrum. Methods Phys. Res. B*, **136/138** (1998), 1028–33.

[41] N. J. Montgomery, D. S. McPhail, R. J. Chater and T. Dingle. *Secondary Mass Spectrometry SIMS X I*, ed. G. Gillen, R. Lareau, J. Bennett and F. Stevie (New York: John Wiley and Sons, 1998), pp. 631–4.

[42] Y. Z. Wang, R. W. Revie, M. W. Phaneuf and J. Li. *Fract. Eng. Mater. Struct.*, **22** (1999), 251–6.

[43] K. Takanashi, H. Wu, N. Ono *et al. Inst. Phys. Conf., Ser.*, **165** (2000), 9–13.

[44] T. Sakamoto, K. Takanashi, Z. H. Cheng *et al. Inst. Phys. Conf. Ser.*, **165** (2001), 9–13.

[45] K. Lohmann, E. D. Gundelfinger, H. Scheich *et al. J. Neurosci. Meth.*, **84** (1998), 143–54.

[46] G. T. Herman, J. Zheng and C. A. Bucholtz. *IEEE Comput. Graphics Appl.*, **12** (1992), 69–79.

[47] S. P. Raya and J. K. Udupa. *IEEE Trans. Med. Imag.*, **9** (1990), 32–42.

[48] M. Levoy. *IEEE Comput. Graphics Appl.*, **8** (1988), 29–37.

[49] R. A. Drebin, L. Carpenter and P. Hanrahan. *Comput. Graphics*, **22** (1988), 65–74.

[50] D. N. Dunn and R. Hull. *Appl. Phys. Lett.*, **75** (1999), s3414–16.

[51] F. G. Rudenauer and W. Steiger. *Ultramicroscopy*, **25** (1988), 115–24.

[52] P. Sigmund. *Phys. Rev.*, **184** (1969), 383–416.

[53] H. E. Schiott. *Radiat. Eff.*, **6** (1970) 107–13.

[54] J. Orloff. *Rev. Sci. Instrum.*, **64** (1993), 1105–29.

[55] S. P. Rayaand and J. Udupa. *IEEE Trans. Medical Imaging*, **9** (1990), 32–42.

[56] G. Herman, J. Zheng and C. A. Bucholtz. *IEEE Comput. Graphics Appli.*, **70** (1992), 69–79.

[57] E. Haines. *Graphics Gems IV* (New York: Academic Press, 1994), pp. 24–46.

[58] J. Russ. *The Image Processing Handbook* (Boca Raton, FL: CRC Press), pp. 161–427.

[59] A. J. Kubis, T. E. Vandervelde, J. C. Bean, D. N. Dunn and R. Hull. *Mater. Res. Soc. Symp. Proc.*, **818** (2004), M14.6.1–M14.6.7.

[60] J. Tersoff, C. Teichert and M. G. Lagally. *Phys. Rev. Lett.*, **76** (1996), 1675–8.

[61] F. Liu, S. E. Davenport, H. M. Evans and M. G. Lagally. *Phys. Rev. Lett.*, **82** (1999), 2528–31.

[62] K. Jousten, K. Bohringer, R. Borret and S. Kalbitzer. *Ultramicroscopy*, **26** (1988), 301.

[63] W. Thompson, A. Armstrong, S. Etchin, R. Percival and A. Saxonis. *Ion–Solid Interactions for Materials Modification and Processing. Mater. Res. Soc. Symp.* (1996), pp. 687–93.

[64] D. N. Dunn, G. J. Shiflet and R. Hull. *Rev. Sci. Instrum.*, **73** (2002), 330.

12

Ion beam implantation of surface layers

DANIEL RECHT AND NAN YAO

Princeton University

12.1 Introduction

Ion implantation is a method for the direct, controlled introduction of impurities into solids. In ion implantation, a beam of dopant ions is aimed at a target material (the substrate) so that the ions are incident with sufficient energy to become permanently embedded. Because ion implantation is an essentially nonequilibrium process, it allows for the creation of concentration profiles that would be impossible to achieve using equilibrium techniques such as diffusion. The advent of focused ion beam (FIB) systems spawned a host of new applications for ion implantation. The ability to create high-resolution (feature sizes of order 10 nm [1,2]) doping configurations without the use of a mask allows not only for rapid prototyping, but also for unique devices whose fabrication would not otherwise be feasible. FIB implantation has been used to make a wide variety of experimental devices including low-dimensional transistors, single photon detectors, subwavelength optics, and quantum computers.

This chapter covers the basics of ion implantation in general and of FIB implantation in particular. Next it considers the challenge of measuring the ion dose introduced into the substrate. It then discusses the major parameters relevant to the FIB implantation process and presents a sample of their complex interrelationships. Finally, it describes the aforementioned applications of FIB implantation to the fabrication of novel devices.

12.2 Basics

In the years since its development in the 1950s [3], ion implantation has become the preferred method for the introduction of impurities into solid

Focused Ion Beam Systems: Basics and Applications, ed. N. Yao.
Published by Cambridge University Press. © Cambridge University Press 2007.

substrates. Because it introduces ions mechanically instead of by thermally activated transport processes, ion implantation is capable of producing concentration profiles that are unachievable via diffusion. Specifically, standard ion implantation can create impurity concentrations far above the solid solubility and concentration gradients along the implantation direction much steeper than could be produced by any equilibrium process. The ability to produce supersaturated solid solutions and sharp junctions was a substantial boon to the semiconductor industry. However, standard ion implantation systems have proven unsuitable for many nanoscale applications because their broad beams prevent the creation of sufficiently small lateral features. Although masking of the beam using a layer of photoresist or other material atop the target can compensate for this to some degree, it is not ideal for many applications. Focused ion beam (FIB) implantation systems represent a significant advance in implantation technology because their narrow beams confer the ability to produce sharp gradients in directions perpendicular to the implantation vector as well as parallel to it.

FIB implantation can be performed on most FIB systems. The standard "point" ion sources such as the liquid-metal and gas field ion sources are adequate for implantation, as are typical column designs, focusing optics, etc. [4]. In general, the only major differences between FIB implantation and other FIB processes come in the handling of the target material. Since the purpose of ion implantation is to inject a precise amount of dopant into a substrate, it is vital that one be able to measure and control that amount (called the dose) with high precision and accuracy. As will be described in the next section, there are many ways to accomplish this.

12.3 Dosimetry

Precise and accurate dosimetry, i.e., measurement of the amount of impurity ions implanted into the substrate, is important in all types of ion implantation. Particular challenges in dosimetry are as varied as the applications of ion implantation. Focused ion beam systems, in particular, require great lateral (horizontal) resolution to match the feature sizes that they can produce. In practice, two sorts of dosimetry are required for successful implantation. First, in what is aptly named dose control, the number of ions introduced into the target must be monitored in real time during the implantation process. Second, after implantation is complete, the three-dimensional distribution of ions in the material must be ascertained. This process will here be referred to as characterization.

Dose control technology used in FIB implantation is quite similar to that applied to standard implantation. Current technology is such that dose control is much more of a solved problem than characterization. Thus, most commercial FIB implanters include dose controllers that rely on the same basic principles. After passing through the column, the ion beam is ideally free of neutral particles and uniform in composition. Standard dose controllers thus begin with a measurement of the instantaneous electric current flowing in the sample. This current is almost entirely due to the flux of ions into the substrate. The current measurements are summed by an integrator and converted to a dose value under the assumption that all the incident ions carry the same charge. This assumption holds well when the beamline (the space through which the ion beam travels) is maintained at a good vacuum pressure. Otherwise, the ions will collide with residual gas molecules in the beamline and either lose or gain electrons, affecting their response to electric and magnetic fields. A net neutralization or charge exchange of ions in the beamline will cause the actual implanted dose to differ from the dose calculated by integration of the current. However, beamline vacuums of 5×10^{-7} torr or less can easily be achieved today, while techniques exist to compensate for dose error at up to 5×10^{-4} torr [5]. Thus, the uniform current assumption is valid for all purposes. Alternatively, the integrator can be excluded to obtain the instantaneous dose rate. These measurements can then be sent into a feedback loop to regulate the dose rate and thus the total dose. Furthermore, if the size and position of the beam are known, these measurements can also be used to produce a map of the dose applied to each beam-sized pixel of the substrate. This is the basis behind patterned implantation using FIBs. In practice, it is often inadvisable to measure the current directly on the substrate. Thus many modern implanters monitor the ion beam using Faraday cups independent of the substrate [6,7].

A complete characterization of the dose in an implanted sample consists of a three-dimensional map of the concentration of impurity ions throughout it. As described above, after implantation is complete the dose controller is capable of outputting a planar map of the total dose applied to each beam-sized pixel of the substrate. In addition to completely ignoring the vertical impurity distribution, this does not take into account the random horizontal scattering of ions. Evidently, the information available from the dose controller is insufficient to determine if an implantation step has met a set of specifications to within appropriate tolerances. Consequently, post-processing for implanted substrates must include a characterization step to determine the actual spatial ion distribution resulting from implantation.

The primary characterization method used to determine the horizontal distribution of ions in conventional implantation is based on the effect of lattice damage on the photomodulation via thermal waves of the reflectance of the substrate. The most common commercial devices using this effect are known as Therma-Probe™ dosimeters. Practically speaking, thermal wave characterization of ion implanted samples operates as follows. Light from a laser, called the pump beam, is focused onto a spot on the substrate, and its intensity is modulated at a high frequency (usually tens of MHz). A second laser beam called the probe, operating at lower power, is focused to an adjacent spot. A photodetector is used to measure the reflected light from the probe beam. Thermal and plasma waves caused by the periodic heating of the modulated pump beam cause the optical parameters of the substrate to also vary periodically [8]. This leads to small but regular oscillations in the reflected power, which increases in magnitude as more damage is introduced into the crystal structure of the substrate [9]. Consequently there is a readily observable (nonlinear) correlation between dose and oscillation amplitude [9]. With appropriate calibration measurements of the reflected probe light can thus be converted to dose. Advances in thermal wave technology since its original application to ion implantation in 1985 have enabled it to make precise measurements over a wide range of doses yielding 2D maps of the type described above much more accurate than those provided by the dose controller [10,11].

Because it relies on light, this technique is not yet fully suitable for use with FIB implantation. The resolution of its measurements is limited by the wavelength of the probe laser. Given the operating wavelength of most lasers today, thermal wave characterization cannot resolve pixels much finer than 0.5 μm in size, whereas FIB implantation was developed to produce features nearly two orders of magnitude smaller. However, a 157 nm wavelength fluorine laser was developed in 2002 [13], Thales Laser is developing 13.5 nm Extreme UV light (EUV) [14], and currently the shortest wavelength reached is 5.9 nm, an achievement matching FIB resolution that was made by the UK X-ray laser consortium. The latter is extremely bulky, but a laser that emits X-rays in the 14–20 nm wavelength range is sufficiently small to fit on a tabletop at the Lawrence Livermore National Laboratory [15]. For now, however, an alternative technique is still necessary to determine the horizontal concentration profiles of FIB implanted materials. The remainder of this section is devoted to examining three possible candidates: secondary ion mass spectrometry (SIMS), Rutherford backscattering spectrometry (RBS), and positron annihilation spectroscopy (PAS).

In SIMS an FIB is used to sputter atoms from the surface of the sample. Some of these atoms will become charged through what is known as secondary ionization. These secondary ions can be analyzed using a standard mass spectrometer to determine the composition of the sample. Because this technique relies on sputtering, it is possible to use a time series of data from the spectrometer to create a depth profile of concentration as the incident ions slowly eat through the substrate. In general, SIMS can produce 3D concentration maps with ~50 nm lateral resolution, which is sufficient for use with FIB implantation. The major drawback to using SIMS is that its reliance on sputtering makes it a destructive characterization process. A sample analyzed with SIMS becomes worthless. Thus SIMS is useful only when naturally occurring sample-to-sample variation can be neglected [12].

RBS is a largely nondestructive technique that is quite similar to SIMS. In RBS an FIB is used to produce a stream of light ions such as He^+ focused onto the substrate. A detector is placed to catch ions that backscatter at an angle of nearly 180°. The energy of these ions depends on the depth of their penetration and the mass of the atom or ion off which they scattered. Thus the energy spectrum can be combined with calibration data and the atomic masses of the implanted and substrate elements to derive a 3D concentration map largely equivalent to that produced by SIMS. The main issue with RBS is that it cannot distinguish between heavy elements with similar masses despite being most useful for detecting heavy ions in a heavy substrate (since there is little risk of the "measuring" ions causing lattice damage). A second nondestructive technique called particle induced X-ray emission (PIXE) analysis needs to be used in conjunction with RBS for accurate mass identification and distinction of such medium to heavy elements [17]. A further complication is that most of the incident ions in RBS do not backscatter but are instead implanted. These combine to limit the usefulness of RBS for the characterization of FIB implanted samples. This is because the small features created by FIB are especially susceptible to the small amounts of damage introduced by RBS of light substrates and implanted species. In addition, the electrical properties of these features are more likely to be degraded by the presence of additional impurities [16,18,19].

Traditionally PAS has been used to characterize the defect distribution in implanted materials, but it is currently being adapted to provide information about dose as well as correlations between the two maps. The basic premise behind the most applicable type of PAS is that the gamma rays produced when a positron and an electron annihilate each other experience Doppler broadening, the extent of which is determined by the momentum of the electron before annihilation. Since free, vacancy-trapped, and surface electrons all have different momenta, the observed gamma ray spectrum can be used to determine the relative concentration of each. This can in turn yield a 2D map of defects in the material with resolution comparable to that of an electron microscope. The energy of the incident positrons can be tuned to modify the depth to which they penetrate, thus extending this map into the third dimension. Limitations arise from the fact that the relatively new technique has not been perfected yet, and the concentration profiles it produces have a vertical resolution that degrades linearly with depth due to scattering of the positron beam after impinging on the sample. However, it was recently shown that for depths of 50–500 nm beam based PAS has a sensitivity of 10 nm [21], a good resolution for FIB characterization. Thus, PAS is well suited for surface or near-surface analysis [20].

In summary it is apparent that although satisfactory methods for dose control exist (as they must for FIB implantation to be possible at all), there is room for improvement in post-implantation characterization. Thermal wave analysis, the standard post-processing dosimetry technology in conventional implantation, will not be well suited to measuring FIB implanted samples until short-wavelength lasers or EUV are successfully employed. While several techniques exist that are capable of providing dose information with the spatial resolution necessary for use with FIBs, none is yet applicable under all circumstances. Thus, it is necessary to choose a dosimetry method based on the specifics of the FIB implantation step being performed. Selecting an appropriate characterization procedure allows implantation to be tuned (via trial and error) to a precision far greater than is achievable using the dose controller alone. As will be discussed in the next section, dose is just one of several parameters that must be tightly controlled in order to fabricate devices using FIB implantation.

12.4 Important parameters

As is described in Chapter 2, the interactions of ions in matter, though mediated by a host of complex relationships, have been modeled with reasonable accuracy. Although these same relationships govern FIB implantation, it is often useful to conceptualize the distribution of normally incident implanted ions using the first-order approximation that the concentration profile as a function of depth is given by a Gaussian distribution with mean R (the range) and standard deviation ΔR_P (the longitudinal straggle) and that the profile as a function of lateral position is a Gaussian surface of revolution centered at the middle of the ion beam with standard deviation ΔR_L (the lateral straggle). Along with the distribution of defects formed during

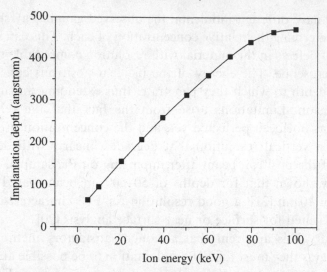

Figure 12.1 Simulation results of ion implantation depth versus ion energy in diamond film [25].

implantation, which is discussed fully in Chapter 2, these are the main variables of FIB implantation that cannot be directly controlled [22].

The objective of a scientist or engineer using FIB implantation to produce a precise concentration profile of a certain impurity species in a particular substrate is to achieve the desired value of R at all points on the target while minimizing ΔR_P and ΔR_L. This minimization is required since ΔR_P and ΔR_L determine the maximum achievable pattern resolution. The free variables over which this three-fold optimization is performed are the implantation energy, implantation angle, dose rate, scan rate (speed at which the beam travels over the target surface), and temperature.

As is well known, R, ΔR_P, and ΔR_L are all increasing functions of the implantation energy, but their precise dependences on it are quite complex [23]. The best understood of the three is R. In fact, to a first approximation, the implantation depth R increases linearly with ion energy [25], as seen in the graph for diamond film (Figure 12.1). The scattering that occurs as the ions enter the substrate limits R in diamond film to less than 500 nm, and leads to a lateral profile that broadens with depth, not unlike that shown in Figure 12.5 [26]. Most often, given a target material and a species to be implanted, experimenters select an implantation energy to provide the desired value of R and then attempt to adjust the other parameters to improve resolution. Jager *et al.* modeled the combined effects of implantation energy and ion mass on the pattern resolution and found, as can be seen in Figure 12.2, that, all

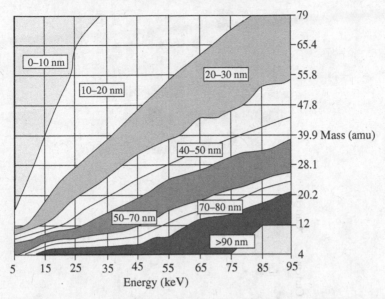

Figure 12.2 Results of Jager's model of the effect on ion mass and implantation energy on the achievable pattern resolution. This graph defines resolution as the lateral range in which 50% of ions are implanted, this corresponds to roughly 0.67 ΔR_{L} on each side [17].

other things being equal, heavy ions can achieve a given resolution at higher energies than light ones [24]. Nakagawa and his colleagues have published several interesting papers on the effect of implantation energy on the ratio of ΔR_{L} to ΔR_{P} [26,27]. They have found through modeling and comparison with data in the literature that the behavior of this ratio with respect to energy in crystalline substrates depends on the implantation angle and the ratio of the masses of the implanted and target species. With an implantation angle of 7°, at which the arrangement of target atoms appears to be random, this ratio increases with energy while, with normal incidence, the ratio remains relatively constant or decreases.

Because focused ion beams have finite resolution, the rastering systems used for FIB implantation must divide the target into an array of discrete pixels, which are implanted one by one. Typically, the ion beam scans meander-like (Figure 12.3) across the sample spending a time, t_{d}, implanting each pixel. This time is called the dwell time and it is inversely related (by the beam diameter) to the scan rate. It is immediately apparent that the dose is equal to the product of the dose rate, dwell time, and number of scans. In other words, for a given dose rate, it is possible to implant the same dose in a large number of fast scans or a small number of slow ones. Hausmann *et al.*

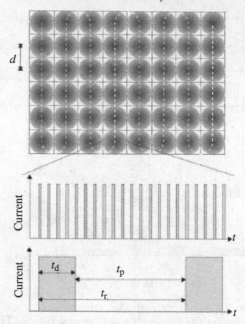

Figure 12.3 Schematic picture of the serial nature of FIB implantation. The upper image gives a top-down view of the implantation area. It is divided into pixels with diameter d. The beam is scanned meander-like over this area. The lower image gives a current vs. time graph as seen by a single pixel. Implantation can be done with a short dwell time, t_d, and thus short pauses between implantations as the beam targets the other pixels, or with a long dwell time and thus long pauses [20].

found in several studies on the formation of cobalt disilicide layers in silicon via FIB implantation of cobalt that dwell time has a significant effect on the results of implantation [28,29]. Specifically, they found that, for dwell times greater than some critical value, the $CoSi_2$ layers that formed after implantation were no longer smooth but instead quite irregular. Furthermore, they demonstrated that the critical dwell time increases exponentially with temperature (Figure 12.4), suggesting that some sort of dynamic annealing process is behind the effect. Dynamic annealing refers to the healing of implant damage even as the implantation process is still occurring, when the heat applied to the substrate makes the point defects more mobile [31]. In a similar vein, Tamura *et al.* observed improved post-anneal electrical activation of boron ions implanted into silicon at low (of order 10^{-2} cm/s instead of order 1) scan speeds [30]. They attributed this to the increased amorphization of the silicon substrate due to the longer dwell time. So while scan rate has not been shown to have a clear effect on the resolution of FIB implantation, it is

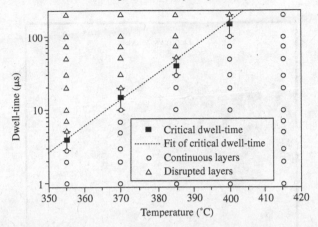

Figure 12.4 Plot showing the quality of $CoSi_2$ layers as a function of dwell time and temperature for FIB implantations of Co^{2+} performed at 70 keV with a current of 0.52 nA and a dose of $1.0 \times 10^{17} cm^{-2}$ [20].

nonetheless an important variable in determining the end result of an implantation step.

Although heating of the substrate may be an inadvertent effect of the ion beam, for FIB lithography with implantation, the substrate with implanted pattern actually needs to be thermally treated by a rapid thermal annealing step (700 °C for 60 s) to prevent ions from spreading continuously in the implanted area (Figure 12.5). This would cause defined lines on the integrated circuit, microoptical, or other device to broaden. Avoiding such loss of lateral resolution is especially crucial for fabrication of Bragg gratings in DFB (distributed feedback) laser diodes, in which the thermal annealing selectively intermixes the quantum wells in the implanted areas. The intermixing increases the band gap of the energy by about 40 meV. The modulation of the band-edge absorption between implanted and nonimplanted areas forms an absorption grating where spacing must be precise and uniform [33].

The dose rate affects FIB implantation in a way similar to the scan rate, though not necessarily through a similar physical mechanism. Dose rate is defined as the number of ions per unit area per unit time incident on the target. As described in the previous section, it is usually obtained by dividing the observed current density by the charge per ion. The narrow beams used in FIB implantation allow for typical dose rates several orders of magnitude larger (order 10^{17} instead of 10^{14} ions/cm²s) than those of unfocused ion beams. In their study, Tamura's group also noted that the effect of increasing the dose rate was quite similar to that of decreasing the scan rate. Both

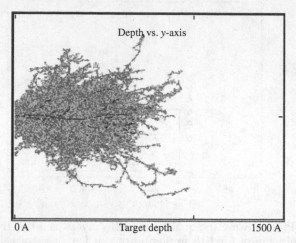

Figure 12.5 Result of Monte Carlo simulation of Ga^+ distribution on surface of fused silica with energy of 50 keV and ion incident angle of 0° simulated by professional software of SRIM2000 [25].

seemed to lead to increased amorphization and thus increased electrical activation of the implanted boron [30]. In a study of Si implantation into GaAs, Lezec *et al.* found a degradation of electrical properties with increasing dose rate. Interestingly, they also found a decrease in longitudinal straggle with increasing dose rate. It is worth note that in Lezec's experiment, the low dose rate was provided by a broad beam system, while the higher rate was provided by an FIB. This means that the effects of focusing might be a confounding factor in their results. Regardless, the reduced longitudinal straggle in the high dose rate FIB implantation is yet another reason that FIB is superior to broad beam technology for nanotechnology applications [32].

Although studies that consider the effect of modifying a single variable are certainly quite useful for isolating dependencies and developing and calibrating models, they tend to overlook some of the more interesting and complex relationships that arise in FIB implantation. In a fairly recent study, Posselt *et al.* considered the effect of dose, dose rate, and temperature on channeling in the implantation of Ge into Si [34]. Channeling, a major source of deviation from the Gaussian approximation discussed above, refers to the increased mean range of ions in crystalline substrates when they are incident along ordered crystallographic planes. Concentration profiles for implantations in which channeling has occurred are strongly skewed toward greater depths. Posselt's experiment and computer modeling revealed a remarkable interdependence between the variables studied. As is apparent from Figure 12.6, the size of the tail of the concentration profile relative to its maximum height, a qualitative measure of the extent to which channeling has occurred, remains relatively constant with increasing dose for a low dose rate at 250 °C, decreases somewhat with dose for a higher dose rate at the same temperature, and decreases the most with dose for the high dose rate at room temperature. In addition, it is quite striking that such a wide variety of concentration profiles could be produced at constant implantation energy and angle of incidence. As can also be seen in Figure 12.6, the experimenters' proposed model of damage accumulation provides a good fit to the observed data.

In summary, this section has tried to make salient the somewhat abstract fact that the parameter space for FIB implantation is significantly larger than that of broad beam technology. This comes about through both extended feasible ranges for parameters such as dose rate and feature size that are common to the two systems, and also through the introduction of parameters such as scan rate that are unique to FIB implantation. Although FIB implantation's greatest advantage is its high resolution, its responsiveness to changes in the variables described above is also fundamentally important. As will be seen in the next section, the creation of novel nanoscale devices using FIB implantation often requires more than just high spatial resolution.

12.5 Device applications

FIB systems are, in many ways, ideal for the fabrication of devices that require nanoscale processing. FIB implantation in particular, is well suited to the creation of high-resolution concentration profiles. In addition, as described in the previous section, the large number of controllable parameters in FIB implantation allow for the tailoring of impurity distributions to precise specifications. In this section, several applications of FIB implantation to nanotechnology will be discussed in an attempt to illustrate the wide variety of uses that have been found for this versatile process.

Just as broad beam ion implantation allowed for the creation of tightly controlled longitudinal dopant distributions in semiconductors, FIB implantation ushered in a new era of precise lateral distributions. FIB implantation's resolution has already permitted the more effective fabrication of binary grating with fine lines and small periods, especially in Bragg grating that feature uniform 100 nm slits, fabricated for use in laser diodes such as DFB lasers and DBR tunable lasers [33]. Taking advantage of such capability, groups have also experimented with several low-dimensional configurations. Two-dimensional field effect transistors (FETs) have been fabricated using the

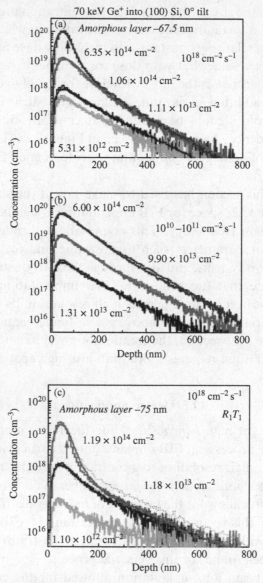

Figure 12.6 Range distributions of Ge in Si. The histograms represent experimental data, while the continuous lines are simulation. The conditions of implantation were (a) 250 °C, high dose rate; (b) 250 °C, low dose rate; (c) room temperature, high dose rate. Note how the tails make up a smaller fraction of the total area under the distributions for lower temperature and higher dose at the higher dose rate.

high dose rate implantation of Ga into AlGaAs/GaAs single heterojunctions to define narrow conducting channels with widths as small as 200 nm (and effective widths an order of magnitude smaller) via the creation of thin, insulating layers of lattice damage [35,36]. The main problem with this design, poor enhancement behavior, was solved by a change in the manufacturing process. Two-dimensional AlGaAs/GaAs FETs with channels defined by the implantation of Si, a p-dopant in this system, showed higher electron mobility near the channel boundaries and thus superior enhancement behavior (1.6 μA with no voltage on the gate, 16 with a gate voltage of 8 V, and no current with −1 V on the gate) [37].

Low-dimensional FETs, are just one of many transistor types that can be made using FIB implantation. The fabrication of junction field effect transistors (JFETs), using this method was first demonstrated more than 20 years ago [38]. In the intervening decades, FIB implantation manufacturing methods for JFETs have matured a great deal [39]. Furthermore, researchers have taken advantage of improvements in FIB technology to devise new uses for these transistors that would not otherwise have been possible. A prime example of this is a fairly recent study by Vitzethum *et al.* [40] demonstrating a highly sensitive photodetector using a structure that combines a low-capacitance p-i-n diode with a JFET. The cross-shaped geometry required for this device, a full discussion of which is beyond the scope of this chapter, requires p-type, insulating, and n-type layers each with one submicrometer spatial dimension and another of order 1 μm. The p-doped layer is created using FIB implantation of Be into GaAs, leaving a natural insulating layer above it on which the n-doped layer is grown by molecular beam epitaxy. The precisely structured nanoscale architecture of this device results in excellent performance characteristics. A preliminary device showed a photoconductive gain of 5×10^6, while it is predicted that a reasonable program of optimization could result in gains $> 10^{12}$, more than sufficient for single photon detection [40]. FIB implantation (also called FIB lithography) also opens the door for entire readily customized integrated circuits. It can be used before, after, or as a step in any lithography process to offer specific customization of other lithographic techniques as a process step or during a prototyping phase, for either one-step fabrication of a device, or partially modifying a template to match a set of specifications [42].

Another application of FIB implantation, which has been explored by Fu and Ann Bryan [25], is fabrication of microoptical waveguides (Figure 12.7). The operating principle of these devices is that the refractive index of an implanted area of a film is greater than that of a nonimplanted area. An optical path for light to follow can be formed by use of FIB implantation

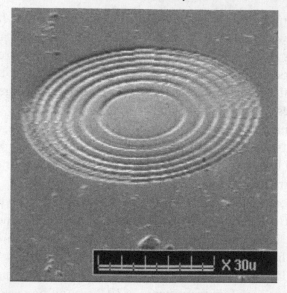

Figure 12.7 SEM micrograph of diffractive optical element (DOE) with six annuli and diameter of 65 μm fabricated by FIB milling with milling sequence from edge to center under an ion energy of 50 keV in diamond film [25].

in commonly used materials such as quartz [25], and this is the basis for the development of complex coupled GaInAsP/InP distributed feedback (DFB) lasers based on maskless FIB lithography. By combining implantation-enhanced wet chemical etching and implantation-induced thermal quantum well intermixing, Konig *et al.* [43] defined a refractive index grating that self-aligned to a gain grating, forming a complex coupled grating lateral to a ridge waveguide. A high single mode yield of over 90% over a large tuning range (88 nm) was achieved. The fabrication of photonic and optoelectronic devices, lateral waveguiding in InP-based materials, the possibility of monolithic integration of bandgap shifted waveguide areas into active devices and the improvement of the lateral carrier confinement in ridge waveguide lasers are all areas of current research [44].

Although the majority of devices fabricated using FIB implantation rely on that process to introduce impurities that change the electrical properties of materials, there are many interesting cases in which FIB implantation has been used for other purposes. For example, Ocelic and Hillenbrand [41] have recently used FIB implantation to construct the first subwavelength phonon photonic structure. Taking advantage of the fact that surface phonon polaritons (couplings of an infrared photon and a phonon) are highly sensitive to crystal structure, the experimenters implanted SiC with Be ions in a

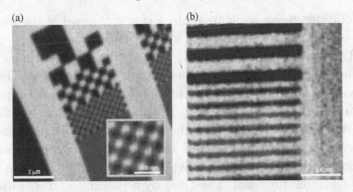

Figure 12.8 Infrared near-field amplitude images of Ga implanted structures. (a) Image of a checkerboard pattern taken at $891\,cm^{-1}$. The inset shows a close-up of the 200 nm squares. (b) Image of implanted stripes 200 and 100 nm wide [31].

checkerboard pattern with squares 1 μm on a side. They found that the implanted areas displayed a significantly damped polariton resonance spectrum (which, incidentally, provides a potential method for measuring lattice damage due to ion implantation). Furthermore, their results suggest that appropriate surface patterning by FIB should be capable of confining surface phonon polaritons. Finally, Ocelic and Hillenbrand [41] used focused Ga implantation into SiC to fabricate a similar checkerboard pattern with squares 200 nm on a side (Figure 12.8). Such structures could potentially be of broad import to photonics as waveguides, negative index materials, or even as read-only memory [41,42].

Just as FIB implantation is useful in the production of technologies such as FETs, which are fairly mature, and DFB lasers and subwavelength optics, which are just emerging, it is also being applied to devices that are barely beyond the realm of theory. For several years now, a sizable group of Australian researchers has been pursuing the possibility of charge based quantum computing using phosphorous ions implanted into Si. Whereas the majority of qubit designs rely on the spin states of electrons or ions, their scheme relies on the position states of an electron shared by a P atom and a P^{+} ion separated by a gap of ~50 nm [46]. The key breakthrough that allowed such a design was the development of single ion implantation in the late 1990s [47] and the resulting progress in the single ion implantation of P ions for the construction of various quantum computing architectures [48–50]. Although, to date, the Australian group has used broad-beam implantation and masking to produce proof-of-concept devices, they have found that this method results in a 50% failure rate because of the possibility of two phosphorous ions being implanted into a single channel in the mask. Their proposed solution to this problem is to use FIB implantation to supply the P^+ ions. By using an ion beam that is roughly the same width as the mask channel, they hope to reduce or eliminate double implantation. They predict that this will significantly reduce the failure rate and allow for the production of the sizable qubit arrays necessary for actual computing [46,51]. In the future, FIB implantation may allow them to dispense with masking entirely.

This section has discussed several examples of the application of FIB implantation to the fabrication of novel and interesting devices at all stages of technological development. These are by no means all of the uses of FIB implantation, nor do they illustrate the exploitation of all of this process's desirable properties. Instead, they are intended to be a sample that is not so much representative as provocative. The above are intended to stand as examples of the creative application of FIB implantation to areas where its usefulness is significant but not obvious; the sources cited in this chapter are presented as evidence of the value that can come from such insight.

12.6 Conclusion

This chapter discussed the basics of focused ion beam (FIB) implantation, considered some techniques for characterizing FIB implanted samples, introduced the major parameters involved in this process and a small sample of their intricate interrelationships, and finally provided some illustrative examples of how FIB implantation is being used to develop new devices. It is customary in chapters of this nature to conclude by speculating on the potential future of the technique under examination. Aside from the obvious point that the resolution of FIB implantation will continue to get better and better and the available parameter space larger and larger as the technology is further refined over time, very little can be said with certainty about the future of FIB implantation. As the previous section has tried to illustrate, FIB implantation is a tool versatile enough to be applied to a great variety of challenges in nano-fabrication. Trying to guess the time and nature of the next creative application of FIB implantation based on the research behind this chapter is equivalent to attempting to predict the publication date and subject matter of the next great work of fiction through a detailed study of the pen. That said, if the history of focused ion beam implantation is any indication, such innovations will be plentiful and the scientific community will be continually and pleasantly surprised by their ingenuity.

References

[1] J. Gierak, C. Vieu, H. Launois, G. Ben Assayag and A. Septier. *Appl. Phys. Lett.*, **70**:15 (1997), 2049–51.

[2] B. D. Huey and R. M. Langford. *Nanotechnology*, **14**: (2003) 409–12.

[3] G. Dearnaley. *Rep. Prog. Phys.*, **32**:2 (1969), 405–91.

[4] J. Orloff, M. Utlaut and L. Swanson. *High Resolution Focused Ion Beams* (New York: Kluwer Academic/Plenum Publishers, 2003).

[5] L. Rubin. Advanced applications of ion implantation and their impacts on vacuum technology. *AVEM International Fall Seminar*, 3 October, 2000, Sheraton Boston Hotel, Boston, MA.

[6] S. Richards, B. Cook, P. Eide, B. Flint and D. Gilbert. A new dose controller for the genus 1510/1520/Kestrel MeV ion implanter. *Proc. 1998 Int. Conf. Ion Implantation Technology* (1998).

[7] J. T. Scheuer, A. Renau, J. C. Olson *et al.* VIISta 810 dosimetry performance. *Conf. Ion Implantation Technology* (2000).

[8] L. Chen *et al. Proc. 10th Int. Conf. Photoacoustic and Photothermal Phenomena, AIP Conf. Proc.*, **463**. (1999), 368–71.

[9] W. L. Smith, A. Rosencwaig and D. L. Willenborg. *Appl. Phys. Lett.*, **47**:6 (1985), 584.

[10] A. Salnick and J. Opsal. *J. Appl. Phys.*, **91**:5 (2002), 2874.

[11] M. Sano, M. Harada, M. Kabasawa, F. Sato and P. Sugitani. Dose monitoring of heavy ion implantation by Therma-Wave signal. *Conf. Ion Implantation Technology* (2002).

[12] S. F. Corcoran, J. L. Hunter and A. Budrevich. SIMS characterization of ion implanted materials: current status and future opportunities. *Conf. Ion Implantation Technology* (1999).

[13] Engineeringtalk Editorial Team. Low-wavelength Laser is Fine for Microstructures. *Laser Lines (Industrial and Medical)*, 27 February (2002), see: www.engineeringtalk.com/news/las/las103.html

[14] EUV generation, see: www.thales-laser.com/appli_euvgeneration.html

[15] J. Dunn, *et al. Phys. Rev. Lett.*, **84** (2000), 4834.

[16] G. W. Arnold and J. A. Borders. *J. Appl. Phys.*, **48**:4 (1977), 1488.

[17] Ion Beam Analysis and Characterization Center, University of Minnesota, see: http://resolution.umn.edu/InstDesc/IBAdesc.html

[18] G. Bourdeault, C. Jeyes, E. Wendler, A. Nejim, R. P. Webb and U. Watjen. *Surf. Interface Anal.*, **33**:6 (2002), 478–86.

[19] M. Takai. *Nucl. Instr. Methods Phys. Res. Section B*, **96**:1–2 (1995), 179.

[20] P. G. Coleman, C. P. Burrows, A. P. Knights *et al.* A new tool for nondestructive monitoring of ion implantation. *Conf. Ion Implantation Technology* (2000).

[21] A. P. Knights and P. G. Coleman. *Mater. Sci. Forum* **445–446** (2004), 123–5.

[22] M. Razeghi. *Fundamentals of Solid State Engineering* (New York: Kluwer Academic Publishers, 2002).

[23] J. F. Ziegler, J. P. Biersack and U. Littmark. *The Stopping and Ranges of Ions in Matter*, Vol. 1. (New York: Pergamon, 1985), p. 321.

[24] P. W. H. D. Jager, C. W. Hagen and P. Kruit. *Microelectron. Eng.*, **30**: (1996), 353–6.

[25] Y. Fu and K. A. N. Bryan. *Opt. Eng.*, **39**:11 (2000), 3008–13.

[26] S. T. Nakagawa, Y. Hada and L. Thome. *Proc. 1998 Int. Conf. Ion Implantation Technology* (1998), pp. 767–70.

[27] S. Nakagawa. *Nucl. Instrum. Methods Phys. Res., Section B*, **153**:1–4 (1999), 446–51.

[28] S. Hausmann, L. Bischoff, J. Teichert, M. Voelskow and W. Moller. *J. Appl. Phys.*, **87**:1 (2000), 57–62.

[29] S. Hausmann, L. Bischoff, J. Teichert *et al. Appl. Phys. Lett.*, **72**:21 (1998), 2719–21.

[30] M. Tamura, S. Shukuri, T. Ishitani, M. Ichikawa and T. Doi. *Jpn. J. Appl. Phys., Part 2*, **23**:6 (1984), L417–L420.

[31] Ion implantation damage annealing, see: www.semiconfareast.com/implant-annealing.htm

[32] H. Lezec, C. Musil, J. Melngailis, L. Mahoney and J. Woodhouse *J. Vac. Sci. Technol., B*, **9**:5 (1991), 2709–13.

[33] A. Orth, *et al. Appl. Phys. Lett.*, **69** (1996), 1906–8.

[34] M. Posselt, J. Teichert, L. Bischoff and S. Hausmann. *Nucl. Instrum. Methods Phys. Res., Section B*, **178** (2001), 170–5.

[35] H. Kim, T. Noda and H. Sakaki. *J. Vac. Sci. Technol., B*, **16**:4 (1998), 2547–50.

[36] U. Dotsch and A. D. Wieck. *Nucl. Instrum. Methods Phys. Res., Section B*, **139**:1–4 (1998), 12–19.

[37] D. Reuter, C. Meier, A. Seekamp and A. D. Wieck. *Physica E*, **13**:2–4 (2002), 938–41.

[38] T. Shiokawa, P. H. Kim, K. Toyoda *et al. J. Vac. Sci. Technol., B*, **1**:4 (1983), 1117–20.

[39] A. J. De Marco and J. Melngailis. *Solid-State Electron.*, **48**:10–11 (2004), 1833–6.

[40] M. Vitzethum, R. Schmidt, P. Kiesel *et al. Physica E*, **12**:1–4 (2002), 570–3.

[41] N. Ocelic and R. Hillenbrand. *Nature Mater.*, **3**:9 (2004), 606–9.

[42] W. K. Barnes, A. Dereux and T. W. Ebbesen. *Nature*, **424**:6950 (2003), 824–30.

[43] H. Konig *et al. Appl. Phys. Lett.*, **75**:11 (1999), 1491–3.

[44] J. P. Reithmaier and A. Forchel. *IEEE J. Sel. Top. Quantum Electron.*, **4**:4 (1998), 595–605.

[45] J. P. Reithmaier, E. Höfling, A. Orth and A. Forchel. *AIP Conf. Proc.*, **392**:1 (1997), 1009–12.

[46] L. C. L. Hollenberg, A. S. Dzurak, C. Wellard *et al. Phys. Rev. B*, **69**:11 (2004), 113301-1–11301-4.

[47] T. Matsukawa, T. Shinada, T. Fukai and I. Ohdomari. *J. Vac. Sci. Technol. B*, **16**:4 (1998), 2479–83.

[48] T. Schenkel, A. Persaud, S. J. Park *et al. J. Vac. Sci. Technol. B*, **20**:6 (2002), 2819–23.

[49] R. P. McKinnon, F. E. Stanley, N. E. Lumpkin *et al. Smart Mater. Struct.*, **11**:5 (2002), 735–40.

[50] T. M. Buehler, R. P. McKinnon, N. E. Lumpkin *et al. Nanotechnology*, **13**:5 (2002), 686–90.

[51] A. S. Dzurak, L. C. L. Hollenberg, D. N. Jamieson *et al. arXiv: cond-mat* (2003), p. 0306265.

13

Applications for biological materials

KIRK HOU AND NAN YAO

Princeton University

13.1 Introduction

Traditional imaging of biological samples has been limited to the use of light microscopes, scanning electron microscopes (SEM), transmission electron microscopes (TEM), and atomic force microscopes (AFM). The information provided by these methods is limited, however, lacking the ability to fully characterize three-dimensional morphology and ultrastructure. Although SEM allows for an analysis of surface morphology, in order, however to study subsurface features complex sectioning must be performed outside of the sample chamber. TEM provides ultra-high resolution, but is unable to offer direct study of three-dimensional morphology. In an opposite manner, AFM provides high resolution in three dimensions, but is unable to reveal information concerning underlying ultrastructure. To overcome these shortcomings, researchers have turned to the focused ion beam (FIB). Traditional use of the FIB has been centered on specimen preparation as well as specimen analysis in the field of semiconductors and microcircuits. Capabilities of the focused ion beam/scanning electron microscope (FIB/SEM) system such as micro-sectioning and in-situ imaging provide an efficient method for failure analysis and repair of defective circuits. Furthermore, gas assisted etching of surface layers can reveal underlying circuitry in localized areas. Development of new techniques for the study of materials by FIB analysis is occurring at a greater frequency as ion beams gain in technical significance.

Despite the well-established use of FIB in the semiconductor field, application of FIB to the study of biological samples remains relatively uncommon [1–10]. Nonetheless, traditional techniques for analysis of traditional

Focused Ion Beam Systems: Basics and Applications, ed. N. Yao.
Published by Cambridge University Press. © Cambridge University Press 2007.

hard samples are also applicable to biological samples. In-situ sample manipulation allows for site-specific morphological and structural analysis through quasi-real time viewing of the sample during processing. The ability to quickly mill cross sections in a variety of directions provides a strong understanding of the relationship between cell structure and morphology. Imaging and sectioning of biological specimens from single yeast cells to small arthropods have provided high-resolution images of biological structures from the subcellular level to the microstructural level of tissue and organs [3,6]. Preliminary results indicate the sample preparation must be of primary concern during the imaging and manipulation of biological samples [3,5,9]. Care must be taken to ensure the fidelity of the biological structures on all size scales and to minimize artefacts due to milling or sample charging.

Novel techniques for the study of ultrastructure and morphology in biological samples include FIB tomography with subsequent three-dimensional reconstruction as well as the proposal of nano-biomachining [5,8,11]. The incorporation of a cooled sample chamber holds the promise for minimizing sample damage while helping to retain integrity of biological structures without artefacts introduced by chemical fixation. This chapter covers the topics important to successful FIB milling and imaging of biological samples.

13.2 Sample preparation

Existing procedures for sample preparation have been developed and optimized for effective TEM and SEM imaging. These techniques exist to strengthen biological structures from bulk tissue down to subcellular nanostructures. Only with extensive processing will these samples be able to retain native configurations with minimal loss of structural components. The general procedure for sample preparation consists of the following steps: chemical fixation, dehydration, embedment, and staining or coating. Other alternatives to chemical fixation are cryotechniques in which structural preservation is accomplished through the "freezing" of cellular components. Recent developments have shown that cryo-dual stage SEM/FIB systems are adequate for imaging unprepared biological samples (Figure 13.1) [31,32]. However, chemical methods are more widely employed as samples prepared through cryotechniques require the availability of a cold stage, and "frozen" cells are thermodynamically unstable at low temperatures. FIB imaging and sectioning of biological samples is a relatively unexplored field and use of varying degrees of sample preservation have been employed in existing literature [1–10,22]. Although there exist some instances in which no preparation was required, chemical fixation and cryotechniques remain popular.

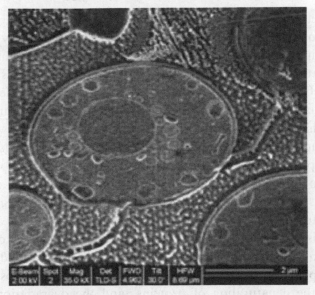

Figure 13.1 Cryo-FIB milled yeast sample originally published in an application note from FEI company [31].

For an in-depth discussion of sample preparation the reader is referred to *Introduction to Biological Electron Microscopy: Theory and Techniques* by Clinton Dawes [12].

13.2.1 Chemical methods for sample preparation
Chemical fixation

By beginning with chemical fixation, cellular structure is preserved in a near *in vivo* state for the subsequent processing. The success of these procedures has been supported by comparison with samples preserved by freezing techniques. Chemicals used for fixation function by forming inter- and intra-molecular cross linkages in proteins to form a gel, preserving cellular support structures and the membrane system of various organelles with little structural change. The quality of a fixation is indicated by the appearance of lipid membranes making up the endoplasmic reticulum and golgi systems. The choice of fixing agent is determined by what cellular constituents are to be preserved with the highest integrity. Common chemicals for fixation include osmium tetroxide, aldehydes, and permanganates. Osmium tetroxide reacts with nearly all cellular components, making it the most popular choice for chemical fixation. The considerations important to successful staining with osmium tetroxide are discussed next.

Osmium tetroxide (OsO_4) is a nonpolar compound, allowing unhindered diffusion through cellular membranes. OsO_4 also has the ability to penetrate hydrophilic lipids as well as aqueous areas of the cell. Osmium has five stable oxidation states, allowing for reactions with most cellular constituents [12]. However, OsO_4 reacts most strongly with carbon–carbon double bonds. As a result, osmium tetroxide reacts most strongly with unsaturated fats and phospholipids. OsO_4 also partially fixes proteins by reaction with phenols and SH groups. This also allows some fixation of nucleic acids by reacting with histone proteins which are involved in DNA packaging. However, OsO_4 is a strong oxidizing agent and may result in the complete oxidation of double bonds to form diols if concentration and fixation duration are not closely monitored [13].

During this procedure, temperature, pH, and osmolarity of the fixing solution are of utmost importance. Buffer solutions are required to maintain a constant pH. As OsO_4 infiltrates the cell, the pH drops drastically, resulting in the denaturation of proteins, and the degradation of protein structures such as microtubules, microfilaments, and intermediate filaments. For animal tissue, the buffer should maintain a pH of 7, while for plant matter a slightly acidic environment of pH 5 to 6 is required. The most common buffers are cacodylate and phosphate due to low reactivity with fixing agents. The fixation solution should also be nearly isotonic with the cellular environment to prevent shrinking and swelling. Also, a hypotonic solution will result in ion leaching, which could destabilize the membrane structures of organelles [12].

As the rate of fixation is directly related to diffusion, temperature begins to play a large role in fixation duration and quality. Room temperature allows for shorter durations, which is less disruptive to the cell; however, leaching of cellular components is highly likely and autolysis of cell membranes will also occur. Temperatures as low as 4 °C prevent these deleterious effects, but require a longer fixation time. If the fixation duration with OsO_4 is too long, however, cellular components may be oxidized too far, resulting in the degradation of cellular structures [12]. Due to the difficulties associated with OsO_4 fixation, aldehydes are often employed for primary fixation of proteins, followed by secondary fixation with OsO_4 to provide electron density and stabilize lipid membranes. By the end of such fixation procedures, electron micrographs will show a granular cytoplasm due to stabilization of cytoplasmic lipo-proteins and dark staining of unsaturated fatty acids. Cell walls will appear darkened along the edges due to low penetration into the dense cell wall. Loss of up to 70% of carbohydrates, 21–50% of lipids, and 12–50% of proteins is common, but these numbers can be reduced by lower temperatures [12].

Post-fixation procedures

For preparation for vacuum, water must be removed to avoid cellular changes upon water evaporation from biological samples. When embedding the sample, dehydration is also required as most embedding media are not miscible with water. Polymerization of plastics will also be hindered by the presence of water. During dehydration, water is slowly replaced by another solvent, usually ethanol or acetone, by slowly increasing the concentration of dehydrant in the infiltrating solution. Acetone is the most popular choice due to miscibility with many plastics, and a lack of reaction with osmium tetroxide. It also will not cause removal of phospholipids. Ethanol is used more sparingly due to cell shrinkage. Special dehydrating agents such as 2,2-dimethoxypropane or 2,2-diethoxypropane can be used to remove water by reaction to produce methanol and acetone or ethanol and acetone [12]. Nonpolar dehydrants can give improved results due to increased retention of lipids and proteins. However, cell compression and membrane expansion will occur during freeze-drying or freeze-substitution.

Although embedding the sample in a plastic or resin is most essential when sectioning in preparation for TEM imaging, sample embedding is also important for the imaging of animal cells without cell walls. Embedment requires the use of a low viscosity plastic to infiltrate the sample. Such plastics can not have volume change upon polymerization, and have low electron scattering effects. Most importantly, the plastic must be stable under ion beam bombardment. To provide the most uniform sectioning, the polymer embedment must be similar to the sample in both density and chemical composition as both parameters affect etching rate. Common embedments include vinyl plastics, methacrylates, epoxy resins, and polyester resins [14]. Drobne *et al.* have used paraplast for the embedding of a crustacean digestive system [8]. Paraffin is another popular choice, although maintaining hot paraffin during preparation is difficult. If the sample is to be imaged in a dual-beam system, stability under electron beam bombardment must also be considered. Epoxies are most stable, while methacrylates are least stable [14].

Staining or coating is required for biological samples in which a large content of carbon, oxygen, and nitrogen prevent high contrast due to low electron density. As a result, coating with heavy metals or post-fixation heavy metal staining is required. Staining refers to either chemical or physical incorporation of heavy atoms into the sample. Common stains include metals such as tungsten, ruthenium, osmium, and lead. These metal ions form coordinate complexes with active groups within the sample, adhering to certain structures to increase contrast. For surface morphology, thin films of gold or palladium are deposited on sample surfaces by sputter coating or high vacuum evaporation. Thin film coatings have the advantage of being an in-situ process that can be performed within the FIB as new sections are generated.

13.2.2 Cryotechniques for sample preparation

Crytotechniques can be used with or without prior chemical fixation and can be performed on living or fixed samples. These techniques are often combined to prepare samples for vacuum by solvent removal. For bulk samples such as arthropods with strong exoskeletons or even yeast cells with thick cell walls, sample embedment is not required; however, removal of liquid within the sample is important [3,9,10]. During ordinary drying, surface tension causes specimen distortion as the liquid volume within the sample decreases [14]. Cryotechniques minimize these effects through increased evaporation or elimination of surface tension effects.

The favored techniques for this task are freeze-drying and critical point drying. Freeze-drying employs vacuum induced sublimation of the solvent to accomplish solvent removal. The largest drawback of freeze-drying is sample shrinkage during freezing. As a result, critical point drying is the more popular choice for solvent removal. During critical point drying, the sample is heated until the solvent reaches its critical point, allowing the solvent to be bled off without surface tension effects or other interactions that could destabilize cellular structures. The critical points of liquid solvents occur at temperatures and pressures too high for practical application, therefore, carbon dioxide or Freon are the preferred solvents. Nevertheless, surface tension damage is not completely avoidable, and a minimal amount of shrinkage is still observed.

Yonehara *et al.* [1] have imaged the small intestine of Wister mice which had been prepared using critical point drying. These samples were first ion beam etched to reveal the small intestine where microvilli were imaged by SEM to obtain high-resolution images after coating with 10 nm of heavy metal [1] (Figures 13.2 and 13.3).

13.2.3 FIB imaging without sample preparation

Ballerini *et al.* [6] have directly imaged and sectioned yeast cells without prior sample preparation (Figure 13.4). Yeast cells on filter paper were directly transferred into the sample chamber. The semipermeable membrane allows for the evaporation of water from within the cell, while the thick cellular wall provided the structure to prevent sample shrinkage and collapse. Despite a

Figure 13.2 Apical view of microvilli of mouse small intestine taken with SEM at 3 kV after ion beam etching to remove surface mucosa [1].

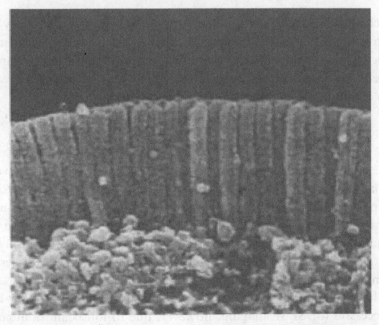

Figure 13.3 Side view of microvilli of mouse small intestine taken with SEM at 3 kV after ion beam etching to remove surface mucosa [1].

Figure 13.4 FIB microsection of yeast cells taken by Ballerini *et al.* with no sample preparation [6].

lack of chemical fixation, ultracellular support structures were clearly visible. They have suggested that imaging of cells such as lymphocytes or chondrocytes will also be possible without much sample preparation (Figure 13.5). Yeast cells may be a special case, however, as they are known to survive extreme conditions including complete dehydration.

13.2.4 Sample mounting

Due to sample embedding, the capsule must be trimmed to reveal the specimen before imaging. Special considerations must be taken to avoid build up of charge on the sample during ion beam or electron beam bombardment depending on the system. Copper tape connecting the sample to the sample holder is an option that is simple and easy to apply. Conductive paints such as silver or carbon have also proven successful in maintaining conduction between nonconducting polymer samples and the sample holder [15]. Complete coatings of iridium, platinum, or gold have been successful for SEM imaging; however, due to ion beam bombardment in the FIB, these methods are less successful as they will be removed. In-situ coatings are possible using ion assisted deposition for small cross

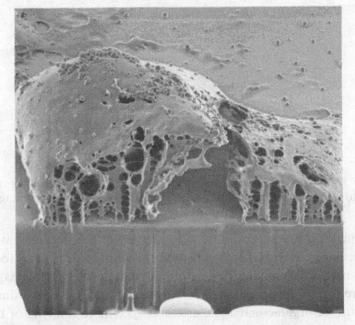

Figure 13.5 Chondrocyte imaged after FIB sectioning by Milani *et al.* [7].

sections, which can increase the effectiveness of dual-beam systems in which SEM imaging is possible.

If serial ultra-thin sectioning is to be performed in a dual-beam system, the angle of the sample face with regard to the ion and electron beams is extremely important. The ion beam must have a grazing angle of incidence to decrease damage to the sample such as amorphous surface layers and ion incorporation into the sample. The configuration of the ion and electron beams must be considered when serial cross sections are examined. Dual-beam machines currently on the market have an FIB at between 48 and 52° from the normal electron beam. Due to the limited tilt of the stage, to obtain a grazing incidence angle of the ion beam, the sample face must be tilted as illustrated in Figure 13.6.

13.3 Ion–sample interactions

Understanding the reactions that occur when energetic ions interact with sample surfaces is essential to preparing high-quality images and uniform cross sections. As with traditional FIB applications within the field of semi-conductors and metals, interactions ranging from ion implantation to sample sputtering and secondary electron emission involve the transfer of energy from

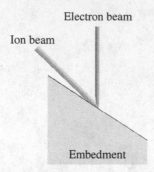

Figure 13.6 Proposed ample embedment angle to achieve grazing surface angle to minimize damage during cross-sectioning with eucentric stage.

the incident ion to the solid surface. Knowledge of these interactions, specifically molecular sputtering theory, enhances the ability to utilize the focused ion beam as an effective milling device. Unfortunately the FIB's imaging capabilities are compromised by artefacts resulting from ion bombardment. Due to the nature of organic materials, artefacts specific to imaging of biological samples become a matter of importance. Careful study of reactions resulting from an incident beam of ions provides a foundation for the analysis and minimization of artefacts generated during FIB milling and sectioning.

With both hard (semiconductors, ceramics, metals, etc.) and soft (organic) materials transfer of momentum from incident ions to the sample surface is the result of elastic collisions and inelastic interactions. The former leads to physical sputtering of surface molecules if the kinetic energy of the impinging ion beam is greater than the bonding energy of those molecules. Complex inelastic interactions with soft materials give similar results to those noted in hard materials such as the generation of secondary electrons and phonons. All ion interactions with the sample result in a decrease in kinetic energy of the impinging ions. If the ions are not backscattered, they will remain implanted within the material. Artefacts such as defects, sample heating, and amorphization typical of hard materials may have even more complex effects in organic materials. Denaturation of proteins, instability of lipid proteins and preferential sputtering may result. Detailed analysis of ion–solid interactions in traditional hard materials are provided in the work by Nastasi *et al.* [16].

13.3.1 Molecular sputtering theory

Physical ejection of atoms from the surface of a target depends on the kinetic energy of the incident electron beam and the energy with which atoms are bound to the surface (sublimation energy). The energy transfer depends on the size, mass, and charge of the ions and atoms. The power potential law proposed by Lindhard *et al.* [17] suggests that the power potential between ions and atoms varies as

$$V(r) = \frac{Z_1 Z_2 e^2 a^{n-1}}{n r^n},$$ (13.1)

where Z_1 and Z_2 are the atomic numbers of the incident ion and target atom, respectively, r is the distance between the two interacting atoms, and e is the charge of an electron. n depends on the type of collision: $n = 1$ for Rutherford type, $n = 2$ for weakly screened collisions, and $n = 5$ for collisions of hard spheres. a is the effective screen radius of the atoms as given by

$$a = 0.8853 \times a_H \times (Z_1^{2/3} + Z_2^{2/3})^{-1/2},$$ (13.2)

where a_H is the Bohr radius for the hydrogen atom.

Kanaya *et al.* (1988) enhance this theory to develop atomic sputtering theory by accounting for the weak-screening effects and hard sphere collisions [2]. Molecular sputtering theory as proposed by Kanaya *et al.* (1992) builds upon atomic sputtering theory to account for bonding energies of molecules [18]. For samples of a pure solid, the sputtering yield is dependent on the nuclear stopping cross section, the number of atoms per unit volume, and the number of primary knock-on atoms per incident particle, as well as the number of atoms that recoil as a result of collision cascade effects.

The proposed semi-empirical relationship for the sputtering yield (ejected atoms per incident ion) as a function of the energy of the incident beam (E) derived by Kanaya *et al.* [2,18] takes the form

$$Y_A = Y_m \left(\frac{E}{E_m} \right)^{1/2 - 1/n}.$$ (13.3)

The maximum theoretical yield Y_m is given by

$$Y_m = \frac{0.45 \times \rho Z_1 Z_2 (1 + M_1/M_2)}{M_2 E_s (Z_1^{2/3} + Z_2^{2/3})^{-1/2}},$$ (13.4)

and E_m is given by

$$E_m = \frac{0.4 Z_1 Z_2 e^2 (M_1 + M_2)}{a M_2},$$ (13.5)

where the coefficients 0.45 and 0.4 have been determined experimentally, M_1 and M_2 are the atomic masses of the incident ion and target atom, respectively,

E_s is the sublimation energy of the atoms, and ρ is the density of the target. It is important to note that the sputtering yield is linearly dependent on target density and inversely proportional to the sublimation energy.

Molecular sputtering theory for organic substances such as polymers and biological specimens is defined by the ratio of removed molecules per incident ion, and is given by a number weighted average of sputtering yields for individual atoms:

$$Y_m = \sum \%_A Y_A, \tag{13.6}$$

where Y_A is given by (13.3). The molecular thinning rate ($nm \cdot cm^2 / mA \cdot min$) in terms of the sputtering yield is given by

$$R_M = 6.23 \times \left(\frac{S_M}{\rho_M}\right) \times G_{-R} \tag{13.7}$$

where G_{-R} is

$$G_{-R} = \sum \frac{A_{-R}}{E_B} \tag{13.8}$$

A_{-R} is the removal mass of a particular molecule and E_B is the bonding energy with which the molecule was bound to the bulk material. Values for removal mass are affected by the packing density factor γ, which accounts for the increased volume within the sample due to vacuoles and lipid vesicles. As a result, the removal mass, A_{-R}, is actually calculated as γA_{-R} for use in (13.8). Once again, an inverse relationship between the mass removal rate and the bonding energy is predicted. The effect of atomic packing is confirmed by comparing the sputtering of enamel and dentine. Due to a decreased packing factor, enamel with its increased hardness shows a lower removal rate despite a similar composition to dentine as demonstrated by Kanaya *et al.* [19]. Enamel shows a predicted thinning rate of 95 $nm \cdot cm^2 / mA \cdot min$ with an experimental value of 86.6 while dentine with a predicted value of 122 $nm \cdot cm^2 / mA \cdot min$ has a measured value of 123 $nm \cdot cm^2 / mA \cdot min$.

13.3.2 Artefacts and considerations

Use of FIB to section and mill biological samples is preferable to ultra-microtomy due to stress-free milling. Samples will not show the effects of shear stress or scratch marks due to the cutting by diamond knives. Despite these advantages, radiation damage as a result of milling or sectioning of a biological specimen using a focused ion beam often results in noticeable artefacts. Such damage includes ion implantation, sample heating, amorphization, destabilization of native structures, and milling artefacts. Despite the unavoidable nature of these phenomena steps can be taken to minimize their occurrence as well as their effects.

Ion implantation is commonplace in any sample that has been milled or imaged with a focused ion beam. The effects of ion implantation range from sample heating due to energy dissipation, defect formation, and amorphization of the surface layer. The penetration depth depends on the stopping ability of the sample, which depends strongly on Lindhard's power potential theory reviewed earlier. Greater penetration depths result in thicker amorphous layers, and therefore increased artefacts when performing surface imaging. Drobne *et al.* [8] noted the presence of such amorphous layers when sectioning the digestive system of a crustacean, although no significant reduction in the quality of the image was reported. Redeposition of sputtered material may resemble amorphization of surface layers; however, these artefacts are unrelated, with redeposition only becoming a significant issue when dealing with deep cross sections or trenches. Ishitani *et al.* [20] suggest that lowered incident ion beam energy will decrease implantation depth and minimize thickness of the amorphous layer, which has been confirmed by Monte Carlo simulations. However, increased damage density at the surface is a likely consequence. Amorphous layers produced by milling with a high energy ion beam can be removed and smoothed by subsequent scanning with a lower energy beam to increase the integrity of imaging.

An important point to consider is the stability of complex cellular structures under ion beam bombardment. High energy ions may result in the denaturation of proteins that form cellular support structures, or destroy cross linkages produce during fixation, which may destabilize cellular features. Stability is also decreased by sample heating indirectly caused by ion implantation. Transfer of kinetic energy from ions to the atoms resulting in the emission of secondary electrons often leads to thermal damage as these electrons are reabsorbed by surrounding atoms. When the electron stopping power of the sample is high, the increase in thermal energy becomes an important factor, especially in frozen samples on a cold stage without chemical fixation. Although Averbeck *et al.* [21] has performed simulations that show that sample heating in hard materials can be considered negligible, these local effects may still have a large impact on biological samples.

On a larger scale, milling artefacts of biological specimen may result in striations or ridge-like filaments in the direction of the ion beam [3,9]. The factors leading to the formation of these structures are poorly understood, however certain measures can be taken to reduce the presence of such artefacts. After initial milling at 5 to 7 nA with a beam energy of 30 keV Ga^+,

Drobne *et al.* [9] performed a final cleaning mill at an aperture corresponding to 0.3 to 1.0 nA to remove surface layers containing milling artefacts. In a similar method, Young *et al.* [3] utilized low energy ions to remove thin surface layers, which revealed surface melting, amorphization, and redeposition. These artefacts that arise as a result of ion–sample interactions are unavoidable. Although their formation is not completely understood, their presence can be minimized through optimization of the FIB for life science applications.

13.4 Life science applications of FIB

As with the focused ion beam applications in the realm of traditional materials such as semiconductors, metals, and ceramics, the most important advantage of FIB is the ability to remove surface layers to reveal underlying microstructures. This has presented scientists with an unprecedented ability to analyze the structure of biological specimens quickly and efficiently without the artefacts resulting from other sectioning methods such as ultramicrotomy or freeze-fracturing. In-situ imaging provided by FIB secondary ion imaging through collection of secondary electrons in a manner similar to that of SEM allows quasi-real time observation of the surface for improved sectioning accuracy. In addition, due to physical effects in electron emission, FIB secondary ion imaging often avoids the charging noted during SEM imaging, as discussed in the previous chapters. Indirect imaging of biological samples is possible when imaging shadowgrams produced by soft X-ray contact microscopy [30]. SIM imaging allows for direct imaging while sectioning of the silicon nitride relief provides a direct height measurement of the relief, which gives a sense of the size and thickness of the organism being imaged. Despite these advantages, FIB utilization within the field of biological studies remains a relatively new field. Thus far, life science applications have been limited to adaptations of existing techniques to fulfill the needs of those studying biological materials. The most popular uses to date are simple sectioning of biological materials and secondary ion mass spectroscopy (SIMS).

13.4.1 Sectioning of biological materials

Generation of cross sections and removal of surface layers are extremely crucial to the study of structural elements in biological specimen ranging from single cells to complete organs and organisms. In the past, sectioning of soft materials required embedment of the sample within a hard plastic matrix

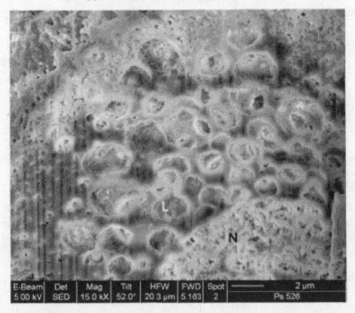

Figure 13.7 SEM micrograph of digestive gland cell by Drobne *et al.* [22] before ion beam bombardment shows charging. N stands for nucleus. L stands for lipid vesicle.

to avoid scratching and extreme deformation due to the applied stresses of diamond knives. Use of FIB provides stress-free sectioning, negating the need for embedding of biological specimen with strong cell walls and exoskeletons as described earlier in this chapter. Through the use of embedments, an increasing range of samples suitable to FIB sectioning are available, such as gland cells, digestive systems, and microvilli of the small intestine of small terrestrial organisms [1,3,8,22]. Sample preparation was performed in accordance with traditional preparation of biological specimens, while FIB sectioning was performed in much the same way as traditional sectioning of hard materials. Charging was noticed in dual-beam systems during imaging with the electron beam; however, pre-bombardment with low current ion beams was able to avoid charging (Figures 13.7 and 13.8) [22].

A natural extension of normal sectioning is three-dimensional reconstruction using serial sectioning. This technique, which has been used extensively to study semiconductors and metallic alloys, is slowly finding its way into the field of life sciences. Three dimensional reconstruction of single cells with the FIB is a key element in improving the knowledge of cell structure and the relationships of different cellular components. Reconstruction using serial sectioning has been slowly developed and refined since the 1960s. Keddie *et al.* [23] and Davison *et al.* [24] utilized serial sectioning

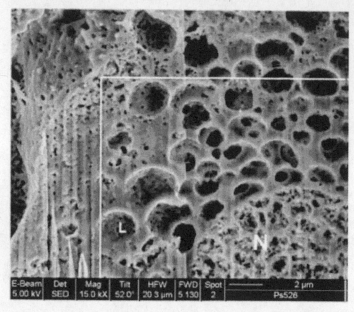

Figure 13.8 SEM micrograph of digestive gland cell by Drobne *et al.* [22] after ion beam bombardment shows no appreciable charging. N stands for nucleus. L stands for lipid vesicle.

by ultramicrotomy to study cell sections by SEM. Models were then generated to study the lipid membrane network within yeast cells. These methods relied on traditional sample preparation, and throughout the years have been slowly refined to include cryotechniques such as freeze-substitution for increased accuracy in quantization of the volume fraction of various cellular components in yeast cells [25]. However, these methods still rely on sectioning by ultramicrotomy. The difficulties of serial sectioning in this manner are presented in work by Fiala *et al.* [26]. Due to random distortions in the sections, such as tilt, rotation, and skewing, digital images must first be processed using complex mathematical transformations to achieve correct alignments. Despite the ability to correct for the nonuniform alignment of serial sections, it is difficult to guarantee section thickness, which increases the difficulty of using shape based interpolation when visualizing three-dimensional volumes and also decreases the accuracy when quantizing cellular components.

The use of FIB sectioning suggests a solution to these problems. Serial sectioning in the FIB provides self-aligned images with stress-free cutting and relatively low amounts of sample surface destruction at grazing angles. Inclusion of pre-milled fiducial marks provides a relatively accurate method for quantizing section depth, eliminating the error involved with inhomogeneous section thickness. Despite the relative lack of published material concerning this technique, the steps required to implement three-dimensional tomography in the FIB are already in place. With the production of conduction polymer embedments by inclusion of carbon nanotubes, sectioning can be performed by FIB, and serial imaging by SEM to provide serial sections with high spatial resolution. New developments in sectioning and etching could lead to nano-biomachining in a similar manner to device preparation in the semiconductor industry [4].

13.4.2 Secondary ion mass spectroscopy (SIMS)

Study of cell composition not only includes the structure analysis but chemical analysis as well. Chemical analysis becomes extremely important when studying the destination of proteins after production, and the uptake of molecules from the extracellular environment. The primary advantage of SIMS is the ability to detect elements with parts per billion sensitivity in a small area. Combined with serial imaging, the ability to generate a three-dimensional elemental map of biological structures becomes a possibility. SIMS has found acceptance in the study of chemical uptake by cancer cells. By analyzing the destination of certain chemicals and the specificity of cancer cells for certain drugs, more effective cancer treatments can be developed. One such method is the preferential uptake of ^{10}B by cancerous brain tissue [27–29]. ^{10}B undergoes fission to give ^{7}Li and ^{4}He when a thermal neutron is absorbed. Release of radiation is enough to kill cells within 10 μm of the reaction site. As a result, the location of ^{10}B within the cancer cell is important, with location closest to the nucleus providing the most effective destruction of cancer cells [28,29]. SIMS has shown concentration of ^{10}B at cellular and nuclear membranes as well as within single cancer cells surrounded by normal brain tissue [27]. These results suggest that the injection of compounds containing ^{10}B will provide an effective cancer treatment, eradicating even single cancer cells. SIMS has found applications in many similar situations, and its usage will continue to grow as high spatial resolution of chemical mapping becomes increasingly important in molecular microbiology.

References

[1] K. Yonehara, N. Baba and K. Kanaya. *J. Electron Microsc. Tech.*, **12** (1989), 71–7.
[2] K. Kanaya, Y. Muranaka, K. Yonehara and K. Adachi. *Micron Microscopica Acta*, **23** (1992), 45–64.

[3] R. J. Young, T. Dingle, K. Robinson and J. A. Pugh. *J. Microsc.*, **172** (1993), 81–8.

[4] M. Ballerini, M. Milani, M. Costato, I. C. Edmond Turcu and F. Squadrini. *Proc. SPIE*, **3260** (1998), 221–30.

[5] M. Milani, M. Ballerini and F. Squadrini. *Proc. SPIE*, **3922** (2000), 212–21.

[6] M. Ballerini, M. Milani, D. Batani and F. Squadrini. *Proc. SPIE*. **4261** (2001), 92–104.

[7] M. Milani, D. Ballerini and D. Batani, *et al. Eur. Phys. J. Appl. Sci.*, **26** (2004), 123–31.

[8] D. Drobne, M. Milani, M. Ballerini, *et al. J. Biomed. Optics*, **9** (2004), 1238–42.

[9] D. Drobne, M. Milani, A. Zrimec, *et al. Scanning*, **27** (2005), 30–4.

[10] M. Milani, D. Drobne, A. Zrimec. *Scanning*, **27** (2005), 60–1.

[11] D. N. Dunn, G. J. Shiflet and R. Hull. *Rev. Sci. Instrum.*, **73** (2002), 330–4.

[12] C. J. Dawes. *Introduction to Biological Electron Microscopy: Theory and Techniques* (Burlington, Vermont: Ladd Research Industries, 1988).

[13] M. Jones. *Organic Chemistry* (New York: Norton, 2005).

[14] L. C. Sawyer and D. Grubb. *Polymer Microscopy*, 2nd edn (Oxford: Chapman & Hall 1996).

[15] Z. R. Li. *Industrial Applications of Electron Microscopy* (New York: Mercel Dekker Inc., 2003).

[16] M. Nastasi, J. W. Mayer and J. K. Hirvonen. *Ion–Solid Interactions: Fundamental and Applications* (Cambridge: Cambridge University Press, 1996).

[17] J. Lindhard, V. Nielsen and P. V. Thomsen. *Mat. Fys. Dan. Vid. Selek.*, **33** (1963), 1–42.

[18] K. Kanaya, Y. Muranaka, K. Yonehara and K. Adachi. *Micron Microscopica Acta*, **23** (1992), 45–64.

[19] K. Kanaya, N. Baba, C. Shinohara and T. Ichijo. *Micron Microscopica Acta*, **15** (1984), 17–35.

[20] T. Ishitani and T. Yaguchi. *Microsc. Res. Tech.*, **35** (1996), 320–33.

[21] R. S. Averbeck and M. Ghaly. *J. Appl. Phys.*, **76** (1994), 3908.

[22] D. Drobne, M. Milani, A. Zrimec, V. Leser and M. Berden Zrimec. *J. Microsc.*, **219** (2005), 29–35.

[23] F. M. Keddie and L. Barajas. *J. Ultrastructure Res.*, **29** (1969), 260–75.

[24] M. T. Davison and P. B. Garland. *J. Gen. Microbiol.*, **98** (1977), 147–53.

[25] S. K. Biswas, M. Yamaguchi, N. Naoe, T. Takashima and K. Takeo. *J. Electron Microsc.*, **52** (2003), 135–43.

[26] J. C. Fiala and K. M. Harris. *J. Am. Med. Informat. Assoc.*, **8** (2001), 1–16.

[27] A. C. Oyedepo, S. L. Brooke, P. J. Heard, *et al. J. Microsc.*, **213** (2004), 39–45.

[28] D. N. Slatkin. *Brain*, **114** (1991), 1609–29.

[29] K. Haselsberger, H. Radner and G. Pendl. *Cancer Lett.*, **131** (1998), 109–11.

[30] M. Milani, D. Drobne, F. Tatti, *et al. Scanning*, **27** (2005), 249–53.

[31] P. Loyd. *3D Cryo-Dual Beam* (Hillsboro; OR: FEI company, 2004).

[32] H. Mulder. *GIT Imaging Microsc.*, **2** (2003), 8–10.

14

Focused ion beam systems as a multifunctional tool for nanotechnology

TOSHIAKI FUJII, TATSUYA ASAHATA, AND TAKASHI KAITO

SII NanoTechnology Inc.

14.1 Introduction

In 1979, Dr. Seliger proposed the concept of the focused ion beam (FIB) using liquid-metal gallium as an ion source [1]. The FIB tool focuses ions generated from an ion source using an electric field, irradiates the ion beam on to specimen surfaces, and observes microscopic specimen surfaces by scanning. The scan region of the ion beam can be selectively sputter etched when ions heavier than electrons are used. A scanning electron microscope (SEM) can be used for observation or analysis, but FIB can be used for both observation and processing. Many research organizations and companies are now involved in FIB development.

The Scientific Instruments Division of Seiko Instruments Inc. (currently SII Nano Technology Inc.) started research and development at the beginning of the 1980s and developed a technology called ion beam induced chemical vapor deposition (CVD). This technology makes it possible to accumulate thin films.

In 1984 SII introduced the world's first FIB photo mask repair tool called the SIR series [2]. White defects, the shaded part of the photo mask used in making integrated circuits that falls off, are filled in and repaired by an ion beam induced CVD of carbon film. Later, there were advances in technical development [3], such as the capability to repair black defects left over from shading material in parts that transmit light, and precision processing that corresponds to a miniaturization of the design rules.

In 1986, SII introduced the world's first multi-purpose commercial FIB tool called the SMI series. It was first put to practical use as a circuit edit

Focused Ion Beam Systems: Basics and Applications, ed. N. Yao.
Published by Cambridge University Press. © Cambridge University Press 2007.

tool for semiconductor integrated circuits [4,5]. New wire formation could be done by localized sputter etching with an FIB, by cutting wire or exposing wiring for drilling beneath the passivation layer, and by depositing tungsten film.

At the same time, US and Japanese semiconductor device makers were competing to develop a 1 Mb DRAM. In particular, the 3D structuring of elements and wires to improve memory integration progressed at the same time as the miniaturization of photo-lithography. To verify the cross-sectional structure of test devices, a method was used to prepare and observe specimens with a microscope. But both specifying and observing a location by this method has proved difficult.

On the other hand, a cross-sectioning and observation function using FIB could observe the topography with a microscope function, determine the cross section position, and realize a previously nonexistent function for specifying the process location [6]. The cross section process observation function is the third application of FIB equipment. This function was able to largely contribute to the development of the semiconductor production process.

Incidentally, the ion beam diameter of the FIB tool at that time was between 50 and 100 nm. Today, the tool is produced with a minimum beam size of 4 nm. This extremely narrow ion beam enables micro-processing at the order of nanometers. This introduces ways in which the latest FIB system can contribute to nanotechnology.

14.2 FIB tools

The FIB tool can observe microscopic areas and perform both etching and deposition processes (Figure 14.1). This section introduces these functions as related to nanotechnology.

14.2.1 Observation

Secondary electrons and secondary ions are produced from the specimen surface by ion beam irradiation. A microscopic image can be obtained through detecting these secondary electrons or secondary ions, converting them to electric signals, and displaying two-dimensional distributions. The idea is the same with SEM. High-resolution observation can be achieved by narrowing the ion beam irradiation area.

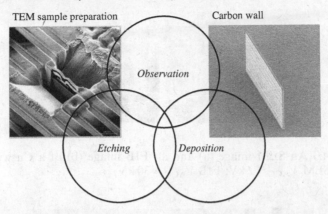

Figure 14.1 The basic functions of FIB.

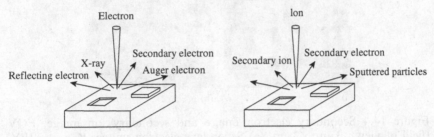

Figure 14.2 The differences between electron beam irradiation and ion beam irradiation.

Figure 14.2 shows the differences between electron beam irradiation and ion beam irradiation. Both SEM and FIB can obtain microscope images using similar principles. However, they use signals of different characteristics to form these images.

It is common to use a secondary electron for microscope observation. The amount of secondary electrons generated by the electron beam depends upon the topography. Imaging by electron beam irradiation gives a topographical contrast image. Consequently, the amount of secondary electrons generated by the ion beam relies not only on the topography but also on material or structural differences in the sample, and thus the image obtained is a structural contrast image.

Figure 14.3 shows a cross section of a Cu wire. Grain observation is possible through the channeling phenomenon of ion beam irradiation. Ion beam irradiation is different to electron beam irradiation. As shown in Figure 14.2, ion beam irradiation generates secondary ions from the specimen surface, detects secondary ions, and enables observation of a secondary ion image.

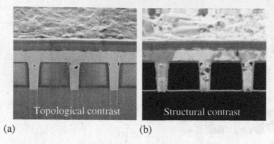

(a)   (b)

Figure 14.3 An SEM image (a) and an FIB image (b) of a Cu wire cross section: SEM $V_{ACC} = 2\,kV$; FIB $V_{ACC} = 30\,kV$.

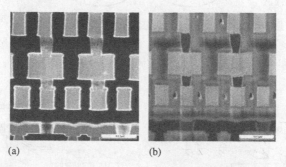

(a)   (b)

Figure 14.4 Secondary electron image and secondary ion image. FOV (field of view): $3\,\mu m \times 3\,\mu m$. (a) Secondary electron image: $V_{ACC} = 30\,kV$, $I_{beam} = 1\,pA$; (b) secondary ion image: $V_{ACC} = 30\,kV$, $I_{beam} = 2\,pA$.

Figure 14.4 shows examples of images of the same cross section. The secondary electron image (Figure 14.4(a)) shows grains of metal, but the secondary ion image (Figure 14.4(b)) shows uniform structure. We can recognize that this portion of the specimen is constructed of one material but its structure is not uniform.

14.2.2 Sputter etching

An FIB is different to an SEM because it irradiates a much heavier gallium ion than an electron. Therefore the sample surface is etched by the collision of heavy gallium ions. Figure 14.5 shows a general diagram of the sputter etching process. Sputter etching advances by irradiating an ion beam continuously in the same domain. Figure 14.6 shows a diagram of how the process area is determined.

At first a domain, including the processing area, is observed in the same ion beam conditions (acceleration voltage, beam current, scan speed, etc.) as used in processing. An ion beam is then irradiated to each pixel as shown in Figure 14.6.

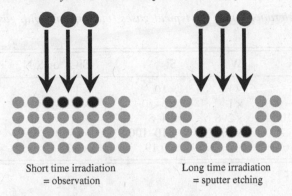

Short time irradiation = observation   Long time irradiation = sputter etching

Figure 14.5 The sputter etching process.

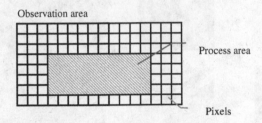

Observation area

Process area

Pixels

Figure 14.6 Choosing the process area.

If the setting of the ion beam irradiation position is the same as the condition of an ion beam, an ion beam is irradiated to the same pixel. Processing happens as shown in Figure 14.5 if an ion beam is continuously irradiated to the same pixel.

The processing area is observed by an ion beam of the same conditions as used for processing, and it is decided by assigning pixels of the observation image.

It is necessary for the ion beam to be focused to a narrow beam. In addition to the stability of the ion beam, irradiation position must be good.

14.2.3 Gas assisted etching (GAE)

GAE uses the reaction between the gas and the work piece. A reactive gas is blown in the proximity of the ion beam irradiation. There are two major types of etching: accelerated etching, in which the etching speed is fast, and decelerated etching, in which the etching speed is slow.

Accelerated etching

Improve the etching rate Generally, the reactive gas exhibits effects in terms of its specific properties. Table 14.1 displays the acceleration ratios for typical

Table 14.1. *Acceleration ratios for typical gases (etching rate with gas/etching rate without gas).*

	Al	W	Si	SiO_2 or SiN_x	Photo resist
Cl_2	×20	×1	×10	×1	×1
Br_2	×20	×1	×6–10	×1	×1
I_2	×8–10	×2–6	×4–5	×1	×1
XeF_2	×1	×10	×10–100	×6–10	×3–5
H_2O	×0.16	–	×0.19	×0.28	×20–30

Figure 14.7 Selective etching using XeF_2. The insulator at the surface of the IC was selectively etched by an ion beam with XeF_2. The wires used in this process can be seen. The etching rate of silicon oxide is large in comparison with aluminum when we use gas assisted etching with XeF_2.

gases. The acceleration rate is expressed as a ratio of the etching rate when the gas is used and the rate when it is not used. For example, with chlorine gas (Cl_2) the etching speed for aluminum (Al) is 20 times greater than that without and the rate for silicon (Si) is 10 times greater. It can be seen from Table 14.1 that no other materials exhibit accelerated etching rates with chlorine.

Selective etching An example of the selective etching of an insulator on an integrated circuit is shown in Figure 14.7. XeF_2 gas was blown onto the sample surface at the same time as ion beam irradiation was used.

Removal of redeposited material Sputter etching causes processed material to accumulate as redeposited material around the process area. As a result, drilling at a high aspect ratio is difficult with only the ion beam. However, when using a gas that makes evaporation reaction products, the reaction

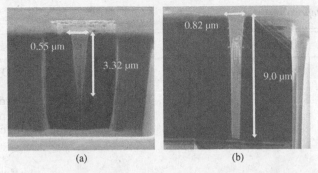

(a)   (b)

Figure 14.8 High aspect drilling of SiO_2 filled by W. (a) Without gas: aspect ratio = 6; (b) with gas: aspect ratio = 11.

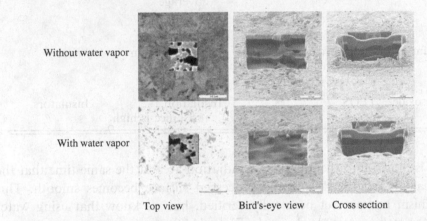

Without water vapor

With water vapor

Top view   Bird's-eye view   Cross section

Figure 14.9 Cu etching with water.

product from the sputter etching in a gas atmosphere is vented by a vacuum pump. As a result, redeposited material will not form in holes, and holes at a high aspect ratio can be formed. Figure 14.8 shows an example of using XeF_2 when etching SiO_2 film or a silicon substrate. Etching without gas has an aspect ratio of 6 for drilling whereas etching with gas improves the aspect ratio to 11 in this example.

Decelerated etching

Cu has been used in recent years in semiconductor IC wires. The etching rate of the ion beam relies largely on the orientation of the crystals. Since the crystal orientation of the grains is not uniform, it is difficult to uniformly etch Cu.

Figure 14.9 shows an example of Cu etched by FIB. It is clear that grains in the direction that is difficult to etch will remain when not using the gas assisted etching. The effect of the grains is remarkably mitigated when water

Table 14.2. *Deposition.*

Film	Material	Properties	Uses
Carbon	Phenanthrene $C_{14}H_{10}$	High deposition rate	3D deposition
		Difficult to etch with ion beam	Protection mask when creating a TEM specimen Resistors
Tungsten	Hexa-carbonyl-tungsten $W(CO)_6$	Low resistivity Good step coverage Low deposition rate	Wires
Platinum	Methylcyclopentadienyl trimethylplatinum $(CH_3C_5H_4)(CH_3)_3Pt$	High deposition rate	Wires
Silicon oxide film	TEOS	Insulation resistance is high	Insulators

vapor is blown into the ion beam irradiation area at the same time that the ion beam is irradiated, and the processed surface becomes smooth. This mechanism has not yet been fully verified, but we know that using water slows the etching rate, as shown in Figure 14.9.

14.2.4 Deposition

The FIB is able to selectively deposit thin films in the ion beam irradiation area if a compound gas is blown to the specimen surface at the same time. Since deposition is induced by the ion beam, it is referred to as ion beam induced CVD.

Secondary electrons are produced when the ion beam irradiates the specimen surface. These secondary electrons decompose the compound gas and gaseous components are vented through the system's vacuum pump. The solid components, on the other hand, remain on the surface of the decomposed specimen and accumulate to form a thin film.

The specimen surface is always being etched due to ion beam irradiation, but the deposition rate looks for conditions that are faster than the etching rate thus enabling formation of a thin film. Gases used today are shown in Table 14.2.

Table 14.3. *SMI3000 series FIB specifications.*

Minimum beam diameter	4 nm @V_{ACC} = 30 kV
Beam current	0.15 pA–20 nA
Beam stability	0.1 μm/10 min
Acceleration voltage	30 kV daily use
	5–30 kV, 5 kV steps
Maximum current density	30 A/cm^2

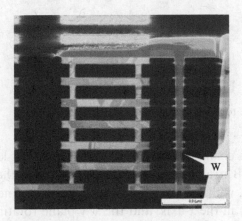

Figure 14.10 W wire formations.

Each deposition film has its own characteristics which must be applied. For example, a tungsten film is able to fill high aspect ratio holes formed by accelerated etching from XeF_2 because the step coverage is excellent. Figure 14.10 shows an example of this process.

14.2.5 System configuration

Figure 14.11 shows a photograph of the SII Nano Technology Inc. SMI3050 FIB tool for small specimens. Its main specifications are given in Table 14.3. This tool can perform cross section process observation in any direction on an entire area of a 50 mm diameter specimen.

The SMI3050 consists of (i) a main console, (ii) an operation table, and (iii) a control cabinet. Also included are accessories including a roughing pump and a power transformer capable of being used in all countries and all regions.

The main console has a specimen chamber and a frame that supports the specimen chamber. An FIB column, secondary electron detector, and gas injector are installed in the specimen chamber. A specimen stage is also installed

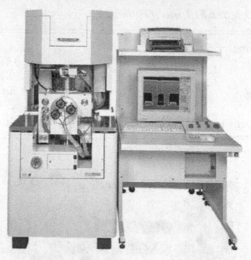

Figure 14.11 Photograph of the SMI3050.

inside the specimen chamber. The specimen stage has at least five drive axes for moving the FIB irradiation angle and position to the specimen. It can move in three axes, with the x–y plane consisting of an incline that tilts with the axis of rotation in the x-axis, and the x–y plane rotating with the axis of rotation of the z-axis.

14.2.6 Scan signal generator

As stated previously, irradiating the ion beam on the specimen surface, processing, and observation are performed by the FIB tool. This is realized by applying a scan signal generated from the scan signal generator to the deflection electrodes of the ion beam column, and controlling the irradiation position of the ion beam. Scan signals are classified as raster scans and vector scans. When performing precision processing, these scan signal properties must be understood and used effectively.

Raster scanning

Figure 14.12 shows a diagram of the raster scanning process. Raster scanning is a typical scan method using a scanning electron microscope or laser microscope. A beam is irradiated one by one along the x columns from one corner of the irradiated area. When the end point of that x column is reached it advances one y column and at the same time returns to the start of the next x column. At this time the blanking signal turns on and prevents the beam

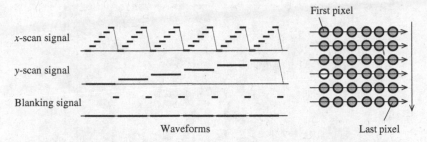

Figure 14.12 The raster scanning process.

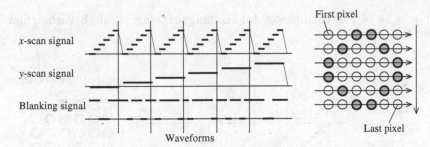

Figure 14.13 The dot blanking scanning process.

from reaching the specimen as it moves from the end point of one x column to the start point of the next x column. A normal rectangular process is a typical result of this scan method (Figure 14.12).

Dot blanking scanning (or bitmap scanning) is a development of raster scanning and is shown in Figure 14.13. Here, the x and y scan signals are the same as for raster scanning, but the blanking signal is controlled at each pixel, and enables the process to form any pattern. Definition of the blanking signal ON/OFF is a general format for displaying picture information on a personal computer, which is usually stored as a .bmp (bitmap) file. The .bmp file can be created using graphics software such as the software that comes standard with Microsoft Windows®.

Figure 14.14 shows two examples of dot blanking scanning. In Figure14.14 (a) Si is etched. In Figure 14.14(b) a carbon film is formed on Si substrate.

Vector scanning

Vector scanning is a highly flexible method of scanning in any order at any pixel, in contrast to raster scanning, which scans periodically in only the x–y direction. Figure 14.15 shows a diagram of a vector scan drawn in the same

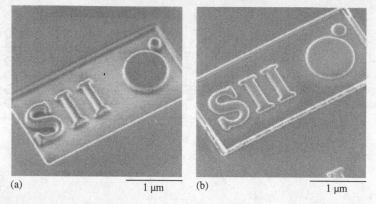

(a)   1 μm   (b)   1 μm

Figure 14.14 Two examples of dot blanking scanning. (a) Si, (b) carbon film on Si.

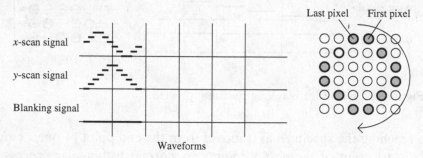

Figure 14.15 The vector scanning process.

shape as Figure 14.13 but which requires less time to scan than dot blanking scanning because:

- beam blanking is not required for each line scan;
- beam irradiation can be at a minimum except toward the process area.

An operator must define the beam trace to use vector scanning. There are two typical methods to define the beam trace. The first is describing using script language. The second is using automatic generation software from picture data. The script language has several standard patterns, e.g., a circle, a rectangle, etc. Operators can describe the detail conditions to draw an ion beam. This method is suitable for precise processing. An example of script language is shown on page 367. An array of 5×5 circles will be drawn by this script (Figure 14.16a). The process field is 40 000×40 000 pixels. Figure 14.16(b) shows results of etching and deposition. These pictures shows us that the results of a vector scan are better than those of a raster scan. An arc of a circle is described at line 5. The direction of the trace is described at line 19.

1. FOV = 40000	/Size of process field
2. p = FOV/800 practical scale on a specimen surface.	/Conversion from script coordination to
3. X0 = 0	/X coordination of the starting point
4. Y0 = 0	/Y coordination of the starting point
5. R1 = 1000	/Size of circle
6. Xp = 2000	/X pitch of circles
7. Yp = 2000	/Y pitch of circles
8. Xa = 5	/X number of repeat
9. Ya = 5	/Y number of repeat
10. px0 = 4*X0/p	/Conversion of X starting point
11. px0 = 4*Y0/p	/Conversion of Y starting point
12. pr = 4*R1/p	/Conversion of size of circle
13. pxp = 4*Xp/p	/Conversion of X pitch
14. pyp = 4*Yp/p	/Conversion of Y pitch
15. for i in xrange(0,ya,1):	/Repeat order fo Y
16. for j in xrange(0,Xa,1):	/Repeat order fo X
17. for k in xrange(1,pr,2):	/Repeat order fo circle
18. blanking(0)	/BLK Off
19. circle(px0 + pxp*i,py0 + pyp*j,k,0,360	/Detail of circle
20. blanking(1)	/BLK On
21. point(px0 + pxp*i + k,py0 + pyp*j)	/End

The method of using graphical software to interpret graphical data is suitable for complicated shapes that are difficult to describe using script language alone. Graphical data are expressed by .bmp format, GDSII format, and so on. The software reads the data and generates a vector for the beam trace. Operators can decide on the direction of the ion beam trace. By this method, complicated shapes are described but it is difficult to define detail.

Scan signals in nanotechnology

Applications in FIB nanotechnology anticipate creating not only flat shapes but also three-dimensional shapes. Here we introduce an example of scan signals for creating 3D shapes. Figure 14.17 is an example of creating 3D shapes with sputter etching. There are several methods for creating scan signals but, as shown in the figure, at the time the 3D shapes are created (not only 2D positions) the ion beam irradiation time for each pixel is controlled and a specified 3D shape is realized. Note that the effects of shaping or redeposition on the processed part must be corrected.

368 *Focused ion beam systems*

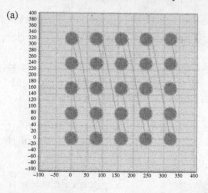

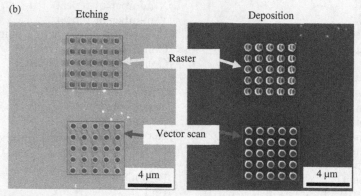

Figure 14.16 (a) Array of 5×5 circles, (b) results of etching and deposition.

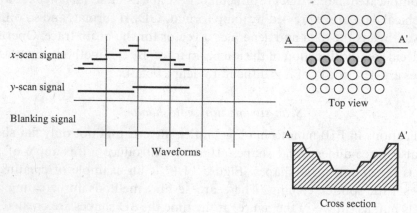

Figure 14.17 3D shaping.

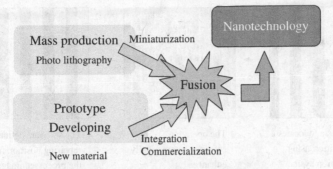

Figure 14.18 Nanotechnology as an industrial product.

14.3 Role of FIB in nanotechnology

14.3.1 Nanotechnology

Figure 14.18 shows that to establish nanotechnology as an industrial product, technology to create structures at the order of nanometers must be effectively fused with mass production technology. For example, information available on Intel's home page shows the realization of manufacturing microscopic devices by applying nanotechnology to integrated circuit manufacturing technology, and by technological trends predicted by Moore's law [7]. In considering the role of the FIB tool in the development of nanotechnology, the FIB tool can create structures of the order of micrometers to nanometers, and since creating a photo mask is not required, the work takes comparatively less time resulting in an extremely useful tool.

14.3.2 Etching process

The biggest features of processing by FIB use the same ion beam in observation and processing. As shown in Figure 14.5, the observation region is sputter etched as observation continues. This function lets you accurately and precisely process extremely microscopic regions.

Microscopic etching process

In recent years, research into the shapes of electrodes with extremely small gaps has greatly advanced. Objectives are many, including trying to make new elements and trying to observe the types of phenomenon that occur in gaps at an order of nanometers.

One experimental result [8] gives an example of a 5 nm width gap formed by applying FIBs. Here the objective is to measure the electric properties of high polymer materials at molecular levels using a manufactured electrode.

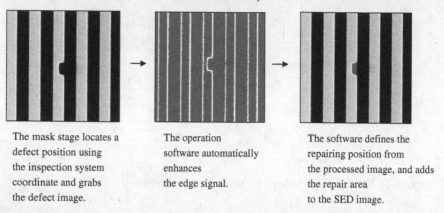

The mask stage locates a defect position using the inspection system coordinate and grabs the defect image.

The operation software automatically enhances the edge signal.

The software defines the repairing position from the processed image, and adds the repair area to the SED image.

Figure 14.19 An example of a mask repair.

At the time the gap is made, the first step is to form a 12 nm wide gap in Ti by sputter etching with the FIB, and the second step is milling of Ti on the mask with Ar and etching the Au.

An SMI9200 (JFIB2300) FIB tool of SII with 7 nm secondary electron resolution was used in this process. The latest tool was improved to have a secondary electron resolution of 4 nm. In the case of the ion beam, the secondary electron resolution is almost the same as the beam diameter. This performance has allowed even narrower gaps to be formed, compared with the SMI9200.

Ordinarily, when expressing process performance with the FIB, there is a lot of dispute over the maximum current and maximum current density because of the desire to increase process throughput. However, if using the FIB tool as an experimental tool for nanotechnology, using a beam current with a beam diameter of less than 10 nm is essential. Processing microscopic areas precisely and rapidly is required by setting a beam current that is optimized for making structures.

Etching using graphic information

Processing must be performed as you monitor process conditions without regard to size.

Figure 14.19 shows an example of a mask repair. The mask is made from a light-penetrating glass board and a shading film such as light-shading chrome film. Figure 14.19 shows an example of white defects where chrome originally should have been, but are now missing. In this case, carbon is deposited at the white defect area using the FIB deposition function [9].

In the case of a mask repair, you must avoid causing damage to the light penetrating part by irradiating it with the ion beam more than is necessary.

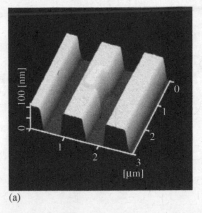 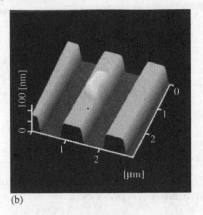

(a)   (b)

Figure 14.20 AFM image of a mask repair. (a) Binary deposition on Cr,
(b) deposition on MoSi.

The ion beam must not irradiate the unnecessary part to prevent damage
being caused to the light penetrating glass. Consequently, graphic information is used. In this example, a secondary ion detector is used in the detector.
Secondary ions are analyzed and the image is made binary from the presence
of chrome. Ion beams irradiate and carbon film is formed at the spot where
chrome should originally exist within the binary image; in other words,
only at the white defect area. Figure 14.20 shows the results of two types of
mask repair observed by an atomic force microscope (AFM). Both were
repaired using the FIB deposition function. A Cr binary mask is a type of
conventional mask, whereas a MoSi mask is a type of phase shift mask.

The technology developed for mask repair is explained in the following. This
technology makes use of graphical information to ensure precise processing.

As shown in Figure 14.9, it is difficult to obtain a flat bottom after Cu
milling. To improve the flatness of the bottom, graphic information was used.

The lower insulating film can be seen as etching of the Cu wire progresses.
Figure 14.21 shows the secondary electron image of Cu during the process.
The bright area is Cu and the dark area is SiO_2.

Observation of the processed region during processing at a fixed period
seeks a binary image that is separated into bright regions and dark regions.
This binary image corresponds to whether there is Cu in the wire. Bitmap and
process data files are created from the binary image and ion beam irradiation
is performed by dot blanking scanning (Figure 14.22).

Observations are successively done during the process, bitmap and process
data files are updated, and the process advances.

Observation performed by secondary electrons has channeling contrast
that relies on grain shape. When creating a binary image based on the image,

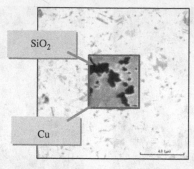

Figure 14.21 Secondary electron image while processing Cu. Only the center area has been processed.

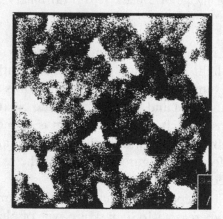

Figure 14.22 Example of bitmap processed data.

areas having Cu can no longer be correctly recognized; therefore, secondary ions are used. Images by secondary ions are contrast images that rely on the quality of the material. Having been affected by channeling contrast, Cu regions and regions without Cu cannot be discriminated (Figure 14.23).

Figure 14.24 shows an example of a processed Cu wire using a secondary ion image. It can be verified that only the Cu is being etched as the process progresses.

Figure 14.25 introduces an example of a process implemented based on bitmap and process data. A process is terminated when the entire processed region becomes bright showing the insulation. Decelerated etching by water is not used in this process. Remarkable improvement is shown compared with the example (without gas) introduced in Figure 14.9.

Finally, the finishing process is done by decelerated etching using water from the condition shown in Figure 14.25. Figure 14.26 shows that the Cu

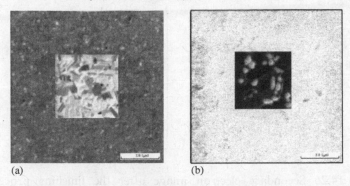

Figure 14.23 (a) Secondary electron and (b) secondary ion images during Cu milling.

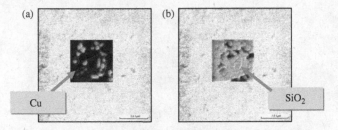

Figure 14.24 Change in secondary ion image during Cu processing. (a) Before etching, (b) after etching. The processed region is the center area, the dark area is Cu, and the bright area is SiO_2.

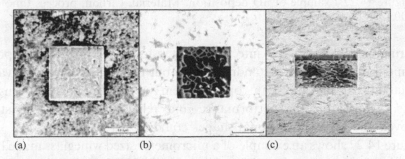

Figure 14.25 Processing using image information, with secondary ion (a), secondary electron (b), and secondary electron tilted (c).

was cleanly removed. Thus, precision processing can be performed by getting feedback from image information.

14.3.3 3D deposition

The first commercialized application of an FIB was mask repair and circuit edit using deposition. Both of them use deposition in two dimensions.

(a)   (b)

Figure 14.26 Secondary electron image after the finishing process by decelerated etching. (a) Top view; (b) bird's-eye view.

2.75 μm

Figure 14.27 Example of 3D deposition. Material: carbon, process time = 600 s.

During the 1990s, research into three-dimensional deposition was reported. The initial report was on a standing pillar, but the objective was to evaluate the stability of the ion beam optic system by making three-dimensional shapes. Afterwards, the performance and reliability of FIBs drastically improved and three-dimensional shapes could be made stably.

Figure 14.27 shows an example of a micrometer sized wineglass made by 3D deposition from carbon deposition developed by NEC, University of Hyogo, and SII [10,11].

This 3D deposition must be performed more uniformly and with higher gas density in the ion beam irradiation region than with 2D deposition. Figure 14.28 shows this process diagrammatically.

A high density, uniform gas is supplied from the gas injector to the area of ion beam irradiation. When forming a pillar, the ion beam continuously irradiated the same position to produce the 3D structure. Overhanging shapes can be achieved by widening the irradiation range in the horizontal direction. The wine

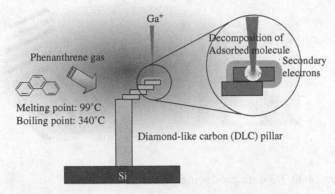

Figure 14.28 The 3D deposition process.

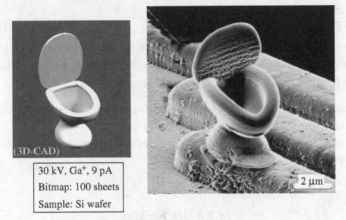

Figure 14.29 Nano-toilet produced using 3D-CAD data for 3D deposition.

glass was made by drawing a circle with a small radius by vector scanning and then extending the radius.

Incidentally, when ion beams are irradiated to the surface of a specimen, the surface is etched by ion beams. The deposition rate must be bigger than the etching rate. Usually, the ion beam current for deposition is set low to achieve a low etching rate.

3D shaping using 3D-CAD data

Figure 14.29 shows an example using 3D-CAD data for 3D deposition. To make this shape, 100 horizontal layers of slice data from a 3D-CAD were created in .bmp format, and 100 carbon deposition film layers were created

Slicing to make 100 sheets of the bitmap data

Figure 14.30 3D data producing process.

in order by the dot blanking scan method, beginning from the lowest layer (Figure 14.30).

We were able to verify from our experimental results that the ion beam mold method could be implemented in the same way as the light mold method, by ion beam induced deposition. In order to realize more precise shapes in the future, the following research needs to be pursued [12]:

1. Technology that corrects shape change from etching that progresses at the same time as deposition film proximity deposition.
2. Establish a deposition data producing algorithm for realizing structures pushed out at locations that don't have a base.

14.3.4 3D Etching

Figure 14.31 shows examples of lens shapes made by etching glass. In these examples, 3D-CAD data were used. In a different way from that used for deposition, space other than the manufactured object at the process space is defined as etching space and creates .bmp files. In the examples in Figure 14.31, sputter etching using dot blanking scanning progresses from the top surface layer. This time, gas accelerated etching is used in order to accurately realize specified shapes. When the redeposited material from sputter etching remains in the process area, the gallium ion for the original shape production is used to remove the redeposited material, and as a result the intended shaped can no longer be realized. Thus, in the actual implementation, we try not to generate redeposited material using XeF_2 gas. The results are that there is no effect from redeposition material and the intended shape can be accurately reproduced.

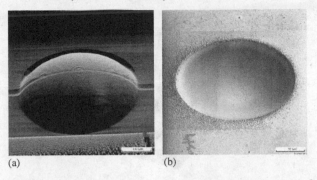

(a)   (b)

Figure 14.31 Lens shaped glass produced using 3D-CAD data. (a) Converging lens, (b) concave lens.

14.4 Unfolding of nanotechnology

This section outlines applications in micro-machining and nanotechnology of microscopic 3D structure production technology that uses all the types of process technology introduced in the previous chapters.

14.4.1 TEM sample preparation

The needs for observation, measurement, and analysis of ultra-microscopic shapes are dramatically increasing together with the development of nano-technology. In particular, there are increasing demands being made of obser-vations by the transmission electron microscope (TEM). TEM is a tool that irradiates accelerated electrons at high voltages of several 100 kV or greater on samples with a thickness of 100 nm, and observes structures within the lamella at atomic levels of resolution by detecting transmitted electrons. TEM is a technology that already has a history going back several decades, but users must make thin lamellae to observe by TEM. It takes more than one day for a skilled technician to make a thin lamella using chemical polishing. With this method, a specified spot on a miniaturized integrated circuit cannot be cut as a specification.

However, Kirk *et al.* [13] developed a practical method of using FIB in TEM specimen production. Automation has progressed, and, in the latest report, TEM specimen production exceeding 100 specimens per day was enabled by an FIB tool and system consisting of one microscope each with a manipulator using the lift-out method (Figure 14.32) [14].

Beam position stability

The thickness of a lamella specimen is commonly 100 nm. Lamellae of thicknesses of 40 nm or less are looked at depending upon the material and

Figure 14.32 Example of TEM specimen production using the lift-out method.

the required TEM observation resolution. In order to realize such micro-processing, the process must be performed by a well-focused and comparatively low current ion beam. For this reason, processing cannot be completed within a short time, and alteration of the ion beam irradiation position caused by environment changes, such as the instrument room temperature, during processing time cannot be ignored. Accordingly, technology to enable the beam irradiation position to be corrected is required.

As shown in Figure 14.33, a mark is formed for position correction within the same visual field as the process region. The position of the position correction mark is periodically verified with the observation field during processing. Moving the mark position in the observation field means moving the beam irradiation position according to the amount of movement, allowing the beam irradiation position to be set at a fixed position. As a result, precision processing can be done even when the process time is long.

Automation

With the design rule miniaturization of integrated circuits, device structures have become miniaturized and complex. As a result, high-resolution observation of device cross-sectional structures by TEM or STEM must increase and multiple productions of TEM samples for observation are required.

In the past, the operator searched for problems and defective spots at the site of device development or production processes, and the main trend was in specimen production.

From this, however, utilization by production control is assumed, and fixed point observation is called for within the routine work of specified points on a

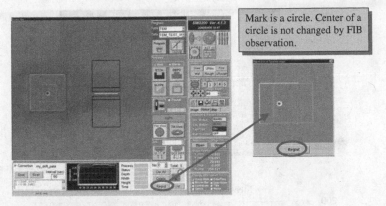

> Mark is a circle. Center of a circle is not changed by FIB observation.

Figure 14.33 Drift correction.

wafer. In this case, a system is required in which the tool automatically detects a targeted pattern on a wafer, and automatically produces a TEM specimen at a set location. The remaining thickness of the specimen is 100 nm.

Figure 14.34 shows an example of evaluating STEM specimen auto production for 200 mm wafers using the SMI3200 FIB tool, which has the same performance as the SMI3050.

The following evaluation method was used:

1. Process target position settings: the process position is set at a distance of 1.7 μm below the wire as shown in Figure 14.34(a). (Note that since the deviation in the *x* direction is sufficiently small for process width, it is not a target for evaluation.)
2. Sample size: width of 10 μm, depth of 7 μm, remaining thickness of 160 nm. The processing time is approximately 20 min for making one sample.
3. Thirty dies on top of the same wafer were selected and automatically processed. The total processing time was approximately 10 hours.

Figure 14.34(b) shows the measured thickness of 30 lamellae. The following results were verified:

- standard deviation of specimen's remaining thickness $(3\sigma) = 15.2$ nm;
- standard deviation of specimen production position $(3\sigma) = 43.6$ nm.

We verified that no large error exists in the image if the specimen thickness has a deviation range of 80 to 150 nm when doing STEM observation of the Si that is used mainly by semiconductor integrated circuits, as shown in Figure 14.35.

Consequently, the specimen thickness deviation obtained by this experiment is permitted to get reasonable STEM images. Technology that automatically produces nanometer level specimens using the FIB tool is now being established.

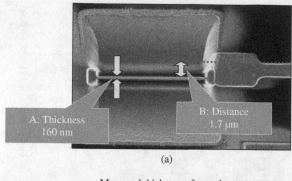

(a)

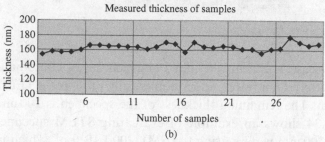

(b)

Figure 14.34 TEM specimen auto production. (a) Process condition, (b) result of evaluation.

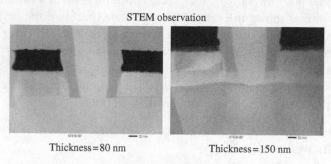

STEM observation

Thickness=80 nm   Thickness=150 nm

$V_{ACC} = 200$ kV

Figure 14.35 Specimen thickness and STEM image observation by JEM2500SE (JEOL).

Low acceleration voltage process

TEM observation is a useful tool for observing the internal structure of specimens and in particular the lattice structure. Nonetheless, the crystal structure may be damaged by FIB processing. Figure 14.36 shows that the amount of damage becomes smaller in proportion to the acceleration voltage of the ion beam used in sputter etching.

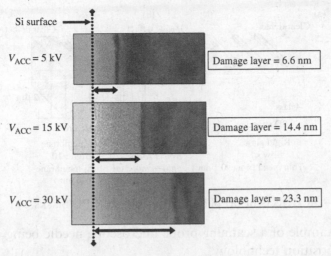

Figure 14.36 Three examples of the damage caused to similar specimens at different acceleration voltages.

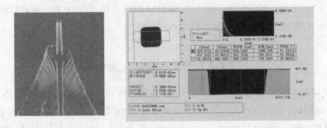

Figure 14.37 Scanning probe microscope needle produced using 3D deposition technology. Length = 1300 nm, diameter = 60 nm.

For this reason, the finishing process in producing a specimen must be performed at a low acceleration voltage when observing lattice images at a high magnification.

The latest research has reported that the acceleration voltages for this purpose can be as low as 1 kV [15].

14.4.2 Probe preparation for the scanning probe microscope

The scanning probe microscope is a tool that scans the surface of specimens with the tip of a needle, detects movement between the needle tip and specimen surface, and observes and measures the shape and physical properties of the specimen surface. Because of its ability to measure various properties, the scanning probe microscope is a powerful tool for nanotechnology.

The performance of the scanning probe microscope cannot be overstated although it is controlled by the properties of the needle itself. Figure 14.37

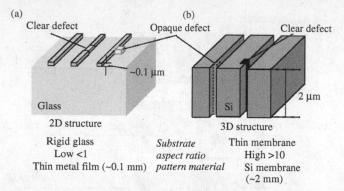

Figure 14.38 Defects of (a) photo masks and (b) EPL masks.

shows an example of a scanning probe microscope needle being made using FIB 3D deposition technology.

The tungsten pillar is formed similarly to that of carbon introduced earlier, and is used as a needle. Depending on the need, the needle can be further sharpened by sputter etching the tip with the FIB. Here we see the start of 3D deposition as applied to an actual product.

14.4.3 EB mask repair

To get smaller structures, a short wavelength light source is needed. A method of making electronic beams a light source electron beam is one of the candidates for new generation lithography. EPL and LEEPLE are known as electron beam lithography technologies. Masks used by electron beam lithography consist of a shading part that does not allow electrons on a silicon substrate to reach the specimen surface and a transparent part that allows electrons to pass through and reach the specimen surface. Figure 14.38 shows defects in the photo and EPL masks.

The masking portion of the conventional photo mask is thin. Its thickness is about 0.1 μm. When white defects and black defects are corrected, operators do not need to think three-dimensionally to fix them.

On the other hand, the EPL mask repair operator must see a 2 μm thick silicon substrate in 3D, grasp the defects, and correct them. These corrections first became possible using FIB 3D shape production technology [16,17].

Several technologies used in correction of masks for electron beam exposure are outlined below.

Taper correction

Ion beam density distribution is a normal distribution. Generally, the "width of half peak" defines the beam diameter and advances the development into

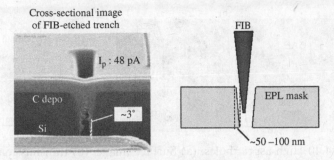

Figure 14.39 Taper correction. The typical taper angle of an FIB etched pattern is roughly 1.5–3°; difference of critical dimension between top and bottom ≈50–100 nm. I_p is the ion beam current at a sample.

narrowing this width. When the FIB tool was first commercialized, the beam diameter was about 100 nm. Today tools with a diameter of 4 nm are being produced. The performance of the FIB has been dramatically improved, but the beam profile relies on physical phenomena and still has a normal distribution.

For this reason, when the sputter etching process is performed, a taper that corresponds to this beam profile in the sidewall of the processed area can be done. This is shown in Figure 14.39. In the case of the EPL mask, you must see through the entrance of the hole of the translucent part to the exit and there must be no obstacles. What is obvious is that the exit slit cannot be narrower than the entrance slit.

In order to conquer this problem, the specimen stage is tilted. As shown by Figure 14.39, a taper of 1.5–3° can be used, and the sidewall can be made vertical to the surface by tilting the specimen to cope with the taper angle. This technology is used even for TEM specimen production. In order to define the thickness of a lamella specimen, the finishing process tilts and the process is performed.

Preparation of high aspect holes

As the mask is for forming microscopic shapes on the surface of specimen, a tool that makes as small a hole as possible is required. Figure 14.40 introduces an example of making a microscopic hole using gas accelerated etching. GAE processing using accelerated etching and normal processing not using gas is performed on a silicon substrate that becomes a mask.

Normal processing can only drill extremely shallow holes because there is no means of transporting the drill cuttings outside the hole. GAE processing is able to open holes of high aspect ratio because the cuttings react with the gas and are carried outside the hole.

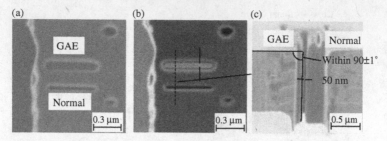

(a)　(b)　(c)

GAE

Normal

0.3 μm

GAE

0.3 μm

GAE　Normal

Within 90±1°

50 nm

0.5 μm

Figure 14.40 High aspect holes. (a) Surface image, (b) transmission image, (c) cross-sectional image. Conditions shown are for halogen gas, $P_a = 3 \times 10^{-3}$ Pa, $I_p = 0.15$ pA, time $= 250$ s. The taper angle is within $90 \pm 1°$. P_a is the pressure of the halogen gas for GAE.

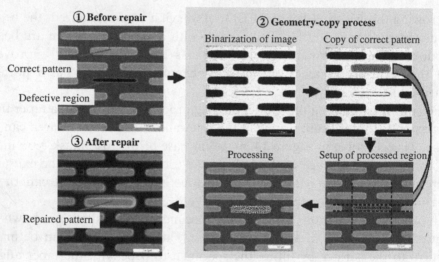

① Before repair

Correct pattern

Defective region

② Geometry-copy process

Binarization of image　　Copy of correct pattern

③ After repair

Repaired pattern

Processing　　Setup of processed region

Corner round: corresponding to correct pattern

Figure 14.41 G-Copy.

Shape reappearance

Correction of defects is based on normal shape data. The best way to obtain normal shape data is from CAD data used in production of the original mask. Nonetheless, in the present state where integration has drastically improved, CAD data typically consist of information of the entire element and the amount of information is large and cannot be easily handled.

Therefore, you can select a normal shape near the defect, create correction information from the image of the normal shape using pattern recognition technology, and perform the process based on the correction information. Figure 14.41 introduces G-Copy, which performs defect correction by this idea. (Note: G-Copy is a function name of the SIR-series.)

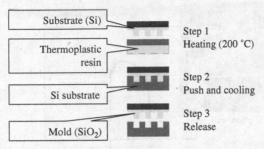

Figure 14.42 Nanoimprint technology.

Nanotechnology is expected to be applied for biotechnology. Cells and genetics are subjects in the field of biotechnology, but when making microscopic structures related to these, the shape information of these must be taken in and cases of production based on the shape information anticipated. From the technology introduced here we can expect a growth in production of structures that correspond to unknown samples.

14.4.4 Micro-mold preparation

Professor Choi proposed nanoimprint technology as a way to mass produce microscopic shapes by pushing a mold in thermoplastic organic matter and transcribing shapes [18]. Figure 14.42 outlines the process of producing nanoimprints.

In this example, the micro-mold was realized by processing a silicon oxide film on a silicon substrate. Thermoplastic organic matter is placed on the silicon substrate then heated at 200 °C. After the thermoplastic organic matter is sufficiently heated, the mold is pushed against the workpiece then cooled. The mold is removed after sufficient cooling, enabling the shape of the mold to be transcribed onto the surface of the workpiece.

This technology is expected to be a technology that complements the current trend of photo lithography. Generally, only 2D shapes can be transcribed with photo lithography but nanoimprinting lets you tilt sidewalls, for example, by making a 3D mold. We believe that making and correcting this mold can be done by using FIB sputter etching and deposition [19]. Making shapes is possible by using technology introduced in Section 14.4.3.

Incidentally, the evaluation of temperature characteristics of microscopic structures made by FIB are not being sufficiently implemented [20]. This is a subject for the future.

Figure 14.43 introduces preparatory experimental results for temperature characteristics of microstructures made by FIB. A carbon wall by carbon

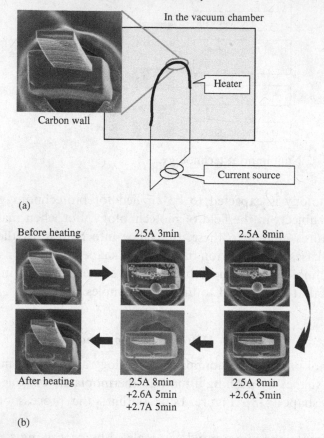

Figure 14.43 Basic experiment with micro molds. (a) Experiment, (b) results.

deposition was made on a heater wire. Changes in the carbon wall are observed when the heater temperature increases.

Because this experiment was preparatory, measurement of temperature was not correctly performed. The heater temperature exceeds 700 °C.

What is clear from the experiment is that gallium is extracted from the surface of the specimen if high-temperature conditions continue, becomes shaped like a water drop, and finally evaporates. Extremely small particles can be seen when the surface of the specimen is observed after high-temperature treatment; however, they can be removed by washing with water.

This experiment shows that heat treatment must be performed in advance and the gallium removed when using a workpiece made by FIB, as a mold, for example, at high temperatures. Since the shape does not change before and after the heat treatment, the possibility that it can be used as a tool for

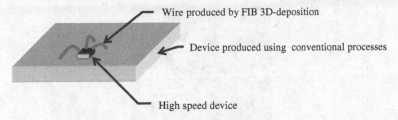

Figure 14.44 Ultra-high speed hybrid device.

making and correcting molds used in nanoimprinting of FIB tools can be verified by gathering more detailed data.

14.4.5 Unfolding of biotechnology

In the latest research, while verifying the behavior of cells one by one, advanced experiments are desired. For this reason, it is desirable that specific conditions be given to cells one by one, and experimental technology that separately verifies how the effects are received is established.

A cell is generally several micrometers across, which is about the size of the nano wine glass introduced in Figure 14.27. In terms of size, production of a container that controls cells one by one is made possible using FIB. If information about the shape of the specimen is precisely obtained in advance, a container of optimum shape can be made from those data.

In addition, a nanoscale manipulator was fabricated by FIB deposition [21]. Cells will be picked up by nanomanipulator, and brought to containers made by FIB deposition.

Today, evaluation has begun on what effect a container made by FIB will have on the organic specimen, such as a cell. These are only the first steps and we look forward to the results of later research.

14.4.6 Ultra-high speed device development

The performance improvement of integrated circuits in which Si is used is remarkable. However, a device made of GaAs is able to get better performance than a device made of Si.

Nevertheless, an ultra-high speed element produced on a semiconductor board made from a chemical compound other than silicon, for example, achieves a high performance not realized by silicon. Accordingly, as shown in Figure 14.44, a hybrid integrated circuit is one of the solutions to achieve a high-speed device.

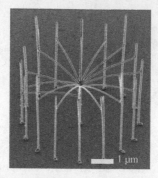

Figure 14.45 Wire basic experimentation in air. (Courtesy of University of Hyogo and Fundamental Research Laboratory of NEC.)

Development engineers of this ultra-high speed hybrid device want to perform device production inside the FIB tool. First, the FIB is used to detach the ultra-high speed element from the substrate and the switch is raised with micro-tweezers. Then the ultra-high speed element is moved to the spot specified in the integrated circuit. This time, both connection between the ultra-high speed element and integrated circuit and production of insulation are possible by selecting the type of deposition film. The electrical circuit connection of the ultra-high speed element and the integrated circuit is performed by FIB by aerial wiring. If this technology was developed, a stray capacitor of wire would be small enough to realize ultra-high speed signal transmission, and will contribute to improved high-frequency signal performance. However, aerial wiring is currently still in the research and development stage.

Figure 14.45 introduces a part of fundamental research which realizes the shape of an aerial wire by mixing carbon and tungsten [22]. Unfortunately, conductivity suitable for this purpose cannot yet be achieved. The realization of a high conductivity rate of the conductive film deposited by an FIB is being sought.

14.4.7 Nano lathe

A precision rotation stage with a small radial run out was installed on the specimen stage of the FIB tool to produce a nano lathe [23]. As shown in Figure 14.46, the rotation center of the precision rotation stage can be set to be parallel or perpendicular to the ion beam.

As shown in both Figures 14.27 and 14.46, an overhung shape with a top that is wider than its bottom or basement can be manufactured using 3D deposition technology. The material used in this method is decided by ion beam induced CVD.

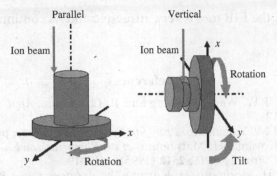

Figure 14.46 Diagram of a nano lathe.

Figure 14.47 A nano wine glass in carbon.

When we use the nano lathe, we can make a wide variety of shapes and use a wide variety of materials. Figure 14.47 shows one example that can be made using this method. The material used is carbon. During milling it was rotated at 120 rpm.

14.5 Conclusions

Listed below are the essential points for the practical use of FIB in the field of nanotechnology:

1. Stable control of sub-picoampere currents for microprocessing.
2. Drift correction for beam irradiation position stabilization.
3. 3D-CAD linkage.
4. Application of pattern recognition technology.
5. Effective use of gas.

Nanotechnology of FIB tools is still leading-edge technology, and there are many items that need research and development. Nevertheless, in the field of nanotechnology, the FIB tool is very attractive and its continued development is desirable.

References

[1] R. L. Seliger, J. W. Ward, V. Wang and R. L. Kubena. *Appl. Phys. Lett.*, **34** (1979), 310–12.

[2] T. Kaito and M. Yamamoto. *Proc. 9th Symp. ISIAT* (1985), pp. 207–10.

[3] K. Aita, Y. Koyama, H. Matsumura, *et al. Photomask and X-Ray Mask Technology II, Proc. of SPIE*, **2512** (1995), 412–19.

[4] Y. Mashiko, H. Morimoto, H. Koyama, *et al. International Reliability Physics Symposium*, (1987), 111–17.

[5] T. Kaito and T. Adachi. *Proc. 1st Micro Process Conference* (1988), 142–3.

[6] K. Nikawa, K. Nasu, M. Murase, *et al. Int. Reliability Phys. Symp.*, (1989), 43–52.

[7] *Moore's Law*, at: http://www.intel.com/technology/mooreslaw/index.htm

[8] T. Nagase, T. Kubota and S. Mashiko. *Thin Solid Films*, **438–439** (2003), 374–7.

[9] K. Aita, A. Yasaka, T. Kitamura, *et al. Photomask and X-Ray Mask Technology II, Proc. of SPIE*, **2793** (1996), s324–35.

[10] S. Matsui, T. Kaito, J. Fujita, *et al. JVST B*, **18** (2000) 3181–4.

[11] J. Fujita, M. Ishida, T. Sakamoto, *et al. JVST B*, **19** (2001), 2834–7.

[12] T. Hoshino, K. Watanabe, R. Kometani, *et al. JVST B*, **21** (2003), 2732–6.

[13] E. C. G. Kirk, D. A. Williams, R. Kometani, *et al. Microsc. Semicond. Mater.*, **100**, Section 7 (1989), 501–6.

[14] H. Suzuki, K. Iwasaki, Y. Ikku, A. Yasaka and T. Adachi. *LSI Testing Symposium 2003*, (2003), 31–5. (In Japanese.)

[15] S. Sadayama, K. Kanda, Y. Yamamoto, *et al. LSI Testing Symposium 2004*, (2004), 133–7. (In Japanese.)

[16] M. Okada, S. Shimizu, S. Kawata and T. Kaito. *JVST B*, **18** (2000), 3254–8.

[17] Y. Yamamoto, M. Hasuda, H. Suzuki, *et al. Photomask and Next-Generation Lithography Mask Technology XI, Proc. SPIE*, **5446** (2004), 348–56.

[18] S. Y. Chou, P. R. Krauss and P. J. Renstrom. *Appl. Phys. Lett.*, **67** (1995), 3114–16.

[19] K Watanabe, T. Morita, R. Kometani, *et al. JVST B*, **22** (2004), 22–6.

[20] T. Fujii and T. Kaito. Fundamental experience for micro mold fabricated by FIB. Oral Presentation at EFUG 2003 (2003).

[21] R. Kometani, T. Morita, K. Watanabe, *et al. JVST B*, **22** (2004), 257–63.

[22] T. Morita, R. Kometani, K. Watanabe, *et al. JVST B*, **21** (2003), 2737–41.

[23] T. Fujii *et al. J. Micromech. Microeng.*, **15** (2005), S286–91.

Index